T0339833

Taxonomy of *Corynoneura* Winnertz (Diptera: Chironomidae)

Taxonomy of
Corynoneura Winnertz
(Diptera: Chironomidae)

YUE FU

College of Biology and Agricultural Resources,
Huanggang Normal University, Hubei Province,
P.R. China

XIANGLING FANG

College of Biology and Agricultural Resources,
Huanggang Normal University, Hubei Province,
P.R. China

XINHUA WANG

College of Life Sciences, Nankai University,
Tianjin, P.R. China

SCIENCE PRESS
Beijing

ELSEVIER

ACADEMIC PRESS

An imprint of Elsevier

Academic Press is an imprint of Elsevier
125 London Wall, London EC2Y 5AS, United Kingdom
525 B Street, Suite 1650, San Diego, CA 92101, United States
50 Hampshire Street, 5th Floor, Cambridge, MA 02139, United States
The Boulevard, Langford Lane, Kidlington, Oxford OX5 1GB, United Kingdom

Notices
Knowledge and best practice in this field are constantly changing. As new research and experience
broaden our understanding, changes in research methods, professional practices, or medical treatment
may become necessary.

Practitioners and researchers must always rely on their own experience and knowledge in evaluating and
using any information, methods, compounds, or experiments described herein. In using such information
or methods they should be mindful of their own safety and the safety of others, including parties for
whom they have a professional responsibility.

To the fullest extent of the law, neither the Publisher nor the authors, contributors, or editors, assume any
liability for any injury and/or damage to persons or property as a matter of products liability, negligence
or otherwise, or from any use or operation of any methods, products, instructions, or ideas contained in
the material herein.

British Library Cataloguing-in-Publication Data
A catalogue record for this book is available from the British Library

Library of Congress Cataloging-in-Publication Data
A catalog record for this book is available from the Library of Congress

ISBN: 978-0-12-815263-8

For Information on all Academic Press publications
visit our website at https://www.elsevier.com/books-and-journals

Publisher: Charlotte Cockle
Acquisition Editor: Anna Valutkevich
Editorial Project Manager: Devlin Person
Production Project Manager: Maria Bernard
Cover Designer: Beijing Tu Yue Sheng Shi Culture and Media
      Co., Ltd., Matthew Limbert

Typeset by MPS Limited, Chennai, India

Working together
to grow libraries in
developing countries

www.elsevier.com • www.bookaid.org

CONTENTS

PREFACE

The *Corynoneura* generic group belongs to the Orthocladiinae, which has recorded 206 species belonging to 9 genera. This group is widely distributed in the major animal geographical areas and has high species diversity. Different species of larvae live in different water bodies and can be indicators of water pollution due to their development of different tolerance values. They are used to evaluate the water quality of rivers and lakes and have an important practical significance for water environment monitoring and protection. This group is the only group in the Chironomidae in which the radial sector is retracted, swollen, and apically fused with the costa in a thick clavus which terminates at, or before, the midpoint of the wing. Because of its morphological characteristics of wing, the group was considered as a separate subfamily of the family Chironomidae, subsequent studies have placed this group in the Tanypodinae and Orthocladiinae, which indicated that the taxon was important in the phylogeny of Chironomidae. In addition, the species of this group is very small (only 1—2 mm in length), which makes research challenging. And in previous descriptions, there are many incorrect identifications and synonyms phenomena in the recorded species. According to the materials in the world or related references, the study was revised the worldwide taxonomy of *Corynoneura*. The results of the study provide the data for the comprehensive determination and classification of the *Corynoneura* generic group and provide the scientific basis for phylogenetic studies of Chironomidae. The book comprises four main sections: an introduction, keys, classification, and zoogeography.

In the introduction, the history of worldwide research and current research on the genus *Corynoneura* are reviewed, materials and methods are described, morphological characters (terminology) are introduced, and diagnostic characters of *Corynoneura* are discussed. The preliminary estimate of the phylogenetic relationships of the sufficiently known males of *Corynoneura* are provided.

In the section on keys, keys to the *Corynoneura* generic group and male species of *Corynoneura* across the world are given.

In the section on classification, 96 species and 7 unnamed species of *Corynoneura* are described in detail with 159 morphological figures. Among the taxa, 41 species are described based on more than two stages (male, female, pupa, or larva). Synonyms of each known taxon are listed in chronological order. The available data on geographical distribution is also reported.

In the section of zoogeography, a preliminary biogeographic analysis of the worldwide fauna is made based on the *Corynoneura* generic group and species distribution data. It is obvious that the species distributed in the Palearctic region are dominant. In addition, the distribution information of *Corynoneura* species research work is listed.

We are much indebted to Prof. Ole A. Sæther (December 9, 1936—January 8, 2013) and Prof. Trond Andersen (Museum of Zoology, University of Bergen, Norway) for inspiration and help by making many valuable suggestions for this research; Dr. A. Shinohara (Department of Zoology, National Science Museum, Tokyo, Japan) for kindly lending many holotypes of Japan; Dr. Michael Bolton (USA), Dr. Ray Pupedis (Yale University), Dr. John H. Epler (Florida, USA), Dr. Bohdan Bilyj (Canada), Dr. Peter H. Langton (University Museum of Zoology, Cambridge), Dr. Torbjørn Ekrem (Norwegian University of Science and Technology, Norway), Dr. M. Spies (Zoologische Staatssammlung München, Germany), and Dr. Hongqu Tang (Jinan University, Guangzhou City, Guangdong Province of China) for sending specimens and research materials; Dr. Endre Willassen (Natural History Collections, Bergen Museum), Dr. T. Kobayashi (Institute for Environmental and Social Welfare Sciences, Japan), Dr. Eugenyi Makarchenko (Institute of Biology and Pedology, Far Eastern Branch of Russian Academy of Sciences, Vladivostok, Russia), Dr. Patrick Ashe (Department of Zoology, Trinity College, Ireland), and Dr. Sofia Wiedenbrug (Universidade Federal de São Carlos, Brazil) for the material and discussions.

The project was supported by the National Natural Science Foundation of China (NSFC) (Grant Nos. 31460572, 31101624), the First Level Disciplines of Forestry of Hubei University for Nationalities.

Yue Fu
September 28, 2016

CHAPTER 1

Introduction

1.1 RESEARCH HISTORY AND CURRENT SITUATION ON *CORYNONEURA* OF THE WORLD

Corynoneura, Thienemanniella, and a few newly described genera are the only genera in Chironomidae in which the radial sector is retracted, swollen, and apically fused with the costa in a thick clavus which terminates at, or before, the midpoint of the wing. The *Corynoneura* group is characterized by having the costa apically fused with R_1 and R_{2+3}, forming a thick clavus (e.g., Sæther & Kristoffersen, 1996). The larvae are also distinctive with modified chaetulae laterales and labral setae, fused meso- and metathoracic segments, ventrobasal spine on the posterior parapods, as well as elongated antennae. The pupae are also distinctive with taeniate setae on the anal lobe, but show some similarities with *Eukiefferiella* Thienemann and related genera, for instance the presence of hooklets on the conjunctives.

Goetghebuer (1919) regarded *Corynoneura* and *Thienemanniella* as a separate tribe, the Corynoneurini; Goetghebuer (1939) regarded the two genera as a separate subfamily, the Corynoneurinae. Brundin (1956), however, regarded the *Corynoneura* group as an integral part of the tribe Metriocnemini and closely related to *Pseudosmittia* (Goetghebuer). Sæther (1977) found support for a tribal system of rearranged Orthocladiini, Metriocnemini, and Corynoneurini except for some basal groups and genera including the *Brillia* group and the *Eukiefferiella* group. Mendes, Andersen, and Sæther (2004), Andersen and Sæther (2005) found strong indications of a close relationship between the *Cardiocladius* (or *Eukiefferiella*) group and the *Corynoneura* group with several synapomorphies. Cranston, Hardy, and Morse (2012), however, found that these groups were not closely related based on molecular phylogeny, but that the similarities were probably based on parallelism or convergence. Their phylogeny of the Orthocladiinae groups resembles that of Sæther (1977, Fig. 36).

The *Corynoneura* group, until relatively recently, included only *Corynoneura* Winnertz, *Thienemanniella* Kieffer, and *Corynoneurella* Brundin. Recently some apparently related genera have been added to the group namely *Physoneura* Ferrington & Sæther, 1995; *Tempisquitoneura*

Epler & de la Rosa, 1995; *Notocladius* Harrison, 1997; *Ichthyocladius* Fittkau, 1974; *Onconeura* Andersen & Sæther, 2005; and *Ubatubaneura* Wiedenbrug & Trivinho-Strixino, 2009 according to the phylogeny in Mendes et al. (2004).

The adult identification of this group is mostly based on genitalic structures, as for many other Chironomidae. Their morphology-based identification is often difficult, especially for nonexperts, and often demands time-consuming genitalic dissections. The identification of adults of the *Corynoneura* group is further complicated by a lack of diagnostic morphological features to distinguish several species. Moreover, identification of some species can be achieved only by individual rearing of larvae and collecting larval and pupal skins to establish the associations between life stages. Despite these drawbacks, reviews of the *Corynoneura* group with descriptions and redescriptions from East Asia and the Afrotropical, Nearctic, and Neotropical regions have been published (e.g., Fu, Hestenes, & Sæther, 2010; Fu & Sæther, 2012; Fu, Sæther, & Wang, 2009, 2010; Wiedenbrug, Lamas, & Trivinho-Strixino, 2012, 2013; Wiedenbrug, Mendes, Pepinelli, & Trivinho-Strixino, 2009; Wiedenbrug & Trivinho-Strixino, 2009, 2011).

With reference to the genus *Corynoneura* Winnertz, 98 species have been described—in the literature, 96 species are reported with the exceptions being *C. imperfecta* (Skuse) and *C. postcinctura* (Tokunaga), and 7 unnamed species are also listed. Forty-six species are recorded from the Palaearctic Region, 19 from the Nearctic Region, 25 from the Neotropical Region, 17 from the Oriental Region, 4 from the Afrotropical Region, and 5 from the Australasian Region (Ashe & O'Connor, 2012; Fu & Sæther, 2012; Fu et al., 2009; Makarchenko & Makarchenko, 2006b, 2010; Makarchenko, Makarchenko, Zorina, & Sergeeva, 2005; Schlee, 1968b; Wang, 2000; Wiedenbrug & Trivinho-Strixino, 2011; Wiedenbrug et al., 2012; Yamamoto, 2004). Recent research work is listed here:

Palaearctic and Oriental Region: Fu et al. (2009), Fu, Hestenes, and Sæther (2010), Fu, Sæther, and Wang (2010) established and redescribed some species, including the sufficiently known males of *Corynoneura* from around the world and gave a preliminary estimate of the phylogenetic relationships of the known males. The range of some measurements given in Cranston, Oliver, and Sæther (1989) was extended. Makarchenko and Makarchenko (2010) and Krasheninnikov (2012) described seven new species from Russia.

Nearctic Region: Roback (1957a, 1957b), Sublette (1970), Boesel and Winner (1980), Bolton (2007), Epler (2001), Hestenes and Sæther (2000) described and gave keys to some species from the Nearctic region. Nine species of *Corynoneura* previously are recorded, namely *C. fittkaui* Schlee, *C. celeripes* Winnertz, *C. lacustris* Edwards, *C. arctica* Kieffer, *C. scutellata* Winnertz, *C. lobata* Edwards, *C. oxfordana* Boesel and Winner, *C. taris* Roback, and *C. diara* Roback. Ten species of *Thienemanniella*, namely *T. cubita* Garrett, *T. majuscula* Edwards, *T. similis* Malloch, *T. elana* Roback, *T. mallochi* Sublette, *T. xena* Roback, *T. boltoni* Hestenes and Sæther, *T. lobapodema* Hestenes and Sæther, *T. obscura* Brundin, and *T. taurocapita* Hestenes *et* Sæther are known from the Nearctic. Among these species, *T. cubita* Garrett and *T. mallochi* Sublette were established on females only; others are still in need of further redescription and should temporarily be regarded as *nomina dubia*. The holotype of *C. oxfordana* Boesel and Winner is reexamined, and found to be a junior synonym of *C. lobata* Edwards, *C. taris* Roback has been considered as a junior synonym of *C. lobata* Edwards (Epler, 2001). The holotypes of the other two Roback species, that is, *T. elana* and *C. diara*, could not be found where they were deposited. Fu and Sæther (2012) gave a review of the Nearctic *Corynoneura* and keys to all stages. The descriptions show that especially the lateral lamellae on the coxosternapodeme and the shape of the labia of female imagines are much better for distinguishing species in *Corynoneura* than different male characters for separating species. Accordingly, including the species mentioned by Oliver, Dillon, and Cranston (1990), there presently are 28 species of these two genera known from the Nearctic region.

Neotropical Region: There are at least 42 species in the Neotropical region. These minute (0.5—1.5 mm) dipterans are found in a broad range of aquatic habitats and are especially diverse in mountain streams in the Atlantic Forest (Wiedenbrug & Trivinho-Strixino, 2011). The larvae seem to be particularly abundant on litter and stones of streams and they are associated with macrophytes from lentic habitats (Henriques-Oliveira, Dorvillé, & Nessimian, 2003; Sanseverino & Nessimian, 2001). The Atlantic Forest has been identified as a biodiversity hotspot of global significance due to its high concentration of endemic taxa and vulnerability to processes that threaten its unique biodiversity (Myers, Mittermeier, Mittermeier, Fonseca, & Kent, 2000). The enormous biodiversity of this biome results, in part from the wide range of latitude it covers, its variations in altitude, its diverse climatic regimes, as well as the geological and climatic history of the whole region (Galindo-Leal & Câmara, 2003). Currently, the coverage of the Atlantic Forest in Southeast Brazil also has been significantly reduced due to a long process of exploitation and degradation (Zaher et al., 2011).

Consequently, species-level keys as well as extensive reference collections are available for most of the genera, making this an ideal group of Chironomidae for investigating molecular/morphological correspondence. The Chironomidae have been increasingly studied from the perspective of DNA barcoding. Most published studies involving the cytochrome c-oxidase subunit I (*CO I*) gene in chironomids have used sequence data to perform phylogenetic analyses (e.g., Allegrucci, Carchini, Todisco, Convey, & Sbordoni, 2006; Martin, Blinov, Alieva, & Hirabayashi, 2007), delimit species (e.g., Carew, Marshal, & Hoffman, 2011; Silva, Ekrem, & Fonseca-Gessner, 2013), document species diversity (e.g., Ekrem, Stur, & Hebert, 2010), biomonitor environmental changes (Brodin, Ejdung, Strandberg, & Lyrholm, 2012), and associated different life stages are based on genetic similarity (e.g., Stur & Ekrem, 2011). DNA barcoding employs a short gene fragment to identify species. It represents a shift from the near-exclusive reliance on morphological features for the identification and delimitation of species to an approach that includes molecular characters in species discrimination (Ekrem et al., 2010).

Orthocladiinae is the second most diverse chironomid subfamily after Chironominae (Ashe & Cranston, 1990). It was established by Edwards (1929) derived from Kieffer's Orthocladiariae group (Cranston, 1995) and its monophyly was supported, with Prodiamesinae as its sister-group. Within the Orthocladiinae, the tribe Corynoneurini is also well-supported with the studied genera *Thienemanniella*, *Corynoneura*, and *Notocladius* (Cranston et al., 2012).

Silva and Wiedenbrug (2014) used DNA barcodes for species delimitation in order to solve taxonomic conflicts in the *Corynoneura* group (Orthocladiinae) in the Atlantic Forest. They used sequences from mitochondrial *CO I* to determine genetic groups in the *Corynoneura* group and examined sequence variation between and within these groups. Their study focused on the genera that occur in the benthic fauna of streams in the Atlantic Forest. They are *Corynoneura*, *Onconeura*, *Thienemanniella*, and *Ubatubaneura*.

1.2 MATERIALS AND METHODS

1.2.1 Collection of Materials

Adults were collected by light trapping, sweep netting, and Malaise trapping. Pharate adults and pupal exuviae were collected with drift nets. Larvae were collected with hand nets or with artificial substrate. The last method used were stones and leaves placed in plastic recipients with small holes. In the case of lakes, the recipients filled with leaves were tied up to buoys in order to be in the same depth as the macrophytes. These collectors were left in the habitat for about a month and sorted out in a laboratory. Living larvae were isolated in small boxes (1 cm × 1 cm) half filled with stream water. Neither substratum nor food was given, except for some detritus carried over with the water. Boxes were checked twice a day for emerged specimens (Fu & Sæther, 2012; Wiedenbrug et al., 2012).

1.2.2 Preservation and Mounting

The material examined was mounted on slides following the procedure outlined by Sæther (1969).

The larval, pupal exuviae, and the adults were preserved in 75%, 85%, and 96% ethanol. When make slide, put the specimen in 70% ethanol, detach the wings (attempt to remove the squama with the wing) and the legs from one side of the body and the antennae using fine forceps into absolute ethanol to dehydrate. Transfer wings, antenna, and legs into clove oil before mounting; Canada balsam or Euparal is used for the final mountant (Fig. 1.1). The wings should be laid out beneath coverslip 1, the legs beneath 2, and the antenna and head beneath 3.

Transfer carcass (abdomen and thorax) into 10% potassium hydroxide (KOH) to clear musculature. Sodium hydroxide or warm KOH can be used and take care to control the degree of clearing.

When cleaning is achieved, transfer carcass to glacial acetic acid for a few minutes, gently removing any bubbles that may form within. Transfer to absolute alcohol or isopropanol to complete dehydration. Then we can use Euparal to cover the abdomen beneath coverslip 4 and the thorax beneath 5. The slide will be naturally dried and then can be preserved for more than 50−60 years.

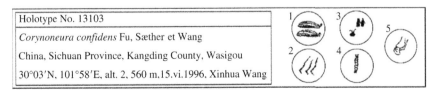

Holotype No. 13103

Corynoneura confidens Fu, Sæther et Wang

China, Sichuan Province, Kangding County, Wasigou

30°03′N, 101°58′E, alt. 2, 560 m.15.vi.1996, Xinhua Wang

Figure 1.1 Slide of specimen.

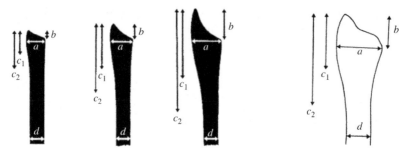

Figure 1.2 Measurements of hind tibia in *Corynoneura* (From Schlee, D. (1968b). Vergleichende Merkmalsanalyse zur Morphologie und Phylogenie der *Corynoneura*-Group (Diptera, Chironomidae). *Stuttgarter Beiträge zur Naturkunde, 180*, 1–150). a = Maximum width. b = Length of ventral elongation. c_1 = Length of strong broadened part, measured from apex. c_2 = Total length of broadening. d = Width of tibia basally to the apical broadening.

1.2.3 Observing and Drawing

The morphological nomenclature follows Sæther (1980). Measurements and ratios of hind tibia follow Schlee (1968b) (Fig. 1.2). Measurements are given as ranges followed by the mean when three or more specimens were measured.

The drawings of adults generally include the hypopygium or genitalia for all species, magnified to fit on an A4 page. The dorsal view is shown to the left and the ventral aspect and the apodemes to the right. In some species, drawings for the wing, antenna, hind tibial apex, tentorium are also given. The drawings of pupa generally include the tergites II—IX, sternites II—IX, and anal segment. The drawings of larva generally include the mandible, mentum, antenna, subbasal seta, of posterior parapod.

1.3 TERMINOLOGY

The morphological terminology follows Sæther (1980). Measurements and ratios of hind tibia follow Schlee (1968b).

1.3.1 Larva

Total length	From anterior margin of head capsule to apex of posterior parapods. Measured in multiple measures and then added up as the larva often is strongly curved in preparation

(Continued)

Head capsule length	Length from anterior margin of head capsule to postoccipital margin
Postmentum length	Length from posterior margin of mentum to postoccipital margin
Antenna	Total length of segments
AR (antennal ratio)	Length of first antennal segment divided by the combined length of remaining segment
First segment width	At the broadest
Blades	Usually at apex of basal antennal segment, with a common base with accessory blade
Mentum	Usually toothed, sclerotized, double-walled medioventral plate of head capsule consisting of dorsomentum and ventromentum with the latter often expanded laterally into ventromental plates
Postmentum	Mentum plus submentum or according to present terminology, nearly all of sclerotized medioventral area of head capsule from apex of mentum to posterior tentorial pits
Mandible length	From posterior base to anterior apex in a longitudinal plane
Mentum width	Width in vertical plan between the sclerotized base of most lateral teeth
Posterior parapod length	From base to apex

1.3.2 Pupa

Total length	From apex of frontal apotome to most posterior part of anal lobes
Anal lobe	Lateral paratergites of anal segment
Anal macrosetae	Usually strong mostly apical, lateral, often hook-shaped seta on each anal lobe
Anterior precorneal setae (PcS_1)	Median setae (or group of setae) on each side of antepronotum
D-setae	Dorsal setae excluding O-setae, normally 5 pairs on tergites II–VII, numbered D_1–D_5
Hooklets	Recurved spines posteriorally on tergite II and sometimes on other conjunctives
Lateral antepronotals	Lateral group of setae on each side of antepronotum
O-setae	Minute setae on extreme anterior margins of tergites and sternites
Pearl row	Row or row of small, blunt tubercles, sometimes with apical pore, along margin of wing sheath

(Continued)

Prealars	Setae in front of wing sheath
Shagreen	Pattern of spinules or minute tubercles on abdominal segments

1.3.3 Adult

Total length	Length of thorax plus length of abdomen. Thorax is measured from anterior apex of scutum to posterior margin of postnotum
Antennal segments	Total length
Cibarial pump	Prominent, well-sclerotized internal rectangular structure of the feeding apparatus forming the floor and partly the roof of the cibarial cavity
Tentorium length	Total longitudinal length
Stipes length	Total longitudinal length
Palpomere lengths	Total length
Hind tibial measurements	In accordance with Schlee (1968b): a = Maximum width. b = Length of ventral elongation. c_1 = Length of strong broadened part, measured from apex. c_2 = Total length of broadening. d = Width of tibia basally to the apical broadening
Wing length	Measured from arculus to apex of wing
Costa length	Measured from arculus to apex of clavus
Media length	Measured from arculus to base of media 1 + 2
Cubitus length	Measured from arculus to cubital fork in a horizontal plane of the wing
Postcubitus length	Measured from arculus to end of vein in a horizontal plane of the wing
Anapleural suture	Total length
Preepisternum width	Measured in a plane parallel to the anapleural suture
Gonostylus length	Length from base to apex
Gonocoxite length	Measured in a longitudinal plan from caudal apex to anterior base
Phallapodeme length	Total length along curve, if bent
Transverse sternapodeme width	Total length along the horizontal plane
Lateral length of Sternapodeme	Length from junction with transverse sternapodeme to caudal apex of coxosternapodeme in a vertical plane
Megaseta	Total length
AR (antennal ratio)	Ratio of length of apical elongated flagellomere plus any flagellomeres distal to it divided by combined length of the more basal flagellomeres
Palpomere 5/3 ratio	Length of palpomere 5 divided by palpomere 3

(Continued)

VR (venarum ratio)	Ratio of length of Cu to length of M
LR (leg ratio)	Ratio of metatarsus to tibia
BR (bristle ratio)	Ratio of longest seta of ta_1 divided by minimum width of ta_1 measured one third from apex
BV (beinverhältnisse)	Combined length of femur, tibia and basitarsus divided by combined length of tarsomeres 2−5
HR (hypopygium ratio)	Ratio of length of gonocoxite to length of gonostylus
HV (hypopygium value)	Ratio of total length to length of gonostylus times 10

1.3.4 Female

Coxosternapodeme	Apodeme of female coxosternite IX nearly or weakly connected to ramus of gonapophysis IX anteriorally and caudolaterally attached to gonocoxite IX
Ramus	Apodeme in oviduct of female genitalia
Length of notum	Length from anterior apex of notum to base
Length of gonapophysis IX	Length from anterior apex of notum posterior base of ramus
Length of seminal capsule	Total length
Width of seminal capsule	Total length
Length of cercus	Total length in a longitudinal plane

1.3.5 Counts

Numbers were counted on the following parts of the specimens.

Imagines: Outer vertical; Dorsocentrals; Prealars; Antepronotals; Scutellars; Tergite and sternite setae; Costal margin setae; Short strong setae on legs; Hind tibial comb setae.

Males: Tergite IX setae; Laterosternite IX setae; Gonocoxite setae.

Females: Tergite IX setae; Gonocoxite IX setae.

Larva: Anal setae.

Pupa: Number of caudal hooklets on abdominal segments; Anal lobe taenia.

1.4 DIAGNOSTIC CHARACTERS OF *CORYNONEURA* WINNERTZ

Generic diagnoses for the male imago was given by Cranston et al. (1989) for the pupa by Coffman, Cranston, Oliver, and Sæther (1986) and for the larvae by Cranston, Oliver, and Sæther (1983). Based on materials from China and references of Boesel and Winner (1980), Brundin (1949), Freeman (1953), Hazra, Nath, and Chaudhuri (2003), Hirvenoja and Hirvenoja (1988), Makarchenko and Makarchenko (2006b), Roback (1957a), Sasa, Kitami, and Suzuki (1999), Sasa and Suzuki (2000b), Sasa, Suzuki, and Sakai (1998), Sublette and Sasa (1994), Singh and Maheshwari (1987), and Tokunaga (1936), the generic diagnosis of the males was corrected or expanded by Fu et al. (2009).

1.4.1 Larva (Figs. 1.3 and 1.4)

Compterosmittia Sæther, 1981; 20; Sæther, 1982.

Small larvae, no more than 3 mm long. Head capsule with surface sculpturing.

Antenna. Subequal to, or longer than head capsule; 4 segmented; segment 3 longer than second; segment 4 minute; segments 2 and 3 frequently darkened. Ring organ distinct with median spine. Blade narrow and weak, shorter than flagellum. Lauterborn organs indistinct or absent.

Labrum. All labral setae weak; S-setae simple; strongly developed setae arising from socket may be either sternite III or sternite I setae. Pecten epipharyngis consisting of 3 spines but with first chaetulae laterales of same size and shape, resembling 5 spines. Two outer chaetulae laterales strongly modified; anterior pair large and strongly plumose; inner, posterior pair large and serrate. Chaetulae basales apparently absent. Premandible with up to 12 minute teeth and strongly developed but translucent brush.

Mandible. Apical tooth ending dorsally of line of 4 inner teeth, scarcely longer than any of inner teeth. Seta subdentalis absent. Seta interna apparently absent in some species, translucent in others.

Mentum. Triangular-shaped with either 2 or 3 median teeth and 5 pairs of lateral teeth. Ventromental plate weak; beard absent.

Maxilla. Small and difficult to interpret. Palp with normal sensillae. Setae maxillaris numerous and apparently all simple. Dorsal side of palpiger with group of fine translucent setae resembling setae interna of overlying mandible.

Body. With 2 thoracic segments; meso-and metathoracic segments fused. Anterior parapods elongate, longer than usual in Orthocladiinae. Posterior parapods also elongate, both parapods bearing claws. Procercus small, bearing 4 apical setae. Anal tubules shorter than posterior parapods, tapering to point. Variably developed seta arising from ventral basal side

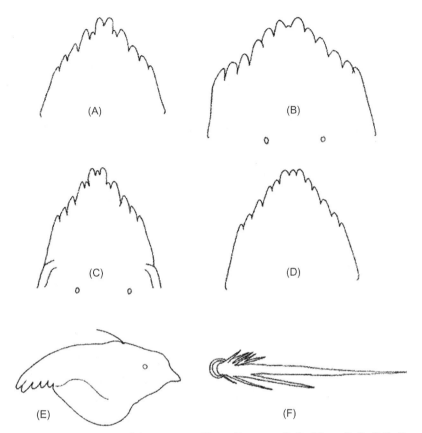

Figure 1.3 (A–D) Larva of *Corynoneura* (From Cranston, P. S., Oliver, D. R., & Sæther, O. A. (1983). The larvae of Orthocladiinae (Diptera: Chironomidae) of the Holarctic region. Keys and diagnoses. In: T. Wiederholm (Ed.), *Chironomidae of the Holarctic Region. Keys and diagnoses. Part 1. Larvae.* Entomol. Scand. Suppl. (*Vol. 19*, pp. 149–292)). (E) Mentum. (F) Mandible. Basiventral spine of posterior parapod.

of posterior parapod, this seta is usually plumose, sometimes dark brown and may be up to half length of posterior parapod.

1.4.2 Pupa (Fig. 1.5)

Corynoneura Winnertz, 1846; 12.

Small pupae, up to 3 mm long, translucent or weakly infuscate.

Cephalothorax. Frontal setae short, arising directly from frontal apotome between antennal bases. Cephalic area smooth or weakly rugulose, without tubercles or warts. Ocular field with 2 fine postorbital setae and without vertical setae.

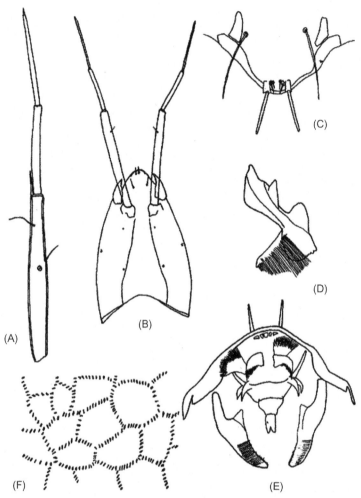

Figure 1.4 Larva of *Corynoneura* (From Cranston, P. S., Oliver, D. R., & Sæther, O. A. (1983). The larvae of Orthocladiinae (Diptera: Chironomidae) of the Holarctic region. Keys and diagnoses. In: T. Wiederholm (Ed.), *Chironomidae of the Holarctic Region. Keys and diagnoses. Part 1. Larvae.* Entomol. Scand. Suppl. (*Vol. 19*, pp. 149−292)) (A) Antenna. (B) Head capsule, dorsal view. (C) Labrum. (D) Premandible. (E) Palatum. (F) Dorsal head capsule sculpturing.

Thorax. With 2 median antepronotals. Thoracic horn absent. Three weak precorneals in a row, 4 weak dorsocentrals, the anterior most displaced ventrally. Thorax with rugulose area near suture. Wing sheaths apically with 1−4 or 5 rows of pearls and with nose sited basally on anterior margin. Leg sheaths lying between wing sheaths, all straight until distally

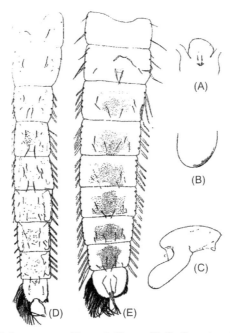

Figure 1.5 Pupa of *Corynoneura* (From Coffman, W. P., Cranston, P. S., Oliver, D. R., & Sæther, O. A. (1986). The pupae of Orthocladiinae (Diptera: Chironomidae) of the Holarctic region. Keys and diagnoses. In: T. Wiederholm (Ed.), *Chironomidae of the Holarctic Region. Keys and Diagnoses. Part 2. Pupae*. Entomol. Scand. Suppl. (Vol. 28, pp. 1–482)) (A) Frontal apotome. (B) Tip of wing sheath. (C) Thorax, lateral view. (D) Tergites. (E) Sternites.

curved around apical margin of wing sheath; posterior leg sheath appreciably longer than anterior leg sheaths.

Abdomen. Shagreen of small points and spinules on tergites (II) III–IX and on sternites (I, II) III–VII. Conjunctives of some tergites and sternites may have single transverse row of some stronger hooklets. Pedes spurii A and B absent. Apophyses weakly delimited or absent.

Abdominal setation. Segment I with 3 D-setae, 0–2 V-setae, and 1–2 L-setae; the posterior L-seta, when present, lamelliform. Segment II with 4 D-setae, 3 V-setae, and 2–4 L-setae of which posterior 1–3 may be lamelliform. Segments III–VII with 5 D-setae, 4 V-setae, and 4 lamelliform L-setae. Segment VIII with 2 D-setae, 3 V-setae, and 4 lamelliform L-setae. O-setae, when present, with 1 pair of dorsal and 1 pair of ventral setae (Coffman, 1979), possibly sometimes absent. Some D-setae and some V-setae may be finely lamelliform.

Anal lobe with fringe of filaments set slightly internal to anal lobe margin. Posterior margin of anal lobes with 1 lamelliform seta and 3 lamelliform macrosetae mounted on tubercles sited at 2/3 length of anal lobe, dorsal to bases of anal lobe fringe setae. Genital sacs of male and female not extending to apex of anal lobe.

1.4.3 Male Adult (Fig. 1.6)

Corynoneura Winnertz, 1846; 12; Schlee, 1968b; Hirvenoja & Hirvenoja, 1988; Fu et al., 2009.

Small species, with wing length 0.35−1.8 mm.

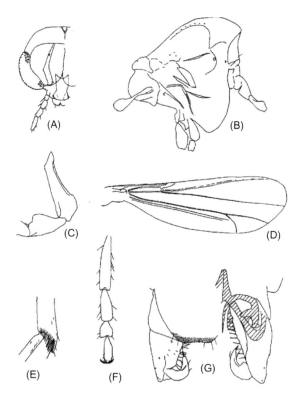

Figure 1.6 Male adult of *Corynoneura* (From Cranston, P. S., Oliver, D. R., & Sæther, O. A. (1989). The adult males of Orthocladiinae (Diptera: Chironomidae) of the Holarctic region. Keys and diagnoses. In: T. Wiederholm (Ed.), *Chironomidae of the Holarctic Region. Part 3. Adult males.* Entomol. Scand. Suppl. (Vol. 34, pp. 165−352)). (A) Head. (B) Thorax. (C) Fore trochanter. (D) Wing. (E) Hind tibial comb. (F) Hind tarsomeres 2−5. (G) Hypopygium.

Antenna. With 6—13 flagellomeres; groove beginning at flagellomere 2; plume absent or moderately developed; apex often with rosette of short hairs; slender sensilla chaetica on flagellomeres (1) 2 and 3. Scape large relative to size of head. Antennal ratio 0.16—1.42.

Head. Eye small, usually bare or rarely pubescent, without dorsomedial extension. Temporal setae absent. Tentorium long and narrow, tapering to blunt point; cibarial pump broad, with variable sized cornua. Palp short, each segment longer than that preceding; often with 1 sensilla clavata on segment 3, without pit.

Thorax. Antepronotum moderately developed, with lobes narrowing dorsally, narrowly in contact anterior to nonextended scutum. Acrostichals absent; few uniserial dorsocentrals, prealars and scutellars present.

Wing. Membrane without seta, with fine punctuation. Anal lobe absent, wing more or less cuneiform. R_1 and R_{4+5} apically fused with costa, forming thick clavus which terminates between 1/4 and 1/2 wing length; weak false vein may continue from RM towards apex of wing, beneath and parallel with anterior margin; Cu_1 weakly sinuous or nearly straight; FCu far distal to RM; postcubitus and anal vein scarcely extending beyond FCu. Veins bare. Squama bare.

Legs. Fore leg trochanter with distinct dorsal keel. Hind tibial apex broadened, sometimes with 1 seta from comb developed as hook. Tarsomere 4 on all legs weakly cordiform, about 1/2 length of tarsomere 5. Tarsal pseudospurs, sensilla chaetica and pulvilli absent.

Abdomen. All setation strongly reduced. Tergite I bare, tergites II—VIII often with 1 or 2 lateral setae and 1 median seta, arising from pale areas. Sternites bare.

Hypopygium. Tergite IX often large, covering much of gonocoxites, with straight or bilobed posterior margin; anal point usually absent or rarely present. Apodemes strongly sclerotized and specifically distinctive. Sternapodeme inverted V- or U-shaped, medially broadened, without oral projections. Phallapodeme relatively straight to strongly curved, sometimes curving anterior of coxapodeme/sternapodeme junction; aedeagal lobe strong to apparently absent. Virga absent. Superior volsella well developed, tongue-shaped to gently curved or absent; inferior volsella variably shaped, sometimes absent. Gonostylus boat-shaped to strongly curved, with medial or baso-medial rounded crista dorsalis in some species; megaseta short and broad to long and fine.

1.5 PHYLOGENY

Schlee (1968a, 1968b) undertook a morphological and phylogenetic study of the *Corynoneura* group.

Adult males of the genus *Corynoneura* have many morphological types (Table 1.1), for example, the shape of male antennal apex, including rounded or pointed; hind tibial apex; the shape and the position of superior volsella, inferior volsella; the shape of phallapodeme, the position relation between phallapodeme, and lateral sternapodeme and so on. In general, the changes between these morphological types are so obvious (discontinuous variation) in most of the species that the morphological classification within the genus *Corynoneura* based on adult males are reliable. If the key includes large numbers of the species, more characters have to be used to distinguish them. But sometimes these morphological types to the same structure are hard to describe when the key is made. For example, phallapodeme have four discontinuous morphological types (Fig. 1.7), type 1 and 2 are scalpel-like, slightly bent, relatively short; type 3 and 4 are not scalpel-like, and obvious curved, relatively long.

Hind tibial apex have three discontinuous morphological types (Fig. 1.8), type 1 and 2 are slightly thickened and elongate, type 3 is distinctly thickened and elongate.

Transverse sternapodeme have four discontinuous morphological types (Fig. 1.9), type 1 developed, straight and wide; type 2/3/4 undeveloped, rounded and narrow, or absent.

Although many species of *Corynoneura* need to be redescribed, a parsimony analysis is warranted in order to obtain at least a preliminary estimation of the phylogenetic relationships, to test the validity of sufficiently described species, and to point out future areas of investigations. This involved the compilation of a data matrix for 44 characters (Table 1.2) in 58 species of *Corynoneura* and for the genera *Corynoneurella*, *Onconeura*, *Tempisquitoneura*, and *Thienemanniella* used combined as an outgroup (Table 1.3). Although the immatures and females show several phylogenetically important characters, they are unknown for most of the species and could not be included in the analysis. However, a few characters synapomorphic for the genus are included. Data were analyzed under parsimony with PAUP 4.0b.10 (Swofford, 2002) operating on a Macintosh G5, and employing 1000 random addition sequence replicates. All characters are unordered.

Table 1.1 Morphological types of adult males of the genus *Corynoneura*

Characters	Type 1	Type 2	Type 3	Type 4
Sensilla chaetica of male antennal apex				
Shape of male antennal apex				
Cibarial pump				
Hind tibial apex				

(Continued)

Table 1.1 (Continued)

Characters	Type 1	Type 2	Type 3	Type 4
Margin of tergite IX				
Transverse sternapodeme				
Phallapodeme				

The position of sternapodeme and phallapodeme

Gonostylus

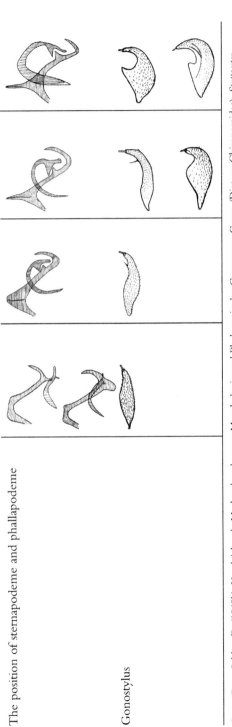

Source: From Schlee, D. (1968b). Vergleichende Merkmalsanalyse zur Morphologie und Phylogenie der *Corynoneura*-Group (Diptera, Chironomidae). *Stuttgarter Beiträge zur Naturkunde, 180,* 1–150.

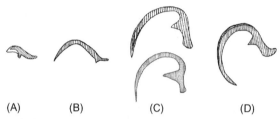

(A)   (B)   (C)   (D)

Figure 1.7 Morphological types of phallapodeme (From Schlee, D. (1968b). Vergleichende Merkmalsanalyse zur Morphologie und Phylogenie der *Corynoneura*-Group (Diptera, Chironomidae). *Stuttgarter Beiträge zur Naturkunde, 180,* 1–150) (A) Type 1. (B) Type 2. (C) Type 3. (D) Type 4.

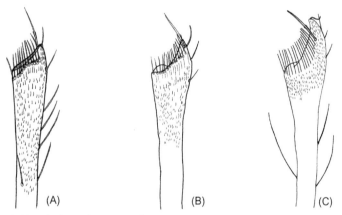

(A)   (B)   (C)

Figure 1.8 Morphological types of hind tibial apex (From Schlee, D. (1968b). Vergleichende Merkmalsanalyse zur Morphologie und Phylogenie der *Corynoneura*-Group (Diptera, Chironomidae). *Stuttgarter Beiträge zur Naturkunde, 180,* 1–150) (A) Type 1. (B) Type 2. (C) Type 3.

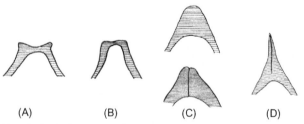

(A)   (B)   (C)   (D)

Figure 1.9 Morphological types of transverse sternapodeme (From Schlee, D. (1968b). Vergleichende Merkmalsanalyse zur Morphologie und Phylogenie der *Corynoneura*-Group (Diptera, Chironomidae). *Stuttgarter Beiträge zur Naturkunde, 180,* 1–150) (A) Type 1. (B) Type 2. (C) Type 3. (D) Type 4.

Table 1.2 Morphological characters and states used in phylogenetic analysis

No.	Morphological characters	States
1	Wing length	(0) >1.00; (1) 0.60−1.00; (2) <0.60
2	Antennal ratio	(0) >1.0; (1) 0.4−1.0; (2) <0.4
3	Number of flagellomeres	(0) 12−13; (1) 11; (2) 9−10; (3) 6−8
4	Sensilla chaetica of male antennal apex	(0) apical; (1) preapical
5	Palpomeres	(0) five palpomeres of normal length; (1) four or less palpomeres or palpomeres strongly reduced in length
6	Sensilla clavata of palpomeres	(0) palpomere 3 with 4−5 sensilla; (1) with less than 4
7	Cibarial pump	(0) not or only slightly incurved; (1) distinctly incurved, often stepwise
8	Temporals	(0) present; (1) absent
9	Dorsocentrals	(0) >5; (1) < 6
10	Prealars	(0) ≥ 3 or more; (1) <3
11	Supraalar	(0) present; (1) absent
12	Wing width/wing length	(0) <0.40; (1) >0.40
13	Clavus length/wing length	(0) >0.30; (1) <0.30
14	Cu/wing length	(0) 0.45−0.50; (1) 0.50−0.57; (2) 0.58−0.71
15	VR	(0) <3.0; (1) 3.0−4.9; (2) >5.0
16	Anal lobe	(0) indicated; (1) absent, wing cuneiform
17	Leg ratio of male (LR_1)	(0) >0.70; (1) 0.55−0.70; (2) <0.55
18	Sensilla chaetica on tarsi	(0) sometimes present; (1) absent
19	Pseudospurs	(0) present and mostly curved; (1) absent or weaker than stiff setae on ventral margin
20	Stiff setae on ventral margin of ta_1	(0) <5; (1) >4
21	Hind tibial apex	(0) at most slightly thickened and elongate [a, b, c_1 and c_2 in Schlee (1968b, Figs. 131−132) as 1.3−1.5, 0.2−1.0, 1−3.5, and 3.5−5 respectively]; (1) at distinctly thickened and elongate [a, b, c_1 and c_2 in Schlee (1968b, Figs. 131−132) as 1.6−2.2, 1.2−2.0, 3−5, and 5−7 respectively]
22	Apical setae of hind tibia	(0) well developed, nearly as long as tibial diameter; (1) shortened, at most half as long as tibial diameter

(Continued)

Table 1.2 (Continued)

No.	Morphological characters	States
23	Apical setae of hind tibia	(0) nearly straight; (1) distinctly curved, S-shaped or angled
24	Anal point	(0) absent; (1) represented by hump-like extension of tergite
25	Tergite IX	(0) with seta; (1) without
26	Tergite IX	(0) evenly curved without posterior humps without indication of median mounds; (1) medially incurved often with indication of pair of low caudal mounds; (2) with posterior humps carrying setae
27	Transverse sternapodeme	(0) present and wide; (1) narrow; (2) absent, lateral sternapodemes meeting in sharp point
28	Lateral sternapodeme	(0) attachment point with phallapodeme placed caudal and caudally directed; (1) attachment point placed caudal and ventrally directed; (2) attachment point placed in caudal third of lateral sternapodeme and directed caudally
29	Phallapodeme	(0) slightly bent, relatively short; (1) strongly curved, long
30	Phallapodeme	(0) not scalpel-like; (1) scalpel-like
31	Phallapodeme	(0) with projection for joint with sternapodeme placed pre-lateral; (1) with lateral apex bifid and enclosing knob joint with sternapodeme
32	Superior volsella	(0) present; (1) absent
33	Superior volsella	(0) joined medially or absent; (1) separate
34	Superior volsella	(0) low or narrow and mostly triangular or absent; (1) conspicuously projecting, often digitiform
35	Inferior volsella	(0) present; (1) absent or barely indicated as long low inner margin of gonocoxite
36	Inferior volsella	(0) well developed, rectangular to triangular; (1) narrow, digitiform to spatulate
37	Gonostylus	(0) straight, nearly equally wide to apex; (1) slightly curved, gradually narrowed near apex, (2) strongly curved
38	Gonostylus	(0) without basal lobe on inner margin; (1) with
39	Megaseta	(0) short and broad; (1) long and fine

(Continued)

Table 1.2 (Continued)

No.	Morphological characters	States
40	Wing sheath	(0) without pearls; (1) with
41	Caudal hooklets on tergal conjunctives	(0) absent or minute; (1) conspicuous
42	Anal macrosetae	(0) three flattened and subequal; (1) absent, 3 bristlelike, or one only
43	Larval antenna	(0) with five segments, no more than 3/4 as long as head; (1) with four segments, subequal to or longer than head
44	Lauterborn organs	(0) moderately large to well developed; (1) weak or absent

Species in which less than 60% of the character alternatives are unknown were excluded from the analyses. This includes *Corynoneura australiensis* Freeman, *Corynoneura centromedia* Hazra, Nath *et* Chaudhuri, *Corynoneura chandertali* Singh *et* Maheshwari, *Corynoneura cristata* Freeman, *C. diara* Roback, *Corynoneura elongata* Freeman, *Corynoneura hirvenojai* Sublette *et* Sasa, *Corynoneura incidera* Hazra, Nath *et* Chaudhuri, *Corynoneura kedrovaya* Makarchenko *et* Makarchenko, *Corynoneura magna* Brundin, *Corynoneura marina* Kieffer, *Corynoneura nasuticeps* Hazra, Nath *et* Chaudhuri, *C. oxfordana* Boesel *et* Winner, *Corynoneura secunda* Makarchenko *et* Makarchenko, and *C. taris* Roback. The constraint that *Corynoneura* is monophyletic was used. Otherwise *Corynoneurella* sometimes is inside the genus.

Fifty-four trees were obtained, each with 328 steps, consistency index (CI) of 0.42, homoplasy index (HI) of 0.83, retention index (RI) of 0.56 and a rescaled consistency index (RC) of 0.24. The strict consensus tree is shown in Fig. 1.10A. When the trees were exposed to successive reweighting according to RC, 165 trees each with 341 steps (when the weights were reset to one), CI of 0.57, HI of 0.57, RI of 0.80, and RC of 0.45 were obtained after two reweightings (Fig. 1.10B).

The characters concerning the apodemes, volsella, and gonostylus (27−39) obviously were more significant than other characters and had been given a weight of 10 in another analysis. This resulted in 396 trees each with 357 steps (when the weights were reset to one), CI of 0.39, HI of 0.80, RI of 0.69, and RC of 0.27. The strict consensus tree is shown in Fig. 1.11A. When the trees were exposed to successive reweighting

Table 1.3 Morphological characters states matrix of *Corynoneura* Kieffer and in the genera *Corynoneurella, Onconeura, Tempisquitoneura,* and *Thienemanniella*

Characters	1	2	3	4	5	6	7	8	9	10	11	12	13	14	15	16	17	18	19	20	21	22
arctica	0	A	2	1	0	1	1	1	1	1	?	?	0	2	0	1	1	1	1	1	1	0
australiensis	1	2	2	0	?	?	?	?	?	?	?	?	1	?	?	1	2	?	?	?	?	?
brundini	0	2	1	1	0	?	?	?	?	?	?	?	0	1	0	1	2	?	?	?	1	1
carinata	A	0	3	1	1	?	0	1	1	?	?	1	?	?	?	1	2	1	0	?	1	1
carriana	1	1	1	1	0	1	1	?	?	?	?	A	0	2	0	1	2	1	?	?	0	1
celeripes	0	1	0	0	0	?	0	?	?	?	?	0	1	2	?	1	?	?	?	?	0	0
celtica	?	C	1	0	1	1	1	1	1	0	0	0	?	2	1	1	1	1	0	0	1	1
centromedia	0	1	3	?	1	1	?	?	A	1	?	?	0	2	1	1	1	1	?	?	?	0
chandertali	1	0	3	?	0	?	?	?	?	?	?	?	1	2	0	1	?	?	?	?	?	?
confidens	0	2	2	0	0	?	0	1	1	0	0	1	?	2	1	1	2	1	0	1	1	0
coronata	1	1	0	0	?	?	1	?	1	1	?	0	0	2	?	1	?	1	?	0	1	0
cristata	?	2	2	0	?	?	?	?	?	?	?	?	?	?	1	1	1	?	?	?	?	?
cuspis	1	1	1	1	1	1	0	1	?	?	?	0	?	2	1	1	1	?	0	?	?	?
cylindricauda	1	1	1	0	0	?	0	?	1	0	0	1	0	2	1	1	2	1	1	0	1	1
dewulfi	0	2	2	0	1	?	?	1	0	1	?	1	1	1	0	1	1	1	?	1	1	1
diara	1	1	1	0	?	?	?	?	?	?	?	?	?	2	?	1	C	?	?	?	?	?
doriceni	?	1	0	1	0	?	1	1	0	1	0	?	1	?	1	1	C	?	1	?	?	0
edwardsi	C	1	1	1	0	?	?	?	A	1	?	0	?	2	0	1	1	1	?	0	1	1
elongata	1	2	2	1	?	?	1	1	?	A	0	?	1	1	?	1	2	?	0	?	?	?
ferelobata	0	2	2	0	0	1	0	1	A	A	?	1	1	2	1	1	2	1	1	0	1	0
fittkaui	1	C	0	0	0	1	1	?	1	1	1	0	0	2	?	1	C	1	1	0	0	0
fujiundecima	0	1	0	0	0	?	1	1	1	?	?	0	?	2	0	1	?	?	?	0	?	0
Characters	**1**	**2**	**3**	**4**	**5**	**6**	**7**	**8**	**9**	**10**	**11**	**12**	**13**	**14**	**15**	**16**	**17**	**18**	**19**	**20**	**21**	**22**
gratias	0	2	3	1	0	1	1	1	0	1	1	0	0	2	A	1	1	1	1	1	1	0
gynocera	1	1	C	1	0	1	?	?	A	1	?	1	0	1	0	1	2	?	?	?	?	?
hirvenojai	1	A	3	?	1	?	?	?	1	1	?	?	?	?	0	1	2	?	?	?	?	0
inanapequea	1	1	2	1	0	1	?	?	1	?	?	0	1	2	1	1	?	?	?	?	?	?

Taxon	1	2	3	4	5	6	7	8	9	10	11	12	13	14	15	16	17	18	19	20	21	22
incidera	2	2	3	?	?	?	?	?	1	0	?	?	?	?	?	1	2	?	?	?	?	0
inefligiata	1	1	2	0	0	1	0	1	1	?	?	1	?	2	2	1	2	1	0	0	1	0
isigaheia	2	2	3	0	0	1	?	1	0	1	?	1	1	1	1	1	2	?	?	1	1	0
kedrovaya	?	1	D	0	1	?	?	?	1	1	?	?	?	?	?	1	2	2	?	?	?	?
kibunelata	1	2	2	0	?	?	1	?	1	?	?	0	1	2	2	1	2	1	?	?	?	?
kibunespinosa	1	2	2	0	0	0	1	?	1	?	?	0	1	1	0	1	?	?	?	?	?	?
kisogawa	1	2	3	0	0	1	1	1	0	1	1	1	1	2	1	1	2	?	1	1	1	0
korema	2	C	A	0	0	1	0	1	1	1	1	1	1	2	1	1	2	1	1	1	1	0
lacustris	A	0	3	1	1	0	0	1	1	1	0	A	0	2	2	1	C	1	1	1	1	1
lahuli	0	2	3	0	0	1	0	1	1	1	0	0	?	C	1	1	1	1	1	1	1	0
latusatra	1	C	C	1	0	?	1	1	1	1	0	1	?	2	1	1	2	1	1	1	1	0
lobata	A	1	1	0	1	1	A	1	1	A	?	A	?	2	A	1	1	1	0	0	1	1
longipennis	0	2	2	1	0	0	?	?	1	0	0	0	?	1	A	1	1	1	?	?	?	?
macdonaldi	1	0	0	0	0	?	1	1	?	?	?	?	A	?	1	1	2	?	?	?	?	1
magna	0	1	1	?	0	?	?	?	1	1	?	?	1	?	1	?	1	1	0	0	1	?
marina	?	2	2	?	?	?	?	1	?	1	0	0	A	2	?	1	?	?	?	?	?	?
medicina	1	1	3	1	?	?	0	?	1	1	?	?	?	?	?	?	?	?	?	?	?	?
nasuticeps	2	2	2	?	0	1	?	1	1	1	0	0	?	2	1	1	1	1	0	0	1	0
nankaiensis	A	2	3	1	0	1	1	1	0	1	1	1	1	?	1	1	2	?	?	?	?	0
Characters	1	2	3	4	5	6	7	8	9	10	11	12	13	14	15	16	17	18	19	20	21	22
oxfordana	A	1	3	0	0	?	?	?	?	?	?	A	?	2	0	1	1	1	1	1	1	0
prima	1	1	3	0	1	1	?	?	1	1	?	1	1	?	1	1	2	1	?	?	?	?
prominens	C	A	C	0	0	?	0	1	1	0	0	1	0	C	1	1	2	1	?	?	?	?
scutellata	A	A	2	1	1	1	1	1	1	1	0	1	?	2	1	1	1	1	1	0	1	0
secunda	1	2	3	0	0	?	?	?	1	1	?	1	1	?	0	1	2	?	?	?	?	1
seiryuresea	1	1	2	0	1	?	1	1	?	?	?	1	0	?	0	1	2	?	?	?	?	?
sorachibecea	0	1	1	1	1	?	1	1	?	?	?	?	?	?	0	?	?	?	?	?	?	?
Taris	1	2	1	0	?	?	?	?	?	?	?	1	1	?	?	1	?	?	?	?	1	?
tenuistyla	1	2	C	1	1	?	?	?	?	1	?	1	1	1	1	1	2	?	?	?	?	?

(Continued)

Table 1.3 (Continued)

Characters	1	2	3	4	5	6	7	8	9	10	11	12	13	14	15	16	17	18	19	20	21	22
tokarapequea	?	1	1	0	0	?	1	?	?	1	?	?	?	?	?	?	2	1	?	?	?	0
tokaraquerea	1	1	1	1	0	?	0	?	?	?	?	1	1	2	1	1	1	?	?	?	?	?
vitalis	0	2	0	?	?	?	?	?	?	?	?	0	1	2	1	1	2	1	?	?	?	?
yoshimurai	C	C	D	0	0	1	?	1	1	1	0	1	1	2	?	1	2	?	0	0	?	0
Corynoneurella	1	0	0	0	?	0	0	?	?	?	A	?	0	1	0	0	0	0	?	0	0	0
Onconeura	1	1	1	0	0	1	1	0	0	0	0	0	1	A	0	0	0	0	0	1	0	?
Tempisquitoneura	1	1	0	0	0	1	0	0	0	0	0	A	0	0	0	0	A	1	0	1	0	?
Thienemanniella	A	C	B	A	0	?	A	1	A	A	0	A	0	0	0	0	1	1	0	1	0	0
Characters	**23**	**24**	**25**	**26**	**27**	**28**	**29**	**30**	**31**	**32**	**33**	**34**	**35**	**36**	**37**	**38**	**39**	**40**	**41**	**42**	**43**	**44**
arctica	1	0	0	C	1	2	1	0	0	0	0	0	1	?	2	0	?	1	0	0	1	1
australiensis	?	0	?	?	1	1	1	0	0	?	?	?	0	0	0	1	1	1	0	0	1	1
brundini	0	0	0	C	2	2	1	0	0	0	1	0	1	0	2	0	1	1	0	0	1	1
carinata	0	0	0	0	2	2	1	0	0	1	0	0	1	?	2	1	1	1	0	0	1	1
carriana	0	0	0	C	2	A	0	1	0	0	1	A	1	?	2	0	A	1	0	0	1	1
celeripes	?	0	?	1	1	2	1	0	1	0	1	?	1	0	1	0	1	1	0	0	1	1
celtica	1	0	0	0	2	2	1	0	0	?	?	?	0	?	2	0	?	1	0	0	1	1
centromedia	0	0	0	0	2	2	1	0	0	?	?	?	0	1	1	0	?	1	0	0	1	1
chandertali	?	0	?	2	2	2	0	0	0	?	0	?	0	?	0	0	0	1	0	0	1	1
confidens	1	0	0	0	2	1	1	0	0	0	1	?	1	1	0	0	1	1	0	0	1	1
coronata	1	0	0	1	1	1	0	0	0	0	?	0	0	?	0	0	?	1	0	0	1	1
Characters	**23**	**24**	**25**	**26**	**27**	**28**	**29**	**30**	**31**	**32**	**33**	**34**	**35**	**36**	**37**	**38**	**39**	**40**	**41**	**42**	**43**	**44**
cristata	?	0	0	?	1	0	0	1	0	1	0	?	0	?	0	0	0	1	0	0	1	1
auspis	?	0	0	?	2	2	1	1	0	1	0	?	0	1	1	1	1	1	0	0	1	1
cylindricauda	1	1	1	1	1	0	0	0	0	0	0	1	0	0	0	0	1	1	0	0	1	1
dewulfi	1	0	1	C	2	2	1	1	0	0	0	0	0	0	0	0	1	1	0	0	1	1
diara	?	0	1	?	2	2	1	0	0	?	?	?	?	?	?	?	?	1	0	0	1	1
doriceni	0	0	?	?	1	1	1	0	0	?	?	0	0	0	0	0	?	?	0	0	1	1
eduardsi	0	0	0	C	2	2	1	0	0	0	1	1	1	?	1	0	1	1	0	0	1	1

	23	24	25	26	27	28	29	30	31	32	33	34	35	36	37	38	39	40	41	42	43	44
elongata	?	0	?	?	2	1	0	0	0	1	0	?	1	?	1	0	0	1	0	0	1	1
ferelobata	1	0	0	C	2	2	1	0	0	0	0	0	0	1	1	0	1	1	0	0	1	1
fittkaui	?	0	0	C	1	0	0	0	1	0	1	1	0	0	2	0	1	1	0	0	1	1
fujiundecima	1	0	0	2	1	0	0	1	0	0	0	0	0	0	0	0	1	1	0	0	1	1
gratias	0	0	0	2	1	1	1	0	0	0	1	0	1	?	2	1	1	1	0	0	1	1
gynocera	0	0	0	C	?	2	?	0	?	?	?	?	?	?	1	?	?	1	0	0	1	1
hirvenojai	?	0	?	0	1	?	?	?	?	?	?	?	?	0	1	0	?	1	0	0	1	1
inauapequea	?	0	1	0	2	2	0	1	0	?	1	0	0	1	1	1	0	1	0	0	1	1
incidera	?	0	?	1	1	1	1	0	0	0	?	?	0	?	1	0	0	1	0	0	1	1
inefligiata	1	0	1	?	1	?	0	1	0	?	?	?	1	0	1	0	?	1	0	0	1	1
isigaheia	0	0	?	0	2	1	1	0	0	0	0	0	0	?	1	0	1	1	0	0	1	1
kedrovaya	?	0	?	2	1	?	0	1	?	?	?	?	1	0	1	1	?	1	0	0	1	1
kibumelata	1	0	1	0	2	1	1	0	0	?	0	?	0	0	1	1	1	1	0	0	1	1
kibunespinosa	?	0	0	0	1	?	0	1	0	0	0	?	?	1	1	0	0	1	0	0	1	1
kisogawa	0	0	1	2	2	2	0	0	0	0	0	0	0	?	1	0	1	1	0	0	1	1
korema	1	0	0	0	1	0	1	0	0	0	1	A	0	1	1	0	0	1	0	0	1	1
lacustris	1	0	1	C	2	2	0	0	0	0	1	1	0	0	1	0	1	1	0	0	1	1
laluli	0	1	0	0	2	2	1	0	0	0	1	0	0	?	1	1	1	1	0	0	1	1
latusatra	1	1	0	C	2	2	1	0	0	0	0	0	1	?	2	0	1	1	0	0	1	1
Characters	23	24	25	26	27	28	29	30	31	32	33	34	35	36	37	38	39	40	41	42	43	44
lobata	?	0	1	1	2	2	1	0	0	0	1	1	0	1	2	0	1	1	0	0	1	1
longipennis	1	?	?	1	2	2	0	1	0	1	0	0	0	1	2	1	1	1	0	0	1	1
macdonaldi	?	0	0	C	1	1	?	?	?	0	0	?	?	0	1	0	0	1	0	0	1	1
magna	?	?	0	1	?	?	?	?	?	?	?	?	?	?	0	0	1	1	0	0	1	1
marina	0	0	0	?	?	?	?	0	0	?	?	0	1	?	?	?	?	1	0	0	1	1
medicina	?	0	?	C	1	2	1	0	0	?	?	?	0	1	1	1	0	1	0	0	1	1
nasuticeps	?	1	?	0	2	2	1	0	0	0	?	0	0	1	1	0	?	1	0	0	1	1
nankaiensis	1	?	0	C	2	2	1	0	0	?	0	?	0	0	2	1	1	?	0	0	1	1
oxfordana	?	0	?	?	?	?	?	?	?	?	?	?	0	0	2	0	?	1	0	0	1	1

(Continued)

Table 1.3 (Continued)

Characters	1	2	3	4	5	6	7	8	9	10	11	12	13	14	15	16	17	18	19	20	21	22
prima	?	0	?	?	1	0	0	1	0	0	0	0	0	0	1	0	1	1	0	0	1	1
prominens	1	1	0	2	1	0	0	1	0	0	1	1	0	0	1	1	1	1	0	0	1	1
scutellata	1	0	0	1	2	2	1	0	0	0	1	A	0	1	2	1	1	1	0	0	1	1
secunda	?	0	?	C	2	2	1	0	?	?	?	?	1	?	1	0	?	1	0	0	1	1
seiryuresea	?	0	0	C	0	0	0	1	0	0	0	?	0	0	1	0	0	1	0	0	1	1
sorachibecea	0	0	1	C	2	1	1	0	0	0	0	0	1	?	1	1	0	1	0	0	1	1
Taris	?	?	?	?	?	?	?	?	?	1	0	0	1	?	2	1	1	1	0	0	1	1
tenuistyla	?	0	?	0	0	0	0	1	0	0	1	0	0	1	1	0	0	1	0	0	1	1
tokarapequea	?	0	1	C	1	0	0	1	0	0	1	0	0	0	1	0	1	1	0	0	1	1
tokaraquerea	?	0	1	0	2	2	1	0	1	?	0	0	1	0	1	1	1	1	0	0	1	1
vittalis	?	0	0	C	0	2	1	0	?	?	?	0	0	?	1	0	?	1	0	0	1	1
yoshimurai	1	0	0	0	0	1	1	0	0	0	1	1	0	0	1	0	0	1	0	0	1	1
Corynoneurella	?	0	0	C	0	0	0	0	0	0	0	0	0	0	C	0	1	0	1	0	?	?
Onconeura	?	0	1	0	0	0	1	0	0	0	0	1	0	0	0	0	0	1	1	1	1	1
Tempisquitoneura	0	1	1	1	0	0	1	0	1	0	?	0	0	0	0	0	0	A	1	1	0	0
Thienemanniella	0	0	0	0	A	0	A	0	0	0	A	A	0	A	0	0	A	0	0	0	0	0

Note: A = 0 & 1, B = 0 & 1 & 2, C = 1 & 2, D = 2 & 3, ? = Unknown.

Strict

Strict

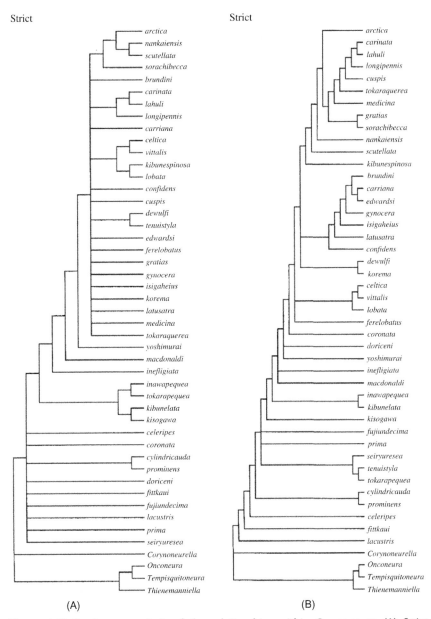

(A)   (B)

Figure 1.10 Parsimony analysis of the relationships within *Corynoneura*. (A) Strict consensus tree. (B) Strict consensus tree after successive reweighting according to the rescaled consistency index.

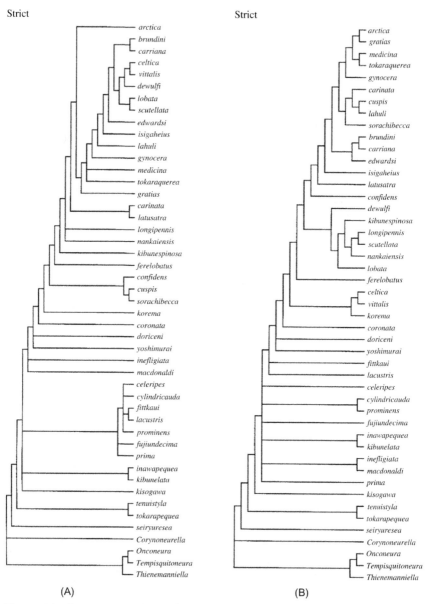

Figure 1.11 Parsimony analysis of the relationships within *Corynoneura*. (A) Strict consensus tree when trends 27–39 are weighted. (B) Strict consensus tree after successive reweighting according to the rescaled consistency index.

according to RC 31 trees each with 350 steps (when the weights were reset to one), CI of 0.58, HI of 0.54, RI of 0.82, and RC of 0.48 were obtained after six reweightings (Fig. 1.11B).

The genus based on Fig. 1.11B can be divided into three main groups, the *scutellata* group (from *C. arctica* Kieffer to *Corynoneura yoshimurai* Tokunaga in the figure), *C. fittkaui* Schlee plus *C. lacustris* Edwards, and the apparently paraphyletic *celeripes* group (from *Corynoneura celeripes* Winnertz to *C. seiryuresea* Sasa, Suzuki *et* Sakai in the figure). Making the *celeripes* group monophyletic does not increase the length of the tree. The scalpel-like phallapodeme [30(1)] is an objective synapomorphy for the apparently paraphyletic *celeripes* group. The long and strongly curved phallapodeme [29(1)] is a synapomorphy for the *scutellata* group. *C. fittkaui* Schlee and *C. lacustris* Edwards have intermediate phallapodeme.

In the *scutellata* group, the attachment point for the phallapodeme is placed in the caudal third of the lateral sternapodeme and directed caudally except for in *Corynoneura gratias* Schlee, *Corynoneura sorachibecea* Sasa *et* Suzuki and the basal three species *Corynoneura coronata* Edwards, *Corynoneura doriceni* Makarchenko *et* Makarchenko, and *C. yoshimurai* Tokunaga which placed caudal and ventrally directed. In the *celeripes* group the attachment point for the phallapodeme is placed caudally and caudally directed except for in *Corynoneura inawapequea* Sasa, Kitami *et* Suzuki, *Corynoneura inefligiata* Fu, Sæther *et* Wang, and *Corynoneura macdonaldi* Fu, Sæther *et* Wang. The absence of a transverse sternapodeme with the lateral sternapodemes meeting in a sharp point (character 27) defines the *scutellata* group minus an apical subgroup which can be called the *gratias* subgroup (*C. arctica* Kieffer to *Corynoneura gynocera* Tuiskunen). Character 31(1), phallapodeme apically bifid and enclosing knob joint with sternapodeme, is found only in a subgroup of the *scutellata* group consisting of *Corynoneura celtica* Edwards, *Corynoneura vittalis* Tokunaga, and *Corynoneura korema* Preapical sensilla chaetica of the male antennal apex is present in the apical group of the *scutellata* group (*C. arctica* Kieffer to *Corynoneura edwardsi* Brundin except *C. gratias* Schlee), in *Corynoneura longipennis* Tokunaga, *C. scutellata* Winnertz, *Corynoneura nankaiensis* Fu, Sæther *et* Wang, and in *C. doriceni* Makarchenko *et* Makarchenko, while in the *celeripes* group it is found only in *C. inawapequea* Sasa, Kitami *et* Suzuki, and in *Corynoneura tenuistyla* Tokunaga. The inferior volsella is reduced in the apical members of the *scutellata* group, *C. arctica* Kieffer to *Corynoneura confidens* Fu, Sæther *et* Wang as well as in *Corynoneura kisogawa* Sasa *et* Kondo of the *celeripes* group. Three groups have gonostylus with a basal lobe on the inner margin: *C. arctica* Kieffer to *C. sorachibecea* Sasa *et* Suzuki, *Corynoneura kibunespinosa* Sasa to *C. nankaiensis* Fu, Sæther *et* Wang of the *scutellata* group, and *C. inawapequea* Sasa, Kitami *et* Suzuki plus *Corynoneura kibunelata* Sasa of the *celeripes* group.

The results are ambiguous. However, the *scutellata* group and some smaller groupings are recognizable in all analyses.

CHAPTER 2

Keys

The *Corynoneura* group is characterized by having costa apically fused with R_1 and R_{2+3}, forming a thick clavus (e.g., Sæther & Kristoffersen, 1996). Nine genera: *Corynoneura* Winnertz, 1846; *Corynoneurella* Brundin, 1949; *Ichthyocladius* Fittkau, 1974; *Notocladius* Harrison, 1997; *Onconeura* Andersen, Sæther, 2005; *Physoneura* Ferrington, Sæther, 1995; *Ubatubaneura* Wiedenbrug, Trivinho-Strixino, 2009; *Tempisquitoneura* Epler & de la Rosa, 1995 and *Thienemanniella* Kieffer, 1911 are included in the group.

2.1 KEY TO *CORYNONEURA* GENERIC GROUP

2.1.1 Key to the Males of the *Corynoneura* Group (From Wiedenbrug & Trivinho-Strixino, 2009)

1. Small size; wing length less than 1.3 mm; wing with clavus where R_1 and R_{2+3} are fused with costa. *Corynoneura* group   2

    Small to large size; wing without clavus   not keyed

2. Anal point weak; median lobes of antepronotum strongly reduced
    ***Physoneura* Ferrington and Sæther**

    Anal point absent; antepronotum usually not reduced   3

3. Eyes hairy; R_{4+5} apically joining costa and continuing as "false vein" to the wing tip (Harrison, 1997, Fig. 3)
    ***Notocladius* Harrison**

    Eyes hairy or bare; R_{4+5} apically not joining costa (Sæther & Kristoffersen, 1996, Figs. 1−2)   4

4. Palp with one palpomere; gonostylus without megaseta (Mendes, Andersen, & Sæther, 2004, Figs. 1 and 9); wing with distinct punctuation
    ***Ichthyocladius* Fittkau**

    Palp with five palpomeres; gonostylus with megaseta; wing with fine punctuation   5

5. Sternapodeme inverted V- or U-shaped; hind tibia usually strongly expanded distally; eyes bare
    ***Corynoneura* Winnertz**

    Transverse sternapodeme not as above; hind tibia not strongly expanded; eyes bare to hairy   6

Taxonomy of Corynoneura *Winnertz (Diptera: Chironomidae)*
DOI: https://doi.org/10.1016/B978-0-12-815263-8.00002-7
33

6. Tergites II − V with raised, median circular group of setae and few lateral setae (Epler & de la Rosa, 1995, Fig. 2E); eyes bare, sternapodeme with strong oral projections
   Tempisquitoneura Epler and de la Rosa
   Tergites II − V not as above; eyes hairy or bare; sternapodeme with or without strong oral projections   7
7. Sternapodeme usually with strong anteriolateral projections; inferior volsella strong; superior volsella low but distinct; eyes bare; tergites with row of setae
   Onconeura Andersen and Sæther
   Other character combination   8
8. Eyes bare; head with right-angled dorsolateral region; terminal flagellomere with triangular apex surrounded by sensilla chaetica; transverse sternapodeme without oral projections
   Ubatubaneura Wiedenbrug and Trivinho-Strixino
   Eyes usually hairy; head shape and apex of antenna not as above; transverse sternapodeme with or without oral projections
   Thienemanniella Kieffer
   Note: The male of *Corynoneurella* is not keyed.

2.1.2 Key to the Females of the *Corynoneura* Group (From Wiedenbrug & Trivinho-Strixino, 2009)

1. Small, wing length less than 1.3 mm; wing with clavus where R_1 and R_{2+3} are fused with costa. *Corynoneura* group   2
   Small to large size; wing without clavus   not keyed
2. Palps with only one palpomere; scutal tubercle present; wing with distinct punctuations
   Ichthyocladius Fittkau
   Palps with more than one palpomere; scutal tubercle absent; wing with fine punctuations   3
3. Tergites II − V with raised, median circular group of setae; eyes bare; seminal capsules about 1/2 the length of cerci; genitalia with sclerotized, spatulate structure between dorsomesal lobes
   Tempisquitoneura Epler and de la Rosa
   Setae on tergites II − V not as above; eyes bare to hairy; seminal capsules larger than 1/2 the length of cerci; genitalia without sclerotized, spatulate structure between dorsomesal lobes   4

4. Eyes bare; all trochanters with dorsal keel; coxosternapodeme usually with few spine-like anterior projections; cerci slightly larger than seminal capsules

Onconeura **Andersen and Sæther**

Eyes bare to hairy; usually not all trochanters with dorsal keel; coxosternapodeme without spine-like anterior projections; cerci usually smaller than seminal capsules   5

5. Eyes bare; head with right-angled dorsolateral region; terminal flagellomere of antenna with triangular apex surrounded by sensilla chaetica

Ubatubaneura **Wiedenbrug and Trivinho-Strixino**

Eyes bare to hairy; head shape and apex of antenna not as above   6

6. Eyes bare; hind tibia usually strongly expanded distally; apodeme lobe and labia large and strongly sclerotized; coxosternapodeme simple or with one to several lateral lamellae

Corynoneura **Winnertz**

7. Eyes usually hairy; hind tibia not strongly expanded distally; apodeme lobe small; labia small to large, but never completely sclerotized; coxosternapodeme simple or with one lateral lamella

Thienemanniella **Kieffer**

Note: Modified from Sæther (1977). The female of *Physoneura* and *Corynoneurella* are unknown; the female of *Notocladius* is not keyed.

2.1.3 Key to the Pupae of the *Corynoneura* Group (From Wiedenbrug & Trivinho-Strixino, 2009)

1. Thoracic horn absent; pedes spurii A and B absent; tergite II without pad of spines or hooks; if hooks present on posterior margin of tergite II, then hooks also present on tergites III and IV; anal lobe fringe usually present, except in *Ichthyocladius* and *Notocladius*. *Corynoneura* group   2

Character combinations different   not keyed

2. Anal lobe strongly reduced, with apical or ventral macrosetae; anal lobe fringe absent; with hooklets on tergites I–IV and sternites IV–VII

Ichthyocladius **Fittkau**

Anal lobe well developed; macrosetae not as above; anal lobe fringe present or absent; hooklets usually absent on sternites   3

3. Anal lobe fringe absent; margin of anal lobe with small, lateral denticles (Harrison, 1997, Fig. 3); with three anal macrosetae

    Notocladius **Harrison**

    Anal lobe fringe present; margin of anal lobe without denticles; anal macrosetae present or absent   4

4. Anal macrosetae absent; inner margin of anal lobe without seta; with humeral callus with broad spine (Epler & de la Rosa, 1995, Fig. 3C); anal lobe with fringe in apical 1/3

    Tempisquitoneura **Epler and de la Rosa**

    Anal macrosetae present; inner margin of anal lobe with seta; humeral callus not as above; anal lobe with full fringe or fringe restricted to apical 1/3   5

5. With single anal macroseta; anal lobe with fringe in apical 1/3, some lateral fringe setae spine-like; median corner of anal lobe with 1−3 spines

    Onconeura **Andersen and Sæther**

    With 3 anal macrosetae; if anal lobe fringe restricted to apical 1/3, then fringe with taeniate setae only; median corner of anal lobe without spines   6

6. Segments III−V each with 2 long, taeniate L-setae; segments VI−VIII without taeniate setae; with anal lobe fringe in distal 1/3; with macroseta on tubercles ventral on anal lobe

    Ubatubaneura **Wiedenbrug and Trivinho-Strixino**

    Segments IV−VIII usually with 4, rarely with 1 taeniate L-setae; anal lobe fringe rarely restricted to distal 1/2; with macrosetae dorsal on anal lobe   7

7. Pearl row on wing sheath present

    Corynoneura **Winnertz**

    Pearl row on wing sheath absent

    Thienemanniella **Kieffer**

    Note: The pupa of *Physoneura* and *Corynoneurella* are unknown.

2.1.4 Key to the Larvae of the *Corynoneura* Group (From Wiedenbrug and Trivinho-Strixino, 2009)

1. Small size, usually less than 4 mm; procercus and posterior parapods distinct; antennal segments evenly sclerotized, with normal, short terminal segment; antennae at least as long as 1/2 head length, or if

shorter than head capsule elongated, at least 1.5 times as long as wide, except in *Ichthyocladius*. *Corynoneura* group   2

    Character combinations different lobe   not keyed

2. Mentum with 16—20 dark, equal sized median and 5 pairs of slightly larger, lateral teeth; mandibular seta interna absent; larvae phoretic on catfishes

      Ichthyocladius **Fittkau**

    Mentum not as above; mandibular seta interna usually present   3

3. Antennae with 4 segments, subequal or longer than head

      Corynoneura **Winnertz**

    Antennae with 5 segments, shorter than head   4

4. Antennal blade slightly longer than flagellum; antenna about 1/4 of head length

      Tempisquitoneura **Epler and de la Rosa**

    Antennal blade much shorter than flagellum; antenna longer than 1/4 of head length   5

5. Antenna about 1/3 of head length; third antennal segment short; lauterborn organs well developed, generally reaching end of third segment; abdominal setae stout, dark and long

      Onconeura **Andersen and Sæther**

    Antenna at least 1/2 of head length; third antennal segment usually long; lauterborn organs usually reduced; abdominal setae usually not as above   6

6. Subbasal seta of posterior parapod apically split

      Ubatubaneura **Wiedenbrug and Trivinho-Strixino**

    Sub-basal seta of posterior parapod simple

      Thienemanniella **Kieffer**

Note: The larva of *Physoneura* and *Corynoneurella* are unknown; the larva *Notocladius* is not keyed.

2.2 KEY TO MALES OF *CORYNONEURA* WINNERTZ

1. Phallapodeme scalpel-like or apical rounded, and slightly bent, relatively short (Schlee, 1968b, Figs. 154—157)   2

    Phallapodeme not scalpel-like, and obvious curved, relatively long (Schlee, 1968b, Figs. 158—161)   51

2. Transverse sternapodeme developed, straight and wide (Schlee, 1968b, Figs. 197—198)   3

    Transverse sternapodeme undeveloped, rounded and narrow, or absent (Schlee, 1968b, Figs. 199—203)   27

3. Antennal apex pointed; megaseta long; hind tibial apex slightly thickened and elongate (Schlee, 1968b, Fig. 132)   4

   Antennal apex rounded; megaseta short; hind tibial apex distinctly thickened and elongate (Schlee, 1968b, Fig. 133)   5

4. Antenna with 11 flagellomeres; gonostylus apically strongly curved and medially unexpanded; superior volsella triangular; inferior volsella small rectangular (Fu & Sæther, 2012, Fig. 5)

   ***Corynoneura capitanea* Fu and Sæther**

   Antenna with 12 flagellomeres; gonostylus apically slightly curved and medially expanded; superior volsella narrow and undeveloped; inferior volsella along the inner margin of gonocoxite (Fu & Sæther, 2012, Fig. 9)

   ***Corynoneura disinflata* Fu and Sæther**

5. Attachment point of lateral sternapodeme with phallapodeme placed in caudal third of lateral sternapodeme   6

   Attachment point of lateral sternapodeme with phallapodeme placed caudal   8

6. Tergite IX with posterior humps and carrying long setae; superior volsella undeveloped and small rounded (Fu & Sæther, 2012, Fig. 6)

   ***Corynoneura caudicula* Fu and Sæther**

   Tergite IX without posterior humps and uncarrying long setae; superior volsella developed and projecting   7

7. Antennal apex distinctly expanded and sensilla chaetica short; gonostylus medially expanded; superior volsella triangular and separate medially (Fu, Sæther, & Wang, 2009, Fig. 7)

   ***Corynoneura inefligiata* Fu, Sæther and Wang**

   Antennal apex slightly expanded and sensilla chaetica long; gonostylus medially unexpanded; superior volsella rounded and joined medially (Makarchenko & Makarchenko, 2010, Figs. 11–14)

   ***Corynoneura collaris* Makarchenko and Makarchenko**

8. Superior volsella bearing long setae   9

   Superior volsella without long setae or bare   12

9. Gonostylus with prominent crista dorsalis   10

   Gonostylus without crista dorsalis   11

10. Inferior volsella rounded and smooth (Wiedenbrug & Triviho-Strixino, 2011, Fig. 19)

    ***Corynoneura unicapsulata* Wiedenbrug and Triviho-Strixino**

    Inferior volsella well defined and rectangular (Wiedenbrug, Lamas, & Trivinho-Strixino, 2012, Fig. 17)

Corynoneura guanacaste **Wiedenbrug, E. Lamas and Triviho-Strixino**

11. Antenna with 9—10 flagellomeres, superior volsella with pointed corner (Wiedenbrug et al., 2012, Fig. 18)

Corynoneura humbertoi **Wiedenbrug, E. Lamas and Triviho-Strixino**

Antenna with 12 flagellomeres, superior volsella with right-angled corner (Makarchenko & Makarchenko, 2010, Figs. 37—42)

Corynoneura schleei **Makarchenko and Makarchenko**

12. Anal point represented by hump-like extension of tergite   13
    Anal point absent   14

13. Antenna with 11 flagellomeres; superior volsella with rounded margin; gonostylus apical strongly bent (Fu et al., 2009, Fig. 4)

Corynoneura cylindricauda **Fu, Sæther and Wang**

Antenna with 8 flagellomeres; superior volsella with pointed corner; gonostylus apical slightly bent (Fu et al., 2009, Fig. 16)

Corynoneura prominens **Fu, Sæther and Wang**

14. Inferior volsella absent (Wiedenbrug et al., 2012, Fig. 25)

Corynoneura salviniatilis **Wiedenbrug, E. Lamas and Triviho-Strixino**

Inferior volsella present   15

15. Superior volsella absent   16
    Superior volsella present   17

16. 1Antenna with 11 flagellomeres; apex of hind tibia without S-shaped seta (Wiedenbrug et al., 2012, Fig. 15)

Corynoneura franciscoi **Wiedenbrug, E. Lamas and Triviho-Strixino**

Antenna with 7 flagellomeres; apex of hind tibia with S-shaped seta (Wiedenbrug et al., 2012, Fig. 23)

Corynoneura renata **Wiedenbrug, E. Lamas and Triviho-Strixino**

17. Inferior volsella with rounded corner   18
    Inferior volsella broad, almost rectangular or posterior margin hooked   21

18. Tergite IX without long setae; superior volsella joined medially (Fu & Sæther, 2012, Fig. 16)

Corynoneura hortonensis **Fu and Sæther**

Tergite IX with long setae; superior volsella separate medially   19

19. Superior volsella digitiform, with many short setae (Wiedenbrug et al., 2012, Fig. 34)

    ***Corynoneura vidiapodeme* Wiedenbrug, E. Lamas and Triviho-Strixino**

    Superior volsella triangular, without seta   20

20. Antenna with 9 flagellomeres, AR 0.65—0.68, antennal apex with a distinct group of about 20 sensilla chaetica (Wiedenbrug et al., 2012, Fig. 7)

    ***Corynoneura diogo* Wiedenbrug, E. Lamas and Triviho-Strixino**

    Antenna with 11 flagellomeres, AR 0.30—0.48, antennal apex with less than 10 sensilla chaetica (Fu & Sæther, 2012, Fig. 14)

    ***Corynoneura floridaensis* Fu and Sæther**

21. Antenna with 12 flagellomeres, inferior volsella posterior margin hooked   22

    Antenna less than 12 flagellomeres, inferior volsella posterior margin unhooked   23

22. AR 0.28—0.4; apex of inferior volsella nearest of gonostylus covered by setae; phallapodeme relatively short and straight (Tokunaga, 1936, Fig. 18; Makarchenko & Makarchenko, 2006b, Figs. 25—26)

    ***Corynoneura tenuistyla* Tokunega**

    AR 0.54—0.74; apex of inferior volsella nearest of gonostylus bare; phallapodeme relatively long and curved (Makarchenko & Makarchenko, 2010, Figs. 47—48)

    ***Corynoneura sundukovi* Makarchenko and Makarchenko**

23. Antenna with 8—9 flagellomeres   24

    Antenna with 10—11 flagellomeres   25

24. Phallapodeme short and straight, superior volsella undeveloped (Fu et al., 2009, Fig. 17)

    ***Corynoneura seiryuresea* Sasa, Suzuki and Sakai**

    Phallapodeme relatively long and curved, superior volsella developed (Makarchenko & Makarchenko, 2010, Figs. 54—58)

    ***Corynoneura tertia* Makarchenko and Makarchenko**

25. Antenna with apical dark brown spot, inferior volsella small rectangular and placed caudally of gonocoxite (Fu & Sæther, 2012, Fig. 18)

    ***Corynoneura macula* Fu and Sæther**

    Antenna without apical dark brown spot, inferior volsella broad rectangular and placed along inner margin of gonocoxite   26

26. Superior volsella with rounded corner, gonostylus straight
    ***Corynoneura ecphora* Fang, Wang and Fu**
    Superior volsella with right-angle corner, gonostylus apical curved
    (Fu et al., 2009, Fig. 19)
    ***Corynoneura tokarapequea* Sasa and Suzuki**
27. Transverse   sternapodeme   almost   absent   (Schlee,   1968b,
    Figs. 202—203)   28
    Transverse sternapodeme narrow and rounded (Schlee, 1968b,
    Figs. 199—201)   29
28. Attachment point of lateral sternapodeme placed in caudal third of
    lateral sternapodeme (Fu et al., 2009, Fig. 6)
    ***Corynoneura inawapequea* Sasa, Kitami and Suzuki**
    Attachment point of lateral sternapodeme placed in caudal
    position of sternapodeme (Fu et al., 2009, Fig. 8)
    ***Corynoneura kibunelata* Sasa**
29. Eyes pubescent   30
    Eyes bare   33
30. Superior volsella absent   31
    Superior volsella present   32
31. Abdominal tergites V−IX brown (Wiedenbrug & Triviho-Strixino,
    2011, Fig. 1)
    ***Corynoneura fortispicula* Wiedenbrug and Triviho-Strixino**
    Abdominal tergite V usually whitish, VI−IX brown (Wiedenbrug
    & Triviho-Strixino, 2011, Fig. 7)
    ***Corynoneura mediaspicula* Wiedenbrug and Triviho-Strixino**
32. Superior volsella rounded with long setae (Wiedenbrug & Triviho-
    Strixino, 2011, Fig. 4)
    ***Corynoneura hermanni* Wiedenbrug and Triviho-Strixino**
    Superior volsella small triangular with short setae (Wiedenbrug &
    Triviho-Strixino, 2011, Fig. 10)
    ***Corynoneura mineira* Wiedenbrug and Triviho-Strixino**
33. Attachment point of lateral sternapodeme with phallapodeme placed
    in caudo-ventral position (Fu et al., 2009, Fig. 13; Schlee, 1968b,
    Fig. 152)
    ***Corynoneura macdonaldi* Fu, Sæther and Wang**
    Attachment point of lateral sternapodeme with phallapodeme
    placed caudally and caudally directed (Schlee, 1968b, Fig. 151)   34
34. Phallapodeme curved, and relatively long (Schlee, 1968b, Figs. 155
    and 157)   5

Phallapodeme almost straight, and relatively short (Schlee, 1968b, Fig. 154)   36

35. Superior volsella conspicuously projecting and with long setae (Schlee, 1968b, Figs. 42—45)

    Corynoneura fittkaui **Schlee**

    Superior volsella unconspicuously projecting and without long setae (Schlee, 1968b, Figs. 46—49)

    Corynoneura lacustris **Schlee**

36. Antenna with 10—12 flagellomeres   37

    Antenna with 7—9 flagellomeres   45

37. Gonostylus straight, medially expanded and relatively short (Fu et al., 2009, Fig. 5)

    Corynoneura fujiundecima **Sasa**

    Gonostylus apical curved, gradually narrow near apex and relatively long   38

38. Inferior volsella undeveloped, barely indicated as long low inner margin of gonocoxite   39

    Inferior volsella developed, conspicuously projecting   41

39. Superior volsella absent (Wiedenbrug & Triviho-Strixino, 2011, Fig. 16)

    Corynoneura sertaodaquina **Wiedenbrug and Triviho-Strixino**

    Superior volsella present   40

40. Antenna with 10 flagellomeres, superior volsella with 1 long seta (Wiedenbrug et al., 2012, Fig. 21)

    Corynoneura longiantenna **Wiedenbrug, E. Lamas and Triviho-Strixino**

    Antenna with 12 flagellomeres, superior volsella without long setae (Fu et al., 2009, Fig. 10)

    Corynoneura kisogawa **Sasa and Kondo**

41. Inferior volsella small rectangular, placed medially on gonocoxite (Makarchenko & Makarchenko, 2010, Figs. 1—5)

    Corynoneura aurora **Makarchenko and Makarchenko**

    Inferior volsella unrectangular, placed caudally on gonocoxite   42

42. Antenna with 10 flagellomeres   43

    Antenna with 11 or 12 flagellomeres   44

43. Superior volsella absent, inferior volsella rounded (Wiedenbrug et al., 2012, Fig. 32)

    Corynoneura trondi **Wiedenbrug, E. Lamas and Triviho-Strixino**

Superior volsella present, inferior volsella broad and the shape like stairs (Fu & Sæther, 2012, Fig. 3)

> *Corynoneura ascensa* **Fu and Sæther**

44. Antenna with 12 flagellomeres, inferior volsella triangular (Schlee, 1968b, Figs. 50—51)

> *Corynoneura celeripes* **Winnertz**

Antenna with 11 flagellomeres, inferior volsella posterior margin hooked (Wiedenbrug & Triviho-Strixino, 2011, Fig. 13)

> *Corynoneura septadentata* **Wiedenbrug and Triviho-Strixino**

45. Superior volsella absent   46

Superior volsella present   48

46. Antenna with 9—10 flagellomeres (Freeman, 1953, Figs. e and f)

> *Corynoneura cristata* **Freeman**

Antenna with 7—8 flagellomeres   47

47. Antenna with 8 flagellomeres, antennal apex with sensilla concentrated apically; laterosternite IX without seta (Wiedenbrug et al., 2012, Fig. 3)

> *Corynoneura boraceia* **Wiedenbrug, E. Lamas and Triviho-Strixino**

Antenna with 7 flagellomeres, antennal apex with sensilla not only apically; laterosternite IX usually with seta (Wiedenbrug et al., 2012, Fig. 1)

> *Corynoneura bodoquena* **Wiedenbrug, E. Lamas and Triviho-Strixino**

48. Superior volsella bearing many setae (Wiedenbrug et al., 2012, Fig. 10)

> *Corynoneura espraiado* **Wiedenbrug, E. Lamas and Triviho-Strixino**

Superior volsella without seta   49

49. Inferior volsella undeveloped and rounded (Makarchenko & Makarchenko, 2006b, Figs. 10—12)

> *Corynoneura prima* **Makarchenko and Makarchenko**

Inferior volsella developed and broad   50

50. Antenna with 9 flagellomeres; inferior volsella rectangular (Fu & Sæther, 2012, Fig. 1)

> *Corynoneura acuminata* **Fu and Sæther**

Antenna with 8 flagellomeres; inferior volsella rounded (Wiedenbrug et al., 2012, Fig. 4)

Corynoneura canchim **Wiedenbrug, E. Lamas and Triviho-Strixino**

51. Phallapodeme curved, relatively short (Schlee, 1968b, Fig. 158)   52
    Phallapodeme strongly curved, relatively long (Schlee, 1968b, Figs. 159 and 160)   53
52. Antenna with 8 flagellomeres, antennal ratio higher than 1.0 (Singh & Maheshwari, 1987, Fig. 2)
    Corynoneura chandertali **Singh and Maheshwari**
    Antenna with 10 flagellomeres, antennal ratio lower than 0.4 (Freeman, 1953, Figs. g and h)
    Corynoneura elongata **Freeman**
53. Sternapodeme inverted U-shaped, transverse sternapodeme present   54
    Sternapodeme inverted V-shaped, transverse sternapodeme absent, lateral sternapodemes meeting in sharp point   64
54. Antenna with 12 flagellomeres   55
    Antenna with 6−11 flagellomeres   57
55. Attachment point of lateral sternapodeme placed caudal (Krasheninnikov, 2012, Figs. 1−5)
    Corynoneura makarchenkorum **Krasheninnikov**
    Attachment point of lateral sternapodeme placed in caudal third of lateral sternapodeme   56
56. Antennal apex with apical sensilla chaetica, inferior volsella not developed (Schlee, 1968b, Figs. 52−53)
    Corynoneura coronata **Edwards**
    Antennal apex with preapical sensilla chaetica, inferior volsella well developed (Fu & Sæther, 2012, Figs. 11−12; Makarchenko & Makarchenko, 2006b, Figs. 2−3; Makarchenko & Makarchenko, 2010, Fig. 59)
    Corynoneura doriceni **Makarchenko and Makarchenko**
57. Attachment point of lateral sternapodeme placed caudal   58
    Attachment point of lateral sternapodeme placed in caudal third of lateral sternapodeme   61
58. Gonostylus with basal lobe on inner margin   59
    Gonostylus without basal lobe on inner margin   60
59. Inferior volsella developed (Freeman, 1961)
    Corynoneura australiensis **Freeman**
    Inferior volsella absent (Hirvenoja & Hirvenoja E, 1988, Fig. 2; Schlee, 1968b, Figs. 54−59)

Corynoneura gratias **Schlee**

60. Transverse sternapodeme wide, inferior volsella with right-angle corner (Tokunega, 1936, Fig. 17)

Corynoneura yoshimurai **Tokunaga**

Transverse sternapodeme narrow, inferior volsella with pointed corner (Makarchenko & Makarchenko, 2006b, Figs. 5–6)

Corynoneura kedrovaya **Makarchenko and Makarchenko**

61. Superior volsella conspicuously projecting, digitiform (Hirvenoja & Hirvenoja, 1988, Fig. 2; Tuiskunen, 1983)

Corynoneura gynocera **Tuiskunen**

Superior volsella not projecting, almost triangular or with rounded corner   62

62. Superior volsella separated, almost triangular (Fu et al., 2009, Fig. 14)

Corynoneura medicina **Fu, Sæther and Wang**

Superior volsella joined medially or absent   63

63. Antenna with 10 flagellomeres, AR c. 1.0; gonostylus strongly curved (Hirvenoja & Hirvenoja, 1988, Fig. 2)

Corynoneura arctica **Kieffer**

Antenna with 11 flagellomeres, AR c. 0.5; gonostylus slightly curved (Fu et al., 2009, Fig. 19)

Corynoneura tokaraquerea **Sasa and Suzuki**

64. Antenna with 12 flagellomeres (Tokunega, 1936)

Corynoneura vittalis **Tokunaga**

Antenna with 6–11 flagellomeres   65

65. Gonostylus with basal lobe on inner margin   66

Gonostylus without basal lobe on inner margin   75

66. Inferior volsella present   67

Inferior volsella absent or barely indicated as long low inner margin of gonocoxite   70

67. Anal point represented by hump-like extension of tergite (Fu et al., 2009, Fig. 15)

Corynoneura nankaiensis **Fu, Sæther and Wang**

Anal point absent   68

68. Antennal ratio 0.3; gonostylus slightly curved (Fu et al., 2009, Fig. 9)

Corynoneura kibunespinosa **Sasa**

Antennal ratio 1.0; gonostylus strongly curved   69

69. Wing very narrow and long, wing width/length < 0.40; superior volsella reduced (Tokunega, 1936, Fig. 21)

Corynoneura longipennis **Tokunaga**

Wing relatively wide and short, wing width/length > 0.40; superior volsella present (Hirvenoja & Hirvenoja, 1988, Fig. 2)

Corynoneura scutellata Winnertz

70. Eyes pubescent (Wiedenbrug et al., 2012, Fig. 35)

Corynoneura zempoala Wiedenbrug, E. Lamas and Triviho-Strixino

Eyes bare   71

71. Attachment point of lateral sternapodeme placed caudally and directed ventrally (Fu et al., 2009, Fig. 18)

Corynoneura sorachibecea Sasa and Suzuki

Attachment point of lateral sternapodeme placed in caudal third of lateral sternapodeme and directed caudally   72

72. Antenna with 11 flagellomeres (Tokunega, 1936, Fig. 19)

Corynoneura cuspis Tokunaga

Antenna with 7–8 flagellomeres   73

73. Antennal ratio lower than 0.4 (Hazra, Nath, & Chaudhuri, 2003)

Corynoneura incidera Hazra, Nath and Chaudhuri

Antennal ratio higher than 1.0   74

74. Superior volsella inconspicuous, joined medially; gonostylus strongly curved (Singh & Maheshwari, 1987, Fig. 1)

Corynoneura carinata Singh and Maheshwari

Superior volsella conspicuously projecting; gonostylus slightly curved (Singh & Maheshwari, 1987, Fig. 3)

Corynoneura lahuli Singh and Maheshwari

75. Inferior volsella present   76

Inferior volsella absent or barely indicated as long low inner margin of gonocoxite   82

76. Phallapodeme with lateral apex bifid, enclosing knob joint with sternapodeme   77

Phallapodeme with projection for joint with sternapodeme placed prelateral   78

77. Antenna with 11 flagellomeres; inferior volsella rounded (Schlee, 1968b, Figs. 84–85)

Corynoneura celtica Edwards, 1924

Antenna with 7 flagellomeres; inferior volsella with digitiform projecting (Fu et al., 2009, Fig. 11)

Corynoneura korema Fu, Sæther and Wang

78. Inferior volsella broad, almost triangular   79

Inferior volsella narrow, digitiform to spatulate   80

79. Superior volsella tongue-shaped; inferior volsella placed basally on gonocoxite, and overlapping with superior volsella (Fu & Sæther, 2012, Fig. 20)

 Corynoneura porrecta **Fu and Sæther**

 Superior volsella not tongue-shaped; inferior volsella placed caudally on gonocoxite, and separated from superior volsella (Harrison, 1992, Fig. 7)

 Corynoneura dewulfi **Goetghebuer**

80. Antenna with 8 flagellomeres; Tergite IX straight, without posterior humps without indication of median mounds (Hazra et al., 2003)

 Corynoneura centromedia **Hazra, Nath and Chaudhuri**

 Antenna with 9 flagellomeres; Tergite IX medially incurved with posterior humps with indication of median mounds   81

81. Antenna with 10 flagellomeres; gonostylus median very expanded and apical hooked (Schlee, 1968b, Figs. 80−83)

 Corynoneura lobata **Edwards**

 Antenna with 9 flagellomeres; gonostylus slender and apical hooked (Sublette & Sasa, 1994, Figs. 44−51; Wiedenbrug et al., 2012, Fig. 13)

 Corynoneura ferelobata **Sublette and Sasa**

82. Antenna with 11 flagellomeres   83

 Antenna with 6−10 flagellomeres   85

83. Superior volsella more or less triangular (Hirvenoja & Hirvenoja, 1988, Fig. 2; Schlee, 1968b, Figs. 60−65)

 Corynoneura carriana **Edwards**

 Superior volsella digitiform, conspicuously projecting   84

84. Antennal ratio 0.4, palpomeres short (Hirvenoja & Hirvenoja, 1988, Fig. 2)

 Corynoneura brundini **Hirvenoja and Hirvenoja**

 Antennal ratio 0.7−0.9, palpomeres relatively long (Hirvenoja & Hirvenoja, 1988, Fig. 2; Schlee, 1968b, Figs. 66−71)

 Corynoneura edwardsi **Brundin**

85. Antennal apex with preapical sensilla chaetica (Makarchenko & Makarchenko, 2010, Figs. 26−27)

 Corynoneura kadalinka **Makarchenko and Makarchenko**

 Antennal apex with apical sensilla chaetica   86

86. Sensilla on antennal apex longer than last flagellomere and curved (Wiedenbrug et al., 2012, Fig. 29)

Corynoneura sisbiota **Wiedenbrug, E. Lamas and Triviho-Strixino**

Sensilla on antennal apex obviously shorter than last flagellomere and uncurved   87

87. Antenna with 6—7 flagellomeres   88

Antenna with 9—10 flagellomeres   89

88. Antenna with 7 flagellomeres, ultimate flagellomere with distal black rosette (Fu et al., 2009, Fig. 12)

Corynoneura latusatra **Fu, Sæther and Wang**

Antenna with 6 flagellomeres, ultimate flagellomere distal without black rosette (Sasa & Suzuki, 2000a)

Corynoneura isigaheia **Sasa and Suzuki**

89. Antenna with 10 flagellomeres; Tergite IX straight, without posterior humps without indication of median mounds; gonostylus straight (Fu et al., 2009, Fig. 3)

Corynoneura confidens **Fu, Sæther and Wang**

Antenna with 9 flagellomeres; Tergite IX medially incurved with posterior humps with indication of median mounds; gonostylus slightly curved (Makarchenko & Makarchenko, 2006b, Figs. 13—15)

Corynoneura secunda **Makarchenko and Makarchenko**

Note: Excluded from the key are: *Corynoneura diara* Roback (Roback, 1957b), *Corynoneura hirvenojai* Sublette *et* Sasa (Sublette & Sasa, 1994), *Corynoneura imperfecta* (Skuse, 1889), *Corynoneura magna* Brundin (Brundin, 1949), *Corynoneura marina* Kieffer (Kieffer, 1924), *Corynoneura nasuticeps* Hazra, Nath *et* Chaudhuri (Hazra et al., 2003), and *Corynoneura postcinctura* (Tokunaga, 1964).

CHAPTER 3

Review of Species

Corynoneura **Winnertz**
Corynoneura Winnertz, 1846: 12.

3.1 *CORYNONEURA ACUMINATA* FU & SÆTHER, 2012 (FIGS. 3.1A—F AND 3.2A—D)

Corynoneura n. sp. 9 Bolton, 2007: 26.
Corynoneura acuminata Fu & Sæther, 2012: 5.

Material examined. Holotype male, with associated larval and pupal exuviae, USA, Ohio, Hocking County, E. Br. Raccoon Creek at Sanner Road, RM 6.6, 3.x.1996, M.J. Bolton (ZMBN Type No. 452). Paratypes: 1 male, 1 pupal and 1 larval exuviae, as holotype; 2 larval exuviae, as holotype except for 21.viii.1995 (ZMBN, MJB).

Diagnostic characters. The adult male is characterized by a strongly concave anterior margin of the cibarial pump; antenna with 9 flagellomeres; superior volsella pointed triangular; inferior volsella broad and placed caudally on gonocoxite; and transverse sternapodeme narrow U-shaped. The pupa has no taeniate L-setae on tergites I−II, tergites III−VII with some thick shagreen, and 14−16 taeniae in fringe of anal lobe (the taeniae are restricted to the posterior half of segment). The larva is separable by having mentum with two large median teeth, lateral tooth adjacent to median tooth very small, antenna slightly shorter than head capsule, and head capsule integument sculptured with fine ridges that are usually in a reticulate pattern.

Male (*n* = 1−2)

Total length 0.87−1.22 mm. Wing length 0.72−0.73 mm. Total length/wing length 1.20−1.72 mm. Wing length/profemur length 3.30−3.42 mm.

Coloration. Head brown. Antenna and legs yellowish. Thorax with upper antepronotum, upper scutum, scutellum, postnotum and ventral part of preepisternum brown, anepisternum and dorsal part of preepisternum yellowish. Abdomen light brown.

Taxonomy of Corynoneura *Winnertz (Diptera: Chironomidae)*
DOI: https://doi.org/10.1016/B978-0-12-815263-8.00003-9

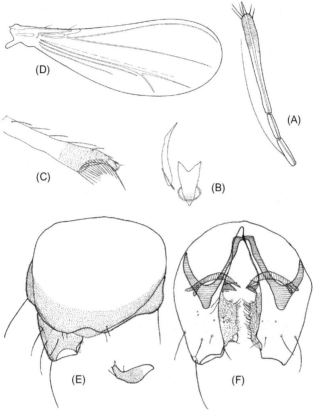

Figure 3.1 *Corynoneura acuminata* Fu & Sæther, 2012, Male imago (From Fu, Y., & Sæther, O. A. (2012). *Corynoneura* Winnertz and *Thienemanniella* Kieffer from the Nearctic region (Diptera: Chironomidae: Orthocladiinae). *Zootaxa, 3536,* 1–61. www.mapress.com/j/zt) (A) Antenna. (B) Tentorium, stipes and cibarial pump. (C) Hind tibial apex. (D) Wing. (E) Hypopygium, dorsal view. (F) Hypopygium, ventral view.

Head. Antenna (Fig. 3.1A) with 9 flagellomeres, AR 0.49−0.52; ultimate flagellomere 135−136 μm, apically expanded, with about 8−10 apical sensilla chaetica; longest antennal seta 135−138 μm. Temporals absent. Tentorium, stipes and cibarial pump as in Fig. 3.1B; tentorium 112−113 μm long, 14 μm wide; stipes 43−55 μm long, 3 μm wide. Anterior margin of cibarial pump strongly concave. Clypeus with 10 setae. Length of palpomeres (in μm): 10−11, 10−12, 15−17, 23−25, 51−52. Palpomere 5/3 ratio 3.1−3.4.

Thorax. Dorsocentrals 4, prealars 2. Scutellum with 2 setae. Anapleural suture 101−105 μm long.

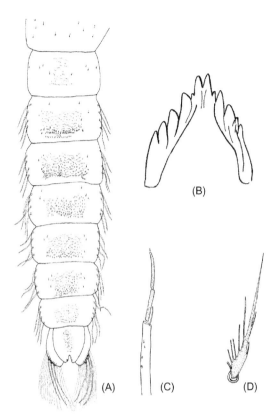

Figure 3.2 *Corynoneura acuminata* Fu & Sæther, 2012, Immatures (From Fu, Y., & Sæther, O. A. (2012). Corynoneura Winnertz and Thienemanniella Kieffer from the Nearctic region (Diptera: Chironomidae: Orthocladiinae). *Zootaxa, 3536,* 1−61. www.mapress.com/j/zt) (A) Pupa: tergites $I-IX$. (B−D) Larva. (B) Mentum. (C) Antenna. (D) Subbasal seta of posterior parapod.

Wing (Fig. 3.1D). VR 3.2−3.5. Cu/wing length 0.57, C 165−168 μm long, Cu 412 μm long, wing width/wing length 0.40−0.41. Brachiolum with 1 seta, costa with 3−4 setae.

Legs. Fore trochanter with dorsal keel. Spurs of fore tibia 19−25 μm and 10 μm long, of mid tibia 11 μm and 14 μm long, of hind tibia 25−29 μm and 11−12 μm long. Width at apex of fore tibia 11−14 μm, of mid tibia 15 μm, of hind tibia (a) 29 μm. Width of hind tibia 1/3 from apex (d) 12 μm, elongation length (b) 28 μm, length of maximum thickening (c_1) 69 μm, total length of thickening (c_2) 124 μm, a/d 2.6, b/d 2.5, c_1/d 6.3, c_2/d 11.3. Hind tibia (Fig. 3.1C) expanded, with comb of 10 setae, 1 seta near spur strongly S-shaped. Lengths and proportions of legs as in Table 3.1.

Table 3.1 Lengths (in μm) and proportions of legs segments of male *Corynoneura acuminata* Fu & Sæther, 2012 ($n = 1-2$)

	fe	ti	ta_1	ta_2	ta_3	ta_4
p_1	210–212	255–256	132–140	71	51	20
p_2	284–296	260–268	168	81	42	20
p_3	232–240	256–260	145	79	36	18

	ta_5	LR	BV	SV	BR
p_1	30	0.52–0.55	3.5	3.3	2.2
p_2	32	0.61	3.2	3.4	2.0
p_3	30	0.56	4.0	3.5	1.9

Source: From Fu, Y., & Sæther, O. A. (2012). *Corynoneura* Winnertz and *Thienemanniella* Kieffer from the Nearctic region (Diptera: Chironomidae: Orthocladiinae). *Zootaxa*, 3536, 1–61. www.mapress.com/j/zt.

Hypopygium (Fig. 3.1E and F). Tergite IX with 4–5 long setae. Laterosternite IX with 1 long seta. Superior volsella pointed triangular, anteriomedially completely separated. Inferior volsella broad, with many glandular setae and strong marginal setae, along the inner margin of gonocoxite and placed caudally. Phallapodeme slightly curved, 28–32 μm long, with lateral apex bifid and joint with sternapodeme placed prelaterally. Transverse sternapodeme narrow, 7–11 μm wide. Large attachment point of lateral sternapodeme with phallapodeme placed and directed caudally. Gonostylus curved, 22 μm long; megaseta 4–6 μm long. HR 3.3–4.5, HV 4.0.

Pupa ($n = 1-2$)

Total length 1.69–1.71 mm. Exuviae yellowish with light brown cephalothorax.

Cephalothorax. Frontal setae 21–26 μm long. Median antepronotals 21 μm and 23 μm long. Lateral antepronotals 22 μm long. Anterior precorneal seta (PcS_1) 8–15 μm long, PcS_2 7–14 μm long, PcS_3 8 μm long. PcS_{1-3} almost in a line, PcS_1 4–7 μm from PcS_2, PcS_2 4–7 μm from PcS_3, PcS_1 10–12 μm from PcS_3, PcS_3 23–28 μm from thoracic horn. Anterior dorsocentral (Dc_1) 6 μm long, Dc_2 4–6 μm long, Dc_3 4 μm long, Dc_4 4–6 μm long. Dc_1 located 17 μm from Dc_2, Dc_2 located 28–30 μm from Dc_3, Dc_3 located 18–29 μm from Dc_4. Wing sheath with 4–5 rows of pearls.

Abdomen (Fig. 3.2A). Shagreen and chaetotaxy as illustrated. No taeniate L-setae on tergites I–II. Anal lobe 111 μm long. Anal lobe fringe with 14–16 setae, 280–300 μm long. Anal macrosetae 141–182 μm long, median setae 80 μm long.

Larva (*n* = 3−4)

Coloration. Head capsule yellowish brown, integument sculptured with fine ridges that are usually in a reticulate pattern. Antenna with basal segment yellowish, other segments brown. Abdomen yellowish.

Head. Capsule length 182−216, 197 μm; width 120−128 μm. Postmentum 154−180, 163 μm long. Sternite II obvious, rising from small tubercle, I and III not visible. Premandible 32−39, 34 μm long. Mentum as in Fig. 3.2B and Bolton (2007). Mandibles 46−59, 51 μm long. Antenna (Fig. 3.2C): AR 0.96−1.04, 1.00. Length of flagellomeres I −IV (in μm): 87−111, 100; 43−53, 47; 48−59, 54; 4−6, 5. Basal segment 14−17 μm wide; length of blade at apex of basal segment 21−23, 22 μm.

Abdomen. Length of anal setae 141−172, 157 μm. Procercus 8−10, 9 μm long; 7−10, 8 μm wide. Subbasal seta of posterior parapods split as Fig. 3.2D: 57−67, 63 μm long.

Distribution. Neoarctic (NE): USA (Ohio State).

3.2 *CORYNONEURA ARCTICA* KIEFFER, 1923 (FIG. 3.3A AND B)

Corynoneura arctica Kieffer: Hirvenoja & Hirvenoja, 1988: 219; Makarchenko & Makarchenko, 2006b: 152; Fu et al., 2009: 6.

Material examined. P.R. China. Tianjin City, Nankai University, 39°8′N, 117°12′E, alt. 2−5 m, sweep net, 5 males, 20.iv.1986, X. Wang;

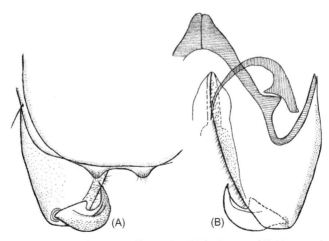

Figure 3.3 *Corynoneura arctica* Kieffer, 1923; Male imago. (A) Hypopygium, dorsal view. (B) Hypopygium, ventral view.

Tianjin City, Wuqing County, Dahuangpu wetland area, 39°3′N, 117° 3′E, alt. 5 m, sweep net, 24 males, 26.iv.2005, X. Wang.

Diagnostic characters. The species can be separated from other members of the genus by having hairy eyes, antenna with 10 flagellomeres, sternapodeme inverted V-shaped, and no inferior volsella.

Male (*n* = 29)

Total length 1.28−1.80, 1.36 mm. Wing length 1.00−1.25, 1.1 mm. Total length/wing length 1.47−1.51, 1.5(24). Wing length/length of profemur 2.4−3.2, 3.0(24).

Coloration. Head dark brown. Antenna and palpomere pale yellow-brown. Thorax dark brown. Abdominal segments dark brown. Legs yellow-brown. Wings transparent and yellowish, with pale yellow clava.

Head. Eyes bare, reniform. Antenna with 10 flagellomeres; ultimate flagellomere 255−350, 290 μm long, apex with short sensilla chaetica which extends a little way back from tip, apically acuate. AR 0.77−1.1, 1.0. Clypeus with 8 setae. Temporal setae lacking. Tentorium 100−163, 143 μm long, 13−25, 20(26) μm wide. Stipes: 62−65, 64 μm long. Palpomeres length (in μm): 12.5−20, 15; 20−25, 22; 25−30, 27; 32.5−42.5, 37; 42.5−75, 63. Palpomere 2 ellipsoid, 3 and 4 rectangular, 5 long and slender. Palpomere 5/3 ratio 1.7−3.0, 2.3.

Thorax. Antepronotum without lateral setae. Dorsocentrals 5−7, 6. Other setae cannot be seen.

Wing. VR 3.0−3.4, 3.3. C length 330−370, 350 μm, C/wing length 0.29−0.37, 0.33. Cu length 650−770, 685 μm. Cu/wing length 0.23−0.37, 0.26. Wing width/wing length 0.36−0.39, 0.37. C with 7−10, 8 setae.

Legs. Fore trochanter with keel. Spur of front tibia 30−35, 32 μm and 13−18, 15 μm long, spurs of middle tibia 15−20, 16 μm long, of hind tibia 35−50, 42 μm long. Width at apex of fore tibia 25−38, 32 μm, of mid tibia 22.5−27.5, 25 μm, of hind tibia 38−50, 45 μm. Tip of hind tibia expanded, with comb of 12−17, 15 setae and 1 seta near spur slightly hooked. Lengths and proportions of legs as in Table 3.2.

Hypopygium (Fig. 3.3A and B). Posterior margin strongly bilobed, and with many short setae, laterosternite IX with 2 long setae. Anal point absent. Inferior volsellae absent, sternapodeme inverted slightly V-shaped, coxapodeme 37.5 μm long. Phallapodeme strongly curved not extending beyond tergite IX, 85−90, 88 μm long. Gonocoxite 80−128, 100 μm long with 4−6 long setae apically. Gonostylus curved

Table 3.2 Lengths (in μm) and proportions of legs segments of male *Corynoneura arctica* Kieffer, 1923 (based on Chinese specimens)

	fe	ti	ta_1	ta_2	ta_3	ta_4
p_1	315–360, 334	355–440, 399	230–265, 241	113–143, 122	68–83, 74	23–33, 31
p_2	450–520, 485	410–475, 438	235–270, 252	100–135, 113	58–70, 63	25–30, 28
p_3	365–425, 396	400–455, 424	225–270, 244	113–150, 131	48–58, 53	23–28, 25

	ta_5	LR	BV	SV	BR
p_1	43–50, 47	0.55–0.66, 0.61	3.5–3.9, 3.7	2.9–3.2, 3.0	1.8–2.3, 2.1
p_2	38–50, 45	0.55–0.61, 058	4.6–5.0, 4.8	3.5–4.0, 3.7	2.5–3.2, 2.8
p_3	38–48, 43	0.55–0.62, 0.58	3.7–4.6, 4.3	3.0–3.6, 3.4	2.1–2.7, 2.5

apically, 25—43, 38 μm long. Megaseta 5—7, 6 μm long. HR 2.1—3.6, 3.0; HV 3.9—5.5, 4.5.

Remarks. This species is close to *Corynoneura edwardsi*, but can be separated by the posterior margin, apex of hind tibia, character of the hypopygium. This species is close to *Corynoneura gratias*, but can be separated easily by the sternapodeme, phallapodeme, gonostylus. This species is also close to *Corynoneura scutellata*, but can be separated by the inferior volsellae absent. When we reviewed the literature regarding this species, we found all previous descriptions and figures too simple to recognize it, so we redescripted and illustrated the species based on Chinese specimens and made other relevant researchers convenient.

Distribution. Palaearctic (PA): Austria, Finland, France, Germany, Great Britain, Norway, Portugal, Spain, Switzerland, Russia including Novaya Zemlya and Russian Far East, Northern China (Tianjin). **NE**: Canada, USA (Oliver et al., 1990; Sæther & Spies, 2004).

3.3 *CORYNONEURA ASCENSA* FU & SÆTHER, 2012 (FIGS. 3.4A—F AND 3.5A—F)

Corynoneura n. sp. 5 Bolton, 2007: 30.
Corynoneura ascensa Fu & Sæther, 2012: 8.

Material examined. Holotype male, with associated larval and pupal exuviae, USA, Ohio, Champaign County, spring at tributary to east branch of Cedar Run south of Woodburn Rd., 15 & 19.iv.1989, M. J. Bolton (ZMBN Type No. 453). Allotype, USA, Ohio, Summit County, spring in Sand Run Metro Park, 18 & 24.iii.1989, M. J. Bolton (ZMBN). Paratypes: 1 female, 1 pupal and 1 larval exuviae, as holotype; 1 male, 1 pupal and 2 larval exuviae, as allotype except for 12.iv.1987, M.J. Bolton (MJB).

Diagnostic characters. The male imago is characterized by the antenna with 10 flagellomeres, AR about 0.35; lateral sternapodeme with very large attachment point and placed caudally; inferior volsella broad and of a particular shape. The female has long hind tibia with a serrated spur, lacks setae on gonocoxite IX, has 4 setae on tergite IX, notum very short, ramus long and broadened medially, and 8 fused lateral lamellae on the coxosternapodeme. The pupa has no taeniate L-setae on tergites I and II; tergites III—VI with few small shagreen anteriorly, shagreen

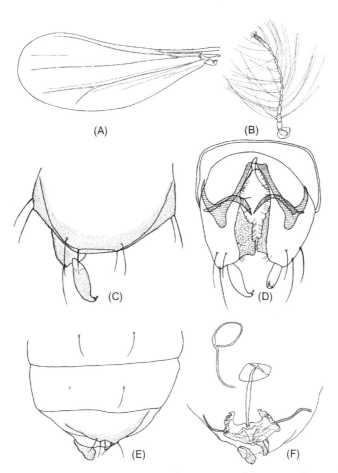

Figure 3.4 *Corynoneura ascensa* Fu & Sæther, 2012; Adult imago (From Fu, Y., & Sæther, O. A. (2012). *Corynoneura* Winnertz and *Thienemanniella* Kieffer from the Nearctic region (Diptera: Chironomidae: Orthocladiinae). *Zootaxa, 3536*, 1–61. www. mapress.com/j/zt). (A–D) Male imago. (A) Wing. (B) Antenna. (C) Hypopygium, dorsal view. (D) Hypopygium, ventral view. (E and F) Female imago. (E) Genitalia, dorsal view. (F) Genitalia, ventral view.

getting thick and larger posteriorly, and there is light brown pigmentation across segments II /III— V /VI in the area of the larger shagreen; and 16 taeniae on the anal lobe that are restricted to the posterior half. The larvae are separable by having smooth head capsule integument, unsculptured; the antenna length is longer than head length (1.1—1.5 times); mentum with trifid median teeth, and lateral tooth adjacent to median tooth very small.

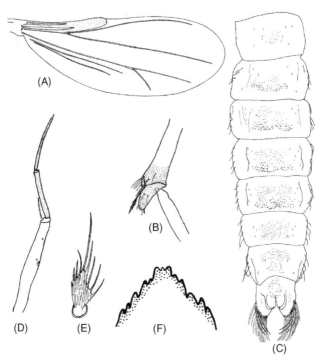

Figure 3.5 *Corynoneura ascensa* Fu & Sæther, 2012 (From Fu, Y., & Sæther, O. A. (2012). *Corynoneura* Winnertz and *Thienemanniella* Kieffer from the Nearctic region (Diptera: Chironomidae: Orthocladiinae). *Zootaxa, 3536*, 1−61. www.mapress.com/j/zt). (A and B) Female imago. (A) Wing. (B) Hind tibial apex. (C) Pupa: Tergites II −IX. (D−F). Larva. (D) Antenna. (E) Subbasal seta of posterior parapod. (F) Mentum.

Male (*n* = 2−3)

Total length 1.33 mm. Wing length 0.82 mm. Total length/wing length 1.63 mm. Wing length/profemur length 3.2 mm.

Coloration. Head and thorax yellowish brown. Legs and abdomen pale yellow.

Head. Antenna (Fig. 3.4B) with 10 flagellomeres, with about 15 apical sensilla chaetica, longest antennal seta 328 µm. AR 0.35. Temporals absent. Tentorium 125 µm long, 20 µm wide; stipes 71 µm long, 2 µm wide. Anterior margin of cibarial pump concave. Clypeus with 8 setae. Length of palpomeres (in µm): 14, 19, 21, 25, 62. Palpomere 5/3 ratio 3.0.

Thorax. Dorsocentrals 4, prealars 2. Scutellum with 2 setae. Anapleural suture 107 µm long.

Wing (Fig. 3.4A). VR 5.5. Cu/wing length 0.62, C 408 µm long, Cu 505 µm long, wing width/wing length 0.44 µm, costa with 4 setae.

Table 3.3 Lengths (in μm) and proportions of legs segments of male *Corynoneura ascensa* Fu & Sæther, 2012 (*n* = 2−3)

	fe	ti	ta$_1$	ta$_2$	ta$_3$	ta$_4$	ta$_5$	LR	BV	SV	BR
p$_1$	252	308	162	91	51	24	32	0.53	3.6	2.5	2.2
p$_2$	360	324	194	91	46	20	32	0.60	4.7	3.5	2.6
p$_3$	288	312	160	99	38	18	32	0.51	4.1	3.8	2.0

Source: From Fu, Y., & Sæther, O. A. (2012). *Corynoneura* Winnertz and *Thienemanniella* Kieffer from the Nearctic region (Diptera: Chironomidae: Orthocladiinae). *Zootaxa*, 3536, 1−61. www.mapress.com/j/zt.

Legs. Fore trochanter with dorsal keel. Spur of fore tibia 23 μm long, spurs of mid tibia 8 μm and 14 μm long, of hind tibia 11 μm and 32 μm long. Width at apex of fore tibia 21 μm, of mid tibia 18 μm, of hind tibia (*a*) 30 μm. Width of hind tibia 1/3 from apex (*d*) 20 μm, elongation length (*b*) 36 μm, length of maximum thickening (*c$_1$*) 57 μm, total length of thickening (*c$_2$*) 111 μm, a/d 1.5, b/d 1.8, c_1/d 2.9, c_2/d 5.6. Hind tibia expanded, with comb of 17 setae, 1 seta near spur strongly S-shaped. Lengths and proportions of legs as in Table 3.3.

Hypopygium (Fig. 3.4C and D). Tergite IX with 4 long setae. Laterosternite IX with 2 long setae. Superior volsella anteriomedially fused, small and rounded. Inferior volsella broad, with many glandular setae and strong marginal setae. Phallapodeme curved, 44 μm long. Transverse sternapodeme 11 μm wide. Large attachment point of lateral sternapodeme with phallapodeme placed caudally and caudally directed. Gonostylus apical curved, 30 μm long; megaseta 5 μm long. HR 2.5, HV 4.4.

Female (*n* = 1)

Total length 1.01 mm. Wing length 0.82 mm. Wing width/wing length 0.46 mm. Total length/wing length 1.24 mm. Wing length/ profemur length 3.6 mm.

Coloration. Head yellowish brown with dark brown eyes. Thorax with yellowish ground color with dorsal part of antepronotum, scutum, ventral part of preepisternum. scutellum, and postnotum light brown. Legs almost hyaline. Abdomen light brown.

Head. Eyes bare. Height of eye/height of head 0.63. AR 0.32. Length of flagellomeres (in μm): 44, 33, 33, 32, 44. Ultimate flagellomere with 10 apical sensilla chaetica, without basal setae. Tentorium 81 μm long, 10 μm wide; stipes 69 μm long, 3 μm wide. Clypeus with 6 setae. Palpomere lengths (in μm): 14, 15, 15, 28, 48. Palpomere 5/3 ratio 3.2; palpomeres 1−3 elliptical, no sensilla clavata.

Thorax. Dorsocentrals 7, uniserial; prealars 2; supraalar 1. Scutellum with 2 setae.

Wing (Fig. 3.5A). Wing broad. VR 3.5; one anal vein present, Cu 495 μm long, Cu/wing length 0.61; C 359 μm long; C/wing length 0.44. Costa with 12 setae.

Legs. Fore trochanter with keel. Spur of fore tibia 12 μm long, spur of mid tibia 12 μm long, and hind tibia with three spurs: a long and serrated spur, 46 μm long; a lateral spur 21 μm long, and a short strong spur, 8 μm long. Width at apex of fore tibia 19 μm wide, of mid tibia 21 μm, of hind tibia (a) 26 μm. Width of hind tibia 1/3 from apex (d) 16 μm, elongation length (b) 32 μm, length of maximum thickening (c_1) 61 μm, total length of thickening (c_2) 121 μm, a/d 1.6, b/d 2.0, c_1/d 3.8, c_2/d 7.5. Hind tibial (Fig. 3.5B) comb with 5 setae, 1 seta near short strong spur S-shaped. Lengths and proportions of legs as in Table 3.4.

Abdomen. Abdomen broad. Number of setae on tergites IV−VIII as: 2, 2, 2, 2, 2. No setae on sternites.

Genitalia (Fig. 3.4E and F). Tergite IX with 4 long caudal setae. Cercus 30 μm long, 22 μm wide. Notum length 30 μm, ramus long and broadened medially. Gonocoxite without long setae. Coxosternapodeme with 8 lateral lamellae. Seminal capsule 67 μm long, neck 10 μm long, 4 μm wide.

Pupa ($n = 1−2$)

Total length 1.80 mm. Exuviae yellowish with light brown cephalothorax.

Cephalothorax. Frontal setae 21 μm long. Median antepronotals 14−21 and 10 μm long. Lateral antepronotals 16 μm long. Thoracic horn length 18 μm, width 4 μm. Anterior precorneal seta (PcS_1) 11−14 μm long, PcS_2 14 μm long, PcS_3 11 μm long. PcS_{1-3} almost in a line, PcS_1 6−10 μm from PcS_2, PcS_2 4 μm from PcS_3, PcS_1 8−12 μm from PcS_3, PcS_3 40−60 μm from thoracic horn. Anterior dorsocentral (Dc_1) 14 μm

Table 3.4 Lengths (in μm) and proportions of legs segments of female *Corynoneura ascensa* Fu & Sæther, 2012 ($n = 1$)

	fe	ti	ta$_1$	ta$_2$	ta$_3$	ta$_4$	ta$_5$	LR	BV	SV	BR
P$_1$	228	280	152	81	46	24	34	0.54	3.6	3.3	2.2
P$_2$	260	300	176	85	40	20	28	0.59	4.3	3.2	2.3
P$_3$	260	288	172	95	36	18	24	0.60	4.2	3.2	1.8

Source: From Fu, Y., & Sæther, O. A. (2012). *Corynoneura* Winnertz and *Thienemanniella* Kieffer from the Nearctic region (Diptera: Chironomidae: Orthocladiinae). *Zootaxa*, 3536, 1−61. www.mapress.com/j/zt.

long, Dc_2 11 µm long, Dc_3 11 µm long, Dc_4 14 µm long. Dc_1 located 6 µm from Dc_2, Dc_2 located 11 µm from Dc_3, Dc_3 located 8 µm from Dc_4. Wing sheath with one row of pearls.

Abdomen (Fig. 3.5C). Shagreen and chaetotaxy as illustrated. No taeniate L-setae on tergite I; O-setae present in all segments. Anal lobe 125 µm long. Anal lobe fringe with 16 setae, 117 µm long, restricted to posterior half. Anal macrosetae 110 µm long, median setae 61 µm long.

Larva ($n = 1-2$)

Coloration. Head light brown; head capsule integument smooth, unsculptured. Antenna with basal segment almost transparent, other segments light brown. Abdomen yellowish.

Head. Capsule length 240−248 µm, width 176 µm. Postmentum 212−220 µm long. Sternite II large, rising from small tubercle, I and III not observed. Premandible 28−30 µm long, brush of premandible not observed. Mentum as Fig. 3.5F. Mandibles 43−48 µm long. Antenna (Fig. 3.5D), AR 1.1−1.5. Length of flagellomeres I−IV (in µm): 131−133, 65−69, 71−73, 8. Basal segment 18−19 µm wide, length of blade at apex of basal segment 35−41 µm, ring organ at 73−76 µm from base of antenna.

Abdomen. Length of anal setae 131−152 µm. Procercus 8−10 µm long, 7−8 µm wide. Subbasal seta of posterior parapods split as Fig. 3.5E.

Distribution. NE: USA (Ohio State).

3.4 *CORYNONEURA AURORA* MAKARCHENKO & MAKARCHENKO, 2010 (FIG. 3.6A−E)

Corynoneura aurora Makarchenko & Makarchenko, 2010: 4.

Diagnostic characters. Total length 0.75−1.20 mm. Wing length 0.64−0.96 mm. Antenna with 10−12 flagellomeres; apex of terminal flagellomere roundish, with some sensitive hairs. t_3 with straight seta in apex; a/d 1.5−1.8; b/d 1.0−1.2. Inferior volsella is situated near middle part of gonocoxite, in distal part become wider, with roundish edge, covered by setae except apical hyaline part. Superior volsella roundish-triangular and bare; under superior volsellae in inner angles of gonocoxites some setae are situated. Sternapodeme inverted U-shaped. Phallapodemes are low and narrow.

Male (Description from Makarchenko & Makarchenko, 2010)

Total length 0.75−1.2 mm. Total length/wing length 1.14−1.33.

Head. Antenna (Fig. 3.6A) with 10−12 flagellomeres, antenna with 10 flagellomeres: AR 0.58−0.62. Antenna with 11 flagellomeres: AR

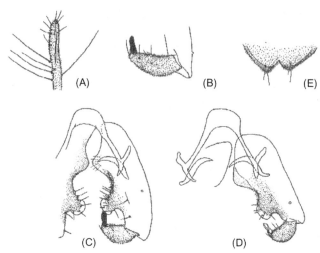

Figure 3.6 *Corynoneura aurora* Makarchenko & Makarchenko, 2010, Male imago (From Makarchenko, E. A., & Makarchenko, M. A. (2010). New data on the fauna and taxonomy of Corynoneura Winnertz (Diptera, Chironomidae, Orthocladiinae) for the Russian Far East and bordering territories. *Euroasian Entomological Journal, 9*(3), 353–370 + II). (A) Distal part of terminal antennal flagellomere. (B) Gonostylus. (C and D) Total view of hypopygium, from above (tergite IX deleted). (E) Posterior margin of tergite IX.

0.36–0.48. Antenna with 12 flagellomeres: AR 0.52–0.60. Clypeus with 6–7 setae. Lengths of palpomeres (in μm): 8–12, 12–18, 18–24, 24–36, 48–60.

Thorax. Brown. Dorsocentrals 4–5, prealars 1–2. Scutellum with 2 setae.

Wing. Wing length 0.64–0.96 mm.

Legs. Fore trochanter with dorsal keel. Spurs of fore tibia 24 μm long, of mid tibia 8 μm and 12–16 μm long. Hind tibia expanded, with comb of 12–13 setae, lengths and proportions of legs as in Table 3.5.

Hypopygium (Fig. 3.6B–E). The posterior margin of tergite IX medially concave, with 2–6 long setae. Laterosternite IX with 1–3 long setae. Gonocoxite 75–110 μm. Superior volsella roundish-triangular. Inferior volsella with roundish edge, placed medially. Phallapodeme curved, 41 μm long, placed in caudal position of sternapodeme. Transverse sternapodeme 20–30 μm wide, curved into U-shaped. Gonostylus slightly convex on the outer edge, 23–35 μm long.

Distribution. PA: Russian Far East.

Table 3.5 Lengths (in μm) and proportions of legs segments of male *Corynoneura aurora* Makarchenko & Makarchenko, 2010

	fe	ti	ta_1	ta_2	ta_3	ta_4
p_1	220–290	250–355	128–190	80–105	38–48	15–21
p_2	270–380	250–370	140–185	73–110	35–53	15–23
p_3	240–320	270–360	160–220	98–133	38–50	15–18

	ta_5	LR	BV	SV	BR	
p_1	30–40	0.51–0.53	3.46–3.92	3.30–3.67	—	
p_2	25–39	0.50–0.56	4.20–4.60	3.71–4.05	—	
p_3	33–45	0.55–0.61	3.62–3.79	3.09–3.62	—	

3.5 *CORYNONEURA AUSTRALIENSIS* FREEMAN, 1961 (FIG. 3.7A AND B)

Corynoneura australiensis Freeman, 1961: 673.

Diagnostic characters. Male antenna with 10 flagellomeres; terminal flagellomere being clubbed and elongate and with traces of a division into 3 segments. Sternapodeme curved into U-shaped. Phallapodeme curved, placed in lateral position of sternapodeme. Gonostylus strongly curved, with a small triangular projecting.

Male (Description from Freeman, 1961)

Total length 1.55 mm. Total length/wing length 1.99.

Coloration. Head blackish, mouthparts and antennae paler. Thorax blackish. Legs brown. Tergites whitish on basal 4 segments.

Head. Eyes bare. Antenna with 10 flagellomeres (Fig. 3.7A); AR about 0.40, the apical one clubbed, with short hair at extremity and showing incipient division into 3 segments; basal two-thirds of apical segment bearing plume hairs.

Thorax. Stripes fused.

Wing. Wing length 1.0 mm. Clavus short, reaching hardly more than one-quarter of wing length; halters pale.

Legs. LR about 0.5.

Hypopygium (Fig. 3.7B). Sternapodeme curved into U-shaped. Phallapodeme curved, placed in lateral position of sternapodeme. Attachment point of lateral sternapodeme with phallapodeme placed and directed caudally. Gonostylus strongly curved, with a small triangular projecting.

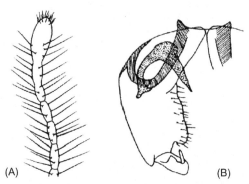

(A)   (B)

Figure 3.7 *Corynoneura australiensis* Freeman, 1961; Male imago (Reproduced from Freeman (1961) with permission from CSIRO Publishing). (A) Apex of male antenna. (B) Ventral view of hypopygium.

Female

Paler than male on thorax where ground color yellowish and stripes separate; wings broader and clavus reaching almost to middle of wing.

Distribution. Australian (AU): Australia (Australian Capital Territory).

3.6 *CORYNONEURA BODOQUENA* WIEDENBRUG ET AL., 2012 (FIGS. 3.8A−D AND 3.9A−D)

Corynoneura bodoquena Wiedenbrug et al., 2012: 5.

Diagnostic characters. The male of *C. bodoquena* is differentiated from the other species which also present phallapodeme attachment placed in the caudal apex of the sternapodeme, phallapodeme sclerotized on posterior margin and apex of hind tibia with a S-shaped seta, by the antenna plumose with 7 flagellomeres, aedeagal lobe with a wide base apically pointed, inferior volsella low and rounded, not apical. The pupa has anal lobe rectangular with posterolateral corner with diagonal sclerotized lines, tergites IV−VIII with 4 taeniate lateral setae of about the same size, tergites II−VI with a posterior row of very strong spines and sternites with shagreen only.

Male (Description from Wiedenbrug et al., 2012)

Total length 0.92−0.93 mm. Wing length 0.46−0.52 mm.

Coloration. Thorax brownish. Abdominal tergites greenish posteriorly brownish. Legs whitish.

Head. AR = 0.84−0.98. Antenna with 7 flagellomeres, apical flagellomere 120−142 μm (Fig. 3.8A). Flagellomeres with more than one row of

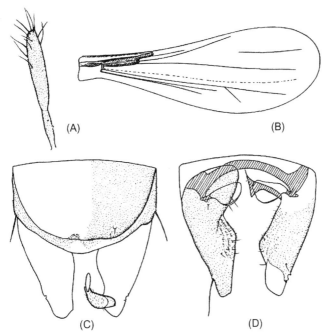

Figure 3.8 *Corynoneura bodoquena* Wiedenbrug et al., 2012, Male imago (From Wiedenbrug, S., Lamas, C. J. E., & Trivinho-Strixino, S. (2012). A review of the genus Corynoneura Winnertz (Diptera: Chironomidae) from the neotropical region. *Zootaxa*, *3574*, 1−61). (A) Terminal flagellomeres. (B) Wing. (C and D) Hypopygium. (C) Tergite IX and gonostylus included. (D) Tergite IX removed, sclerites hatched; right is dorsal view; left is ventral view.

setae each. Clypeus with 6−8 setae. Palpomere lengths (in μm): —, —, 10−12, 12−15, 37. Eyes pubescent.

Thorax. Antepronotal lobes dorsally tapering. Dorsocentrals 3, prealars 2. Scutellars 2.

Wing (Fig. 3.8B). Clavus/wing length 0.23. Anal lobe absent (Fig. 3.8B).

Legs. Hind tibial scale 25 μm long, with one small S-seta.

Hypopygium (Fig. 3.8C and D). Tergite IX with 4 setae. Laterosternite IX with 1 seta. Superior volsella absent. Inferior volsella low and rounded, not apical. Aedeagal lobe with a wide base, apically pointed. Sternapodeme rounded, phallapodeme caudal attached. Phallapodeme 20 μm long, posteriorly sclerotized and rounded. Gonocoxite 52−57 μm long. Gonostylus 15−20 μm long.

Pupa (Description from Wiedenbrug et al., 2012)

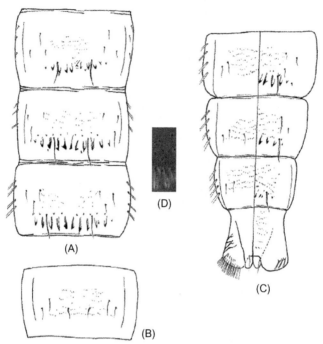

(D)

(A)

(C)

(B)

Figure 3.9 *Corynoneura bodoquena* Wiedenbrug et al., 2012; Pupa imago (From Wiedenbrug, S., Lamas, C. J. E., & Trivinho-Strixino, S. (2012). A review of the genus *Corynoneura* Winnertz (Diptera: Chironomidae) from the neotropical region. *Zootaxa, 3574,* 1−61). (A) Tergites II−IV. (B) Sternite IV. (C) Segments VI−IX and anal lobe; left is ventral view, right is dorsal view. (D) Detail of posterior shagreen of tergite IV.

Total length 1.21−1.27 mm.

Coloration (exuviae). Cephalothorax light brown, abdomen transparent except grayish muscle markings.

Cephalothorax. Frontal apotome granulated. Thorax suture almost smooth except scutal tubercle region with small wide spinules. All Dc-setae thin taeniate, except Dc_3 taeniate 70−82 μm long. Dc_1 displaced ventrally. Wing sheaths with two to five rows of pearls.

Abdomen (Fig. 3.9A−D). Tergite I and sternites I and II almost bare, tergites II−VIII with fine shagreen and posterior row of very strong spines, spines at tergite II and VIII smaller. Sternite III with very fine shagreen, IV−VIII with fine shagreen. Sternite VII with posterior shagreen slightly stronger. Conjunctives tergite II/III and tergite III/IV with few spinules. Segment I with 1, II with 3 L-setae, III with 3 taeniate L-setae and IV−VIII with 4 long taeniate L-setae. Anal lobe rectangular, posterolateral corner with diagonal sclerotized lines (Fig. 3.9C). Anal lobe with fringe restricted to posterior margin, 3 taeniate macrosetae and inner setae taeniate.

Remarks. Pupal exuviae of *C. bodoquena* were collected in small streams with stony bottom and coarse sand of a Semideciduous Forest in Mato Grosso do Sul, Brazil.

Distribution. Neotropical (NT): Brazil (Bodoquena).

3.7 *CORYNONEURA BORACEIA* WIEDENBRUG ET AL., 2012 (FIG. 3.10A−C)

Corynoneura boraceia Wiedenbrug et al., 2012: 9.

Diagnostic characters. The male of *Corynoneura boraceia* can be differentiated from others species with a distinct group of sensilla chaetica on the antennal apex by the lateral sternapodeme concave and transverse sternapodeme rounded.

Male (Description from Wiedenbrug et al., 2012)

Wing length 0.61 mm.

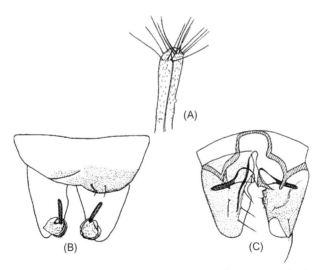

Figure 3.10 *Corynoneura boraceia* Wiedenbrug et al., 2012; Male imago (From Wiedenbrug, S., Lamas, C. J. E., & Trivinho-Strixino, S. (2012). A review of the genus *Corynoneura* Winnertz (Diptera: Chironomidae) from the neotropical region. *Zootaxa*, *3574*, 1−61). (A) Terminal flagellomere. (B and C) Hypopygium. (B) Tergite IX and gonostylus included; left is ventral view of tergite IX, right is dorsal view. (C) Tergite IX and gonostylus removed, sclerites hatched; right is dorsal view, left is ventral view.

Coloration. Thorax brownish. Abdominal tergites I–IV whitish, other tergites and genitalia brownish.

Head. AR = 0.32. Antenna with 8 flagellomeres, apical flagellomere 150 μm. Flagellomeres with more than one row of setae each. Antennal apex with 15 sensilla (Fig. 3.10A). Eyes pubescent.

Thorax. Antepronotal lobes dorsally tapering. Dorsocentrals 3, prealars 2. Scutellars 2.

Wing. Clavus/wing length 0.19. Anal lobe absent.

Legs. Missing.

Hypopygium (Fig. 3.10B and C). Tergite IX with 4 setae. Laterosternite without seta. Superior volsella absent. Inferior volsella slightly rounded at some distance before the apex of gonocoxite. Aedeagal lobe short and triangular. Transverse sternapodeme rounded, lateral sternapodeme concave. Phallapodeme 25 μm long, oral margin sclerotized, attached in the caudal apex of the sternapodeme. Gonocoxite 52 μm long. Megaseta long.

Remarks. *C. boraceia* was collected with a Malaise trap in the Atlantic Forest of São Paulo State, Brazil. See remarks of *Corynoneura diogo*.

Distribution. NT: Brazil (Salesópolis).

3.8 *CORYNONEURA BRUNDINI* HIRVENOJA & HIRVENOJA, 1988 (FIG. 3.11A–E)

Corynoneura brundini Hirvenoja & Hirvenoja, 1988: 223.

Diagnostic characters. Male antenna with 10 flagellomeres; AR 0.39–0.47; tip of hind without S-shaped. Inferior volsellae absent, sternapodeme inverted V-shaped, phallapodeme and gonostylus strongly curved.

Male (Description from Hirvenoja & Hirvenoja, 1988)

Head. Antenna with 10 flagellomeres, the last flagellomere with many long bristles (Fig. 3.11A), as long as the previous 3–3.5 flagellomeres together; AR 0.39–0.47. Lengths of palpomeres (in μm) (Fig. 3.11B): —, 24–26, 29–31, 29–34, 34–43.

Wing (Fig. 3.11C). Wing length 1.0–1.1 mm.

Legs. Tip of hind tibia expanded, without S-shaped (Fig. 3.11D). Lengths and proportions of legs as in Table 3.6.

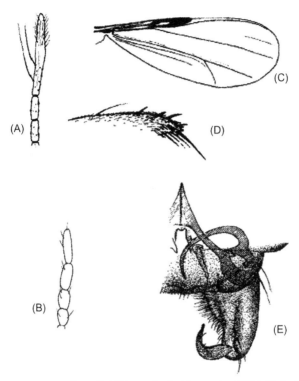

Figure 3.11 *Corynoneura brundini* Hirvenoja & Hirvenoja, 1988, Male imago (From Hirvenoja, M., & Hirvenoja, E. (1988). *Corynoneura brundini* spec. nov. Ein Beitrag zur Systematik der Gattung *Corynoneura* (Diptera, Chironomidae). *Spixiana Supplement, 14*, 213–232). (A) Terminal flagellomere. (B) Palpomeres. (C) Wing. (D) Apex of Hind tibia. (E) Hypopygium.

Hypopygium (Fig. 3.11E). Posterior margin weakly bilobed, and with many short setae, each of laterosternite IX with 2 long setae. Inferior volsellae absent, sternapodeme inverted V-shaped, phallapodeme strongly curved not extending beyond tergite IX, gonostylus curved apically.

Female (Description from Hirvenoja & Hirvenoja, 1988)

Head. Antenna with 6 flagellomeres. Lengths of palpomeres (in μm): —, 17–24, 22–24, 24–32, 34–38.

Wing. Wing length 0.9–1.2 mm.

Legs. Lengths and proportions of legs as in Table 3.7.

Distribution. PA: Finland.

Table 3.6 Lengths (in μm) and proportions of legs segments of male *Corynoneura brundini* Hirvenoja & Hirvenoja, 1988

	fe	ti	ta_1	ta_2	ta_3	ta_4	ta_5	LR
p_1	276–333	305–371	124–143	52–57	33–38	24–29	38–43	0.38–0.41
p_2	381–429	352–381	190–210	86–95	48–52	29	38–48	0.53–0.57
p_3	295–343	386–436	190–218	67–76	29–38	24–29	38	0.49–0.54

Table 3.7 Lengths (in μm) and proportions of legs segments of female *Corynoneura brundini* Hirvenoja & Hirvenoja, 1988

	fe	ti	ta_1	ta_2	ta_3	ta_4	ta_5	LR
p_1	229–276	276–324	114–133	48–52	29–33	24–29	29–38	0.40–0.45
p_2	333–371	305–357	171–190	67–76	38	24–29	29–38	0.51–0.59
p_3	276–333	370–420	200–230	62–71	29–34	24–29	38	0.51–0.56

3.9 *CORYNONEURA CANCHIM* WIEDENBRUG ET AL., 2012 (FIGS. 3.12A–C, 3.13A–D, AND 3.14A–G)

Corynoneura canchim Wiedenbrug et al., 2012: 10.

Diagnostic characters. The male of *Corynoneura canchim* can be differentiated from other species with phallapodeme attachment placed in the caudal apex of the sternapodeme, phallapodeme sclerotized on posterior margin and apex of hind tibia with a S-shaped seta, by the antenna plumose with 8 flagellomeres, tergite X apparent, superior volsella low without long seta, inferior volsella low and broad, gonostylus curved. The female of *C. canchim* is characterized by the seminal capsules 32–37 μm long, both similarly sclerotized; copulatory bursa semi-circled, orally median invaginated. The pupa has rounded anal lobe, tergites with homogeneous shagreen and sternite II with long and colorless spinules. The larva has long antennae with the first segment longer than the

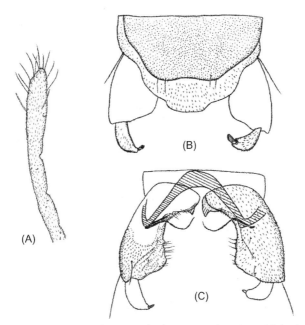

Figure 3.12 *Corynoneura canchim* Wiedenbrug et al., 2012; Male imago (From Wiedenbrug, S., Lamas, C. J. E., & Trivinho-Strixino, S. (2012). A review of the genus *Corynoneura* Winnertz (Diptera: Chironomidae) from the neotropical region. *Zootaxa*, *3574*, 1–61). (A) Terminal flagellomeres. (B and C) Hypopygium. (B) Tergite IX and gonostylus included. (C) Tergite IX removed, sclerites hatched; right is ventral view, left is dorsal view.

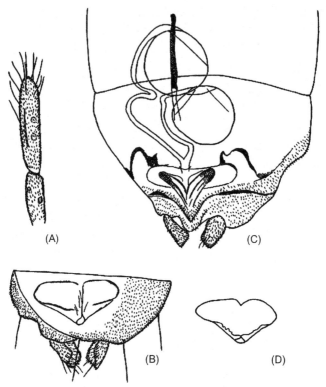

(A)   (C)

(B)   (D)

Figure 3.13 *Corynoneura canchim* Wiedenbrug et al., 2012; Female imago (From Wiedenbrug, S., Lamas, C. J. E., & Trivinho-Strixino, S. (2012). A review of the genus *Corynoneura* Winnertz (Diptera: Chironomidae) from the neotropical region. *Zootaxa, 3574*, 1−61). (A) Terminal flagellomeres. (B−D) Genitalia. (B) Dorsal view with view of labia. (C) Ventral view of genitalia. (D) Ventral view of labia.

postmentum length, mentum with two median teeth and head integument with dorsolateral sclerotized fold.

Male (Description from Wiedenbrug et al., 2012)

Total length 1.00−1.32 mm. Wing length 0.54−0.56 mm.

Coloration. Thorax brownish. Abdominal tergites I−IV whitish, other tergites and genitalia brownish. Legs whitish.

Head. AR = 0.70−0.76. Antenna with 8 flagellomeres, apical flagellomere 120−150 μm (Fig. 3.12A). Flagellomeres with more than one row of setae each. Clypeus with 8 setae. Palpomere lengths (in μm): —, —, 15, 17, 45. Eyes pubescent.

Thorax. Antepronotal lobes dorsally tapering. Dorsocentrals 3, prealars 2. Scutellars 2.

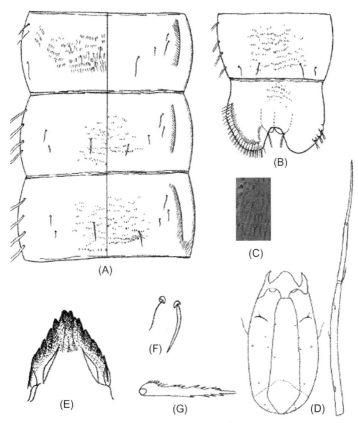

Figure 3.14 *Corynoneura canchim* Wiedenbrug et al., 2012; Immatures imago (From Wiedenbrug, S., Lamas, C. J. E., & Trivinho-Strixino, S. (2012). A review of the genus *Corynoneura* Winnertz (Diptera: Chironomidae) from the neotropical region. *Zootaxa, 3574,* 1–61). (A–C) Pupa imago. (A) Segments II–IV; left is ventral view, right is dorsal view. (B) Tergites VIII, IX and anal lobe; left is ventral view of anal lobe, right is dorsal view of anal lobe. (C) Detail of shagreen of sternite II. (D–G) Larva imago. (D) Head, dorsal view and separated antenna. (E) Mentum. (F) Abdominal setae. (G) Subbasal seta of posterior parapod.

Wing. Clavus/wing length 0.21. Anal lobe absent.

Legs. Hind tibial scale 35–37 μm long, with one small S-seta.

Hypopygium (Fig. 3.12B and C). Tergite X exposed beyond tergite IX. Tergite IX with 4 setae. Laterosternite IX with 1 seta. Superior volsella low without long setae. Inferior volsella broad, low and rounded. Aedeagal lobe small. Transverse sternapodeme rounded. Phallapodeme 27 μm long, with posterior margin sclerotized, rounded, attached in the caudal apex of the sternapodeme. Gonocoxite 52−57 μm long. Gonostylus 12−17 μm long.

Female (Description from Wiedenbrug et al., 2012)

Total length 0.95 mm. Wing length 0.57 mm.

Coloration. Thorax brownish. Abdominal tergites I−III whitish, other tergites and genitalia brownish. Legs whitish.

Head. AR = 0.48−0.49. Antenna with 5 flagellomeres, apical flagellomere 50−53 μm long (Fig. 3.13A). Flagellomeres with more than one row of setae each. Eyes pubescent.

Thorax. Antepronotal lobes dorsally tapering.

Wing. Clavus/wing length 0.41. Anal lobe absent.

Legs. Hind tibial scale 30−37 μm long, with one small S-seta.

Genitalia (Fig. 3.13B−D). Tergite IX with 2 setae. Laterosternite IX with 1 seta. Two seminal capsules respectively 32 μm and 37 μm long; one spermathecal duct with a loop, second shorter and straighter, both ducts join together shortly before seminal eminence, which has sclerotized outer borders. Notum 45 μm long. Apodeme lobe well sclerotized, apically pointed. Coxosternapodeme, strongly curved, with one end at roof of copulatory bursa, last semicircle-shaped, orally median invaginated. Labia membranous, bare, funnel shaped, apically building the accessory gonopore (Fig. 3.13D). Gonocoxapodeme straight, gonapophyses median smoothly pointed. Postgenital plate rounded with few microtrichia. Cercus 20−27 μm long.

Pupa (Description from Wiedenbrug et al., 2012)

Total length 1.21−1.35 mm.

Coloration (exuviae). Cephalothorax light brown, abdomen transparent except brownish muscle markings, lateral margin and anal lobe.

Cephalothorax. Frontal apotome rugose. Thorax suture smooth. All Dc-setae thin taeniate. Dc_1 displaced ventrally. Wing sheaths with two rows of pearls.

Abdomen (Fig. 3.14A−C). Tergite I almost bare, tergite II with few small shagreen points, shagreen on tergites III−IX quite homogeneous, on tergites III−VI posterior points slightly larger, on III−V posterolateral points slightly larger than posteromedian. Sternite II with long colorless spinules (Fig. 3.14C). Other sternites with fine shagreen. Conjunctives sternites IV / V −VII/VIII with small spinules. Segment I with 1, II with 3 L-setae and IV−VIII with 4 long taeniate L-setae. Anal lobe rounded (Fig. 3.14B). Anal lobe with fringe not complete, 3 taeniate macrosetae and inner setae taeniate.

Larva (Description from Wiedenbrug et al., 2012)

Head. Postmentum 145−162 µm long. Head capsule integument smooth, with dorsolateral sclerotized fold (Fig. 3.14D). Mentum with two median teeth, first lateral tooth small and adpressed to median, five additional lateral teeth (Fig. 3.14E). Antenna 404−425 µm long, segments two and three brown (not drawn). First segment longer than postmentum length.

Abdomen. Ventral setae modified, slightly wider than dorsal setae (Fig. 3.14F). Subbasal seta on posterior parapod serrated at the base and mostly at one margin (Fig. 3.14G).

Remarks. Larvae of *C. canchim* were collected in artificial substrate of stones and leaves at slow flowing, lowland streams in São Paulo State, Brazil.

Distribution. NT: Brazil (São Carlos).

3.10 *CORYNONEURA CAPITANEA* FU & SÆTHER, 2012 (FIG. 3.15A AND B)

Corynoneura capitanea Fu & Sæther, 2012: 12.

Material examined. Holotype male, USA, Fliclay County, unnamed creek, 1 pharate male, 13.i.1995, J. H. Epler (ZMBN Type No. 454).

Diagnostic characters. The pharate male is characterized by antenna with 11 flagellomeres, AR 0.61, apically acute; hind tibia without S-shaped seta; gonostylus with well-developed megaseta.

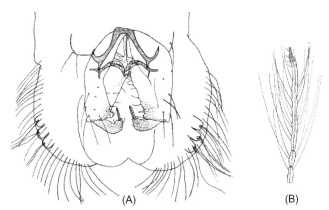

(A)   (B)

Figure 3.15 *Corynoneura capitanea* Fu & Sæther, 2012; Pharate male imago (From Fu, Y., & Sæther, O. A. (2012). *Corynoneura* Winnertz and *Thienemanniella* Kieffer from the Nearctic region (Diptera: Chironomidae: Orthocladiinae). *Zootaxa, 3536*, 1−61. www.mapress.com/j/zt). (A) Hypopygium, ventral view and anal lobe. (B) Antenna.

Pharate male (*n* = 1)

Coloration. Head brown, other parts yellowish brown.

The limited characters known: antenna (Fig. 3.15B) with 11 flagello-meres, ultimate flagellomere 212 μm, AR 0.61, apically acute, with about 8 apical sensilla chaetica. Hind tibia slightly expanded, with comb of 11 setae, without S-shaped seta.

Laterosternite IX with 2 long setae. Superior volsella triangular, anterome-dially separated. Inferior volsella small, rectangular, placed caudally. Phallapodeme slightly curved, 44 μm long, in caudal position of sternapodeme. Transverse sternapodeme 20 μm wide. Gonocoxite 95 μm long. Gonostylus apically curved, 34 μm long; megaseta 14 μm long. HR 2.8 (Fig. 3.15A).

Distribution. NE: USA.

3.11 *CORYNONEURA CARINATA* SINGH & MAHESHWARI, 1987 (FIG. 3.16A−F)

Corynoneura carinata Singh & Maheshwari, 1987: 11.

Diagnostic characters. Male antenna with 8 flagellomeres. Gonostylus curved. This species comes near to *Corynoneura brevinervis* Kieffer but differs in flagellum being seven segmented, differences in antennal and leg ratios, venarum ratio and the ratio of gonocoxite and gonostylus. It can easily be differentiated by the presence of keel or carina on gonostylus.

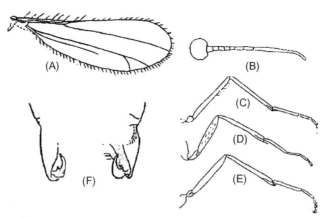

Figure 3.16 *Corynoneura carinata* Singh & Maheshwari, 1987, Male imago (From Singh, S. & Maheshwari, G. (1987). Chironomidae (Diptera) of Chandertal Lake, Lahul Valley (Northwest Himalaya). Annals of Entomology, 5(2), 11−20). (A) Wing. (B) Antenna. (C−E) Fore, mid and hind leg. (F) Hypopygium.

Table 3.8 Lengths (in μm) and proportions of legs segments of male *Corynoneura carinata* Singh & Maheshwari, 1987

	fe	ti	ta₁	ta₂	ta₃	ta₄	ta₅	LR
p₁	290	390	180	—	60	20	30	0.50
p₂	420	400	200	80	40	20	20	0.50
p₃	370	390	190	100	50	30	40	0.40

Male (Description from Singh & Maheshwari, 1987)

Total length 2.0−2.1 mm. Wing length 1.29 mm.

Coloration. Pedicel dark brown, flagellum light brown, head brown, eyes black, palpi light brown. Scutum, scutellum and postnotum blackish brown. Haltere light brown. Legs light brown. Abdomen brown.

Head. Ratio of maximum head width and width between eyes 2.13. Antenna (Fig. 3.16B) with 8 flagellomeres, ultimate flagellomere elongated, AR 1.16. Lengths of palpomeres (in μm): 7, 16, 13, 11.

Thorax. Dorsocentrals 5, prealars 2. Scutellum with 2 setae. Anapleural suture 91 μm long.

Wing (Fig. 3.16A). Costa with 9 setae, R$_{4+5}$ fused with costa and forms clavus. Costa ends about one-third of the length of wing. VR 2.60.

Legs (Fig. 3.16C−E). Fore trochanter with dorsal keel, fore femur narrow at both ends with six setae distally; apex of fore tibia with two spurs, outer spur not reaching up to half of inner spur; middle tibial apex with two spurs, outer spur not reaching up to half of inner spur; fourth tarsal segment smaller than the fifth one. Apex of hind tibia enlarged and with one comb and two spurs, size of enlarged tibial apex 0.06 mm. Lengths and proportions of legs as in Table 3.8.

Hypopygium (Fig. 3.16F). Setae present on the inner margin of gonocoxa. Gonocoxa 73 μm long. Gonostylus folded inwards, 44 μm long. Ratio of gonocoxa and gonostylus 1.90.

Distribution. Oriental (OR): India (Himachal Pradesh).

3.12 *CORYNONEURA CARRIANA* EDWARDS, 1924 (FIG. 3.17A−H)

Corynoneura carriana Edwards, 1924: 188; Hirvenoja & Hirvenoja, 1988: 231; Langton & Visser, 2003: 319; Langton & Pinder, 2007: 88; Makarchenko & Makarchenko, 2010: 358.

Corynoneura heterocera Kieffer, 1915: 87.

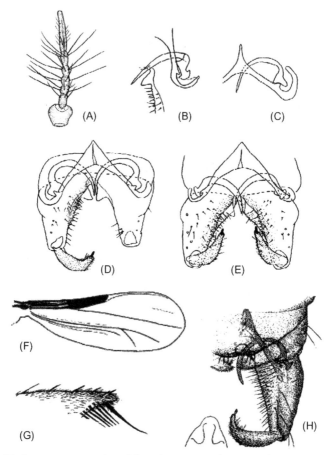

Figure 3.17 *Corynoneura carriana* Edwards, 1924, Male imago (From Hirvenoja, M., & Hirvenoja, E. (1988). *Corynoneura brundini* spec. nov. Ein Beitrag zur Systematik der Gattung *Corynoneura* (Diptera, Chironomidae). *Spixiana Supplement, 14*, 213–232). (A) Antenna. (B) Connection of phallapodeme with sternapodeme. (C) Sternapodema and phallapodeme. (D and E) Total view of hypopygium, from above (tergite IX deleted) (From Makarchenko, E. A., & Makarchenko, M. A. (2010). New data on the fauna and taxonomy of *Corynoneura* Winnertz (Diptera, Chironomidae, Orthocladiinae) for the Russian Far East and bordering territories. *Euroasian Entomological Journal, 9*(3), 353–370 + II). (F) Wing. (G) Apex of hind tibia. (H) Hypopygium.

Corynoneura crassipes Kieffer, 1925: 564.

Diagnostic characters. The male imago is characterized by antenna with 10 flagellomeres (except the species recorded in Russia which antenna with 6 flagellomeres), superior volsella projecting triangular; inferior volsella absent; sternapodeme curved into V-shape.

Table 3.9 Lengths (in μm) and proportions of legs segments of male *Corynoneura carriana* Edwards

	fe	ti	ta$_1$	ta$_2$	ta$_3$	ta$_4$
p$_1$	240−272	260−292	120−148	60−80	40	16−20
p$_2$	308−328	280−300	148−168	72−80	38−40	20
p$_3$	256−280	240−280	132−152	72−88	28−32	16−20

	ta$_5$	LR	BV	SV	BR
p$_1$	36	0.46−0.51	3.97−4.24	3.70−4.17	—
p$_2$	36	0.53−0.56	4.43−4.52	3.74−3.97	—
p$_3$	32−36	0.53−0.55	4.13−4.21	3.68−3.78	—

Source: From Makarchenko, E. A., & Makarchenko, M. A. (2010). New data on the fauna and taxonomy of Corynoneura Winnertz (Diptera, Chironomidae, Orthocladiinae) for the Russian Far East and bordering territories. *Euroasian Entomological Journal*, *9*(3), 353−370 + II.

Male (Description from Hirvenoja & Hirvenoja, 1988; Makarchenko & Makarchenko, 2010; Schlee, 1968b)

Total length 1.1−1.3 mm. Total length/wing length 1.59−1.97.

Head. Antenna (Fig. 3.17A) with 6 flagellomeres (Russian species), antenna with 10 flagellomeres (other areas recorded this species), AR 0.62−0.73. Clypeus with 6−7 setae. Lengths of palpomeres (in μm): —, 12−18, 20−24, 24−28, 36−40.

Thorax. Pronotum dark yellow. Dorsocentrals 5, prealars 2.

Wing (Fig. 3.17F). Wing length 0.64−0.69 mm. Cu/wing length 0.61−1.62 mm, costa length/wing length 0.40−0.44 mm, wing width/ wing length 0.38−0.42 mm.

Legs. Fore trochanter with dorsal keel. Spurs of fore tibia 12−16 μm long, of mid tibia 8 μm and 8 μm long, of hind tibia 24−28 μm long. Hind tibia expanded, with comb of 10−13 setae (Fig. 3.17G). Lengths and proportions of legs as in Table 3.9.

Hypopygium (Fig. 3.17B−E, H). The posterior margin of tergite IX medially concave, with 4−8 long setae. Laterosternite IX with 2−4 long setae. Gonocoxite 80−88 μm, superior volsella hyaline triangular, inferior volsella absent. Sternapodeme curved into V-shaped. Gonostylus narrow, slightly curved, 40−56 μm long.

Distribution. PA: Algeria, Austria, Belgium, Corsica, Czech Republic, Denmark, Finland, France, Germany, Great Britain, Ireland, Italy, Kaliningrad, Morocco, Netherlands, Norway, Poland, Russia (Far East), Slovakia, Spain, Sweden, Switzerland.

3.13 *CORYNONEURA CAUDICULA* FU & SÆTHER, 2012 (FIGS. 3.18A—G AND 3.19A—E)

Corynoneura n. sp. 1 Bolton, 2007: 29;

Corynoneura caudicula Fu & Sæther, 2012: 12.

Material examined. Holotype male, USA, Ohio, Highland County, Evans Road, Clear Creek, 30.ix.1986, M. J. Bolton (ZMBN Type No. 455). Paratypes: Ohio, Highland County, Rocky Fork, 30.ix.1986, 4 males, 4 pupal and 4 larval exuviae, M. J. Bolton; Ohio, Delaware County, unnamed tributary to Olentangy River north of winter road,

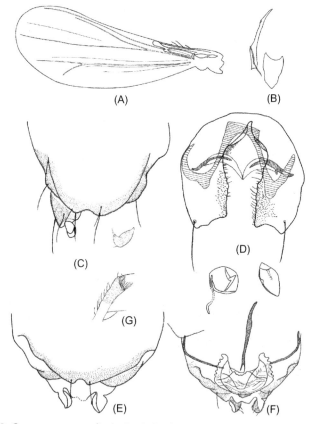

Figure 3.18 *Corynoneura caudicula* Fu & Sæther, 2012 (From Fu, Y., & Sæther, O. A. (2012). *Corynoneura* Winnertz and *Thienemanniella* Kieffer from the Nearctic region (Diptera: Chironomidae: Orthocladiinae). *Zootaxa, 3536,* 1—61. www.mapress.com/j/zt). (A—D) Male imago. (A) Wing. (B) Tentorium, stipes and cibarial pump. (C) Hypopygium, dorsal view. (D) Hypopygium, ventral view. (E and F) Female imago. (E) Genitalia, dorsal view. (F) Genitalia, ventral view. (G) Apex of hind tibia.

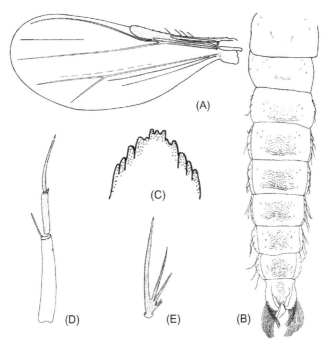

Figure 3.19 *Corynoneura caudicula* Fu & Sæther, 2012 (From Fu, Y., & Sæther, O. A. (2012). *Corynoneura* Winnertz and *Thienemanniella* Kieffer from the Nearctic region (Diptera: Chironomidae: Orthocladiinae). *Zootaxa, 3536,* 1−61. www.mapress.com/j/zt). (A) Female imago: wing. (B) Pupa: Tergites I −IX. (C−E). Larva. (C) Mentum. (D) Antenna. (E) Subbasal seta of posterior parapod.

15.x.1987, 1 female, 1 pupal and 1 larval exuviae, M. J. Bolton (ZMBN, MJB).

Diagnostic characters. The male imago is characterized by a strongly concave anterior margin of cibarial pump, palpomere 3 with 1 sensillum clavatum; superior volsella not developed; inferior volsella broad, placed caudally on gonocoxite; sternapodeme curved into U-shaped, and lateral sternapodeme with large caudal attachment point. The female has 6 transparent lateral lamellae on coxosternapodeme, tergite IX with 2 large setae, and gonocoxite with 1 long seta. The pupa have no taeniate L-setae and shagreen on tergite I , tergites IV / V −VI/VII with somewhat thick shagreen, and 17−21 taeniae in fringe of anal lobe restricted to posterior half of lobe. The larvae are separable by the antenna shorter than head (AR 0.85−0.91), mentum with 3 median teeth, lateral tooth adjacent to median tooth very small, apex of second antennal segment with small blade, and head capsule integument smooth.

Male (*n* = 2−3)

Total length 1.16−1.34, 1.24 mm. Wing length 0.67−0.73, 0.71 mm. Total length/wing length 1.69−1.83, 1.75 mm. Wing length/profemur length 2.35−3.10, 2.85 mm.

Coloration. Head brown. Thorax with upper antepronotum, upper scutum, scutellum, postnotum and preepisternum brown, median yellowish. Legs and abdomen light brown.

Head. Antenna with 9−10 flagellomeres, flagellomeres 9 and 10 partly to fully separated; when 9 flagellomeres, ultimate flagellomere 131−149 µm long (AR 0.57−0.63); when 10 flagellomeres, ultimate flagellomere 101−119 µm long (AR 0.36−0.41); ultimate flagellomere distinctly expanded apically, with about 10−15, 12 apical sensilla chaetica; longest antennal seta 210−220, 214 µm. Temporals absent. Tentorium, stipes and cibarial pump as in Fig. 3.18B, tentorium 117−131, 123 µm long; 14−16, 15 µm wide; stipes 44−63 µm long, 3 µm wide. Anterior margin of cibarial pump strongly concave. Clypeus with 8−10, 9 setae. Palpomere 3 with 1 sensilla clavata. Length of palpomeres (in µm): 14−18, 16; 11−14, 12; 17−19, 18; 21−26, 24; 48−58, 53. Palpomere 5/3 ratio 2.8−3.4, 3.0.

Thorax. Antepronotals 5. Dorsocentrals 3−4, prealars 2. Scutellum with 2 setae. Anapleural suture 111−131, 121 µm long.

Wing (Fig. 3.18A). VR 3.4−3.6, 3.5. Cu/wing length 0.57−0.61, 0.59; C 168−180, 173 µm long; Cu 384−440, 419 µm long; wing width/wing length 0.40−0.43, 0.41. Brachiolum with 1 seta, costa with 5−7 setae.

Legs. Fore trochanter with dorsal keel. Spur of fore tibia 21−28, 25 µm long, of mid tibia 8−11, 10 µm long, and of hind tibia 30−36, 32 µm long. Width at apex of fore tibia 14−24, 18 µm, of mid tibia 15−19, 17 µm, of hind tibia (*a*) 32−35, 33 µm. Width of hind tibia 1/3 from apex (*d*) 12−19, 14 µm, elongation length (*b*) 36−39, 38 µm, length of maximum thickening (c_1) 62−69, 66 µm, total length of thickening (c_2) 90−116, 100 µm; *a*/*d* 1.7−2.9, 2.4; *b*/*d* 2.1−3.3, 2.8; c_1/*d* 3.6−5.6, 4.8; c_2/*d* 6.1−7.9, 7.3. Hind tibia expanded, with comb of 14−16, 15 setae, 1 seta near spur strongly S-shaped (Fig. 3.18G). Lengths and proportions of legs as in Table 3.10.

Hypopygium (Fig. 3.18C and D). Tergite IX with 4 long setae. Laterosternite IX with 1 long setae. Superior volsella anteriomedially almost separated, reduced. Inferior volsella broad, with many glandular setae and strong marginal setae, along the inner margin of gonocoxite and

Table 3.10 Lengths (in μm) and proportions of legs segments of male *Corynoneura caudicula* Fu & Sæther, 2012 (*n* = 3)

	fe	ti	ta_1	ta_2	ta_3	ta_4
p_1	216−232, 225	252−280, 269	131−145, 138	67−74, 72	44−48, 45	18−20, 19
p_2	300−320, 306	256−280, 272	123−176, 158	73−81, 78	40	18−20, 19
p_3	232−260, 249	256−280, 272	136−160, 152	75−85, 80	30−36, 34	16−20, 19

	ta_5	LR	BV	SV	BR
p_1	20−34, 26	0.49−0.55, 0.52	3.9−4.2, 4.0	3.6−4.1, 3.8	2.0−2.2
p_2	28−32, 30	0.48−0.63, 058	4.1−4.5, 4.3	3.4−4.5, 3.8	2.4−2.9, 2.6
p_3	26−31, 28	0.53−0.57, 0.56	4.1−4.2	3.4−3.6, 3.5	2.3−2.9, 2.5

placed caudally. Phallapodeme curved, 30−41, 35 μm long, with projection for joint with sternapodeme placed prelateral. Transverse sternapodeme 14−19 μm wide. Large attachment point of lateral sternapodeme with phallapodeme placed and directed caudally. Gonostylus curved, 68−83, 76 μm long; megaseta 4−6, 5 μm long. HR 2.4−3.2, 2.8; HV 4.2−5.5, 4.8.

Female ($n = 1−2$)

Total length 1.20 mm. Wing length 0.73−0.76 mm. Total length/wing length 1.58 mm. Wing length/profemur length 3.42−3.79 mm.

Coloration. Head brown. Thorax and legs yellowish except upper antepronotum, upper scutum, scutellum, postnotum and preepisternum light brown. Abdomen light brown.

Head. Height of eye/height of head 0.76. AR 0.35−0.38. Length of flagellomeres (in μm): 41−67, 28−43, 28−44, 28−41, 35−50. Ultimate flagellomere with 13−15 apical sensilla chaetica, without basal setae. Tentorium 83 μm long, 7 μm wide; stipes 50 μm long, 2 μm wide. Clypeus with 8 setae. Palpomere lengths (in μm): 11−15, 12−15, 18−19, 22−23, 48−50. Palpomere 5/3 ratio 2.6−2.7; palpomeres 1−3 elliptical, sensilla clavata not visible.

Thorax. Dorsocentrals 5−7; prealars 2−4; supraalar 1. Scutellum with 2 setae.

Wing (Fig. 3.19A). Wing wider than in male. VR 2.7, one anal vein present, Cu 304 μm long, Cu/wing length 0.40, C 326 μm long, C/wing length 0.43, wing width/wing length 0.45. Costa with 13 setae.

Legs. Fore trochanter with small keel. Spur of fore tibia 12−14 μm long, spurs of mid tibia 7 μm long, of hind tibia 35−38 μm long and 10 μm long. Width at apex of fore tibia 18−20 μm wide, of mid tibia 18 μm, of hind tibia (a) 23−26 μm. Width of hind tibia 1/3 from apex (d) 18−20 μm, elongation length (b) 34−36 μm, length of maximum thickening (c_1) 60−62 μm, total length of thickening (c_2) 105−110 μm, a/d 1.8−2.0, b/d 1.9−2.0, c_1/d 3.2−3.4, c_2/d 6.0−6.1. Hind tibial comb with 14−18 setae, 1 seta near strong spur S-shaped. Lengths and proportions of legs as in Table 3.11.

Abdomen. Number of setae on tergites V −VIII as: 3, 2, 2, 2.

Genitalia (Fig. 3.18E and F). Tergite IX with 2 long caudal setae. Cercus 28−36 μm long, 21−22 μm wide. Notum length 97−101 μm. Gonocoxite with 1 long seta. Coxosternapodeme with 6 lateral lamellae. Seminal capsule 44−65 μm long; neck 8 μm long, 4 μm wide.

Table 3.11 Lengths (in μm) and proportions of legs segments of female *Corynoneura caudicula* Fu & Sæther, 2012 ($n = 1-2$)

	fe	ti	ta$_1$	ta$_2$	ta$_3$	ta$_4$
p$_1$	200–212	260–264	127–132	67–72	40	20
p$_2$	280	276	162	77	40	18
p$_3$	252–256	264–268	137–141	79–81	32–34	16–18

	ta$_5$	LR	BV	SV	BR	
p$_1$	32	0.49–0.50	3.5–4.3	3.5–3.7	1.0–1.2	
p$_2$	32	0.59	4.3	3.4	2.0	
p$_3$	28–32	0.52–0.53	4.0–4.2	3.7–3.8	2.0–2.3	

Pupa ($n = 1-3$)

Total length 1.53 mm. Exuviae yellowish with light brown cephalothorax.

Cephalothorax. Frontal setae 14 μm long. Median antepronotals 16, 18, 20 μm and 22 μm long. Lateral antepronotals 14 μm long. Thoracic horn length 32 μm, width 6 μm. PcS$_1$ 14–15 μm long, PcS$_2$ 11–17 μm long, PcS$_3$ 14–17 μm long. PcS$_{1-3}$ in a line; PcS$_1$ 4–6 μm from PcS$_2$; PcS$_2$ 4–7, 5 μm from PcS$_3$; PcS$_1$ 10–12 μm from PcS$_3$; PcS$_3$ 28–32 μm from thoracic horn. Anterior dorsocentral (Dc$_1$) 20 μm long, Dc$_2$ 20 μm long; Dc$_3$ 8 μm long; Dc$_4$ 8 μm long. Dc$_1$ located 10 μm from Dc$_2$, Dc$_2$ located 7 μm from Dc$_3$, Dc$_3$ located 7 μm from Dc$_4$. Wing sheath with several rows of pearls.

Abdomen (Fig. 3.19B). Shagreen and chaetotaxy as illustrated. No taeniate L-setae on tergite I; Anal lobe 91–121 μm long. Anal lobe fringe with 17–21 setae, 220–260 μm long. Anal macrosetae 110 μm long; median setae 76–90 μm long.

Larva ($n = 1-3$)

Coloration. Head yellowish. Antenna with basal segment almost transparent, other segments light brown. Abdomen yellowish.

Head. Capsule length 220–240, 236 μm; width 150–154, 152 μm. Postmentum 184–212, 200 μm long. Sternite II large, arising from small tubercle, I simple, III not visible. Premandible 32–36, 34 μm long, brush of premandible not observed. Mentum as in Fig. 3.19C. Mandibles 48–61, 53 μm long. Antenna (Fig. 3.19D), AR 0.85–0.91, 0.88. Lengths of flagellomeres I–IV (in μm): 86–108, 97; 40–65, 52; 52–63, 59; 3–6, 4. Basal segment width 15–18, 16 μm; length of blade at apex of basal segment 28–32, 30 μm; ring organ at 44–58, 52 μm from the base of antenna.

Abdomen. Length of anal setae 121–162 µm. Procercus 11 µm long, 8–10 µm wide. Subbasal seta of posterior parapods split as in Fig. 3.19E and Bolton (2007).

Distribution. NE: USA(Ohio State).

3.14 *CORYNONEURA CELERIPES* WINNERTZ, 1852

Corynoneura celeripes Winnertz, 1852: 50; Schlee, 1968b: 18.

Diagnostic characters. Male antenna with 13 flagellomeres; AR 0.59–0.70. Inferior volsella large and triangular. Gonostylus slim, almost straight.

Male (Description from Schlee, 1968b)

Wing length 0.90 mm.

Wing. Wing width/Wing length 0.37. Costa length to wing length from 0.25 to 0.26. Stick length FCU to wing length from 0.58 to 0.59.

Hypopygium. Basal joint. The posterior margin of tergite IX medially concave, with two very weakly projecting bumps, each with 3–5 short setae. Laterosternite IX with 2–3 long setae. Inferior volsella large and triangular. Gonostylus slim, almost straight.

Distribution. NE: Canada (Ontario, Saskatchewan), Greenland, USA (Illinois, Indiana, Maine, New York, Ohio). PA: Austria, Balearic Islands, Belarus, Belgium, Croatia, Czech Republic, Finland, France, Germany, Great Britain, Hungary, Ireland, Moldova, Mongolia, Netherlands, Norway, Poland, Romania, Russia (CET, NET, SET, East Siberia, West Siberia), Slovakia, Spain, Sweden.

3.15 *CORYNONEURA CELTICA* EDWARDS, 1924

Corynoneura celtica Edwards, 1924: 186; Schlee, 1968b: 47.

Diagnostic characters. Male antenna with 10 flagellomeres, AR 0.46, the last flagellomere as long as the previous three together.

Male (Description from Schlee, 1968b)

Wing. Wing length 1.1 mm. Cu/wing length 0.61, costa length/wing length 0.24, wing width/wing length 0.40.

Legs. Fore trochanter with dorsal keel. hind tibia with comb of 18 setae and one S-shaped seta.

Distribution. PA: Algeria, Austria, Balearic Island, Belgium, Corsica, Croatia, Czech Republic, Estonia, Finland, France, Germany, Great

Britain, Hungary, Ireland, Japan, Lebanon, Morocco, Norway, Romania, Russia (CET, NET), Slovakia, Spain, **Sweden, Switzerland**.

3.16 *CORYNONEURA CENTROMEDIA* HAZRA, NATH, CHAUDHURI, 2003 (FIGS. 3.20A−D, 3.21A−D, AND 3.22A−F)

Corynoneura centromedia Hazra et al., 2003: 69.

Diagnostic characters. The adult male is characterized by antenna with 8 flagellomeres; AR 0.40−0.43; tergite IX covering much of the gonocoxites with straight posterior margin; sternapodeme inverted V-shaped, phallapodeme strongly curved extending beyond tergite IX. The pupa has sternite I medially with few points; shagreen of small points on tergites II, VIII and IX; tergites III−VII with anteriorly located shagreen of small points, and posteriorly located spinules; tergites IV−VII with longitudinal rows of spinules; anal lobe with 25−26 filaments in fringe covering the lobes completely. The larva is separable by having mentum triangular with a small tooth present in between the bases of median teeth.

Male (Description from Hazra et al., 2003)

Total length 1.27 mm.

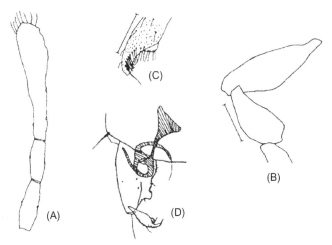

Figure 3.20 *Corynoneura centromedia* Hazra et al., 2003, Male imago (From Hazra, N., Nath, S., & Chaudhuri, P. K. (2003). The genus *Corynoneura* Winnertz (Diptera. Chironomidae) from the Darjeeling-Sikkim Himalayas of India, with description of three new species. *Entomologist's Monthly Magazine, 139*, 69−82). (A) The last three flagellomeres of antenna. (B) Fore trochanter. (C) Hind tibial comb. (D) Hypopygium: ventral view.

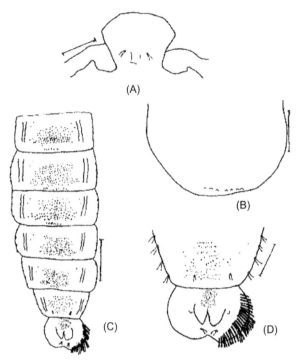

Figure 3.21 *Corynoneura centromedia* Hazra et al., 2003, Pupa imago (From Hazra, N., Nath, S., & Chaudhuri, P. K. (2003). The genus *Corynoneura* Winnertz (Diptera. Chironomidae) from the Darjeeling-Sikkim Himalayas of India, with description of three new species. *Entomologist's Monthly Magazine, 139,* 69—82). (A) Frontal apotome. (B) Tip of wing sheath. (C) Tergite. (D) Tergite VIII, anal lobe and male genital sac.

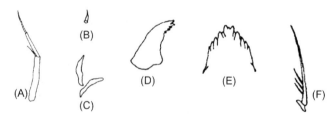

Figure 3.22 *Corynoneura centromedia* Hazra et al., 2003, Larva imago (From Hazra, N., Nath, S., & Chaudhuri, P. K. (2003). The genus *Corynoneura* Winnertz (Diptera. Chironomidae) from the Darjeeling-Sikkim Himalayas of India, with description of three new species. *Entomologist's Monthly Magazine, 139,* 69—82). (A) Antenna. (B) S I of labrum. (C) Premandible. (D) Mandible. (E) Mentum. (F) Basiventral spine of posterior parapod.

Head. Antenna with 8 flagellomeres; AR 0.40−0.43, length of flagellomeres (I−VIII): 26, 22, 33, 37, 37, 37, 33, 93; little clubbed apex with rosette of short hairs and a few plume hairs basally (Fig. 3.20A). Clypeus with 6−7 setae. Tentorium 115−119 μm long. Lengths of palpomeres (in μm): 10−12, 10−12, 15−17, 18−20, 28−30. Cibarial pump 52−56 μm long.

Thorax. Antepronotum with lobes narrowing dorsally without lateral setae; dorsocentrals 5−6, prealars 2. Scutellum with 2 setae.

Wing. Damaged in all specimens.

Legs. Fore trochanter as in Fig. 3.20B. Spur of fore tibia 16−18 μm long, of mid tibia 10−12 μm long, of hind tibia 28−30 μm long. Width of the apex of fore tibia 13−15 μm, mid tibia 13−15 μm and of hind tibia 26−28 μm. Hind tibia expanded, with comb of 15−16 setae and hind apex as in Fig. 3.20C, lengths and proportions of legs as in Table 3.12.

Abdomen. Tergites II−VIII with 1 lateral seta on each side and 1 median seta.

Hypopygium (Fig. 3.20D). Tergite IX covering much of the gonocoxites with straight posterior margin. Gonocoxite 52−56 μm long with well-developed inferior volsella; gonostylus 18−26 μm long tapered to a finely pointed apex with a pointed megaseta; sternapodeme inverted V-shaped, phallapodeme strongly curved extending beyond tergite IX; HR 2.15−2.88.

Pupa (Description from Hazra et al., 2003)

Total length of exuviae 1.69 mm.

Cephalothorax. Frontal seta 203 μm long. Lamelliform as in Fig. 3.21A. Postorbital seta 20 μm long, lamelliform; 2 median antepronotals 18 μm long each, distance between two 10; anterior precorneal 14 μm long, median and posterior precorneals 11 μm long each, distance between

Table 3.12 Lengths (in μm) and proportions of legs segments of male *Corynoneura centromedia* Hazra et al., 2003

	fe	ti	ta_1	ta_2	ta_3	ta_4
p_1	185−190	230−234	127−132	74−78	43−45	20−22
p_2	256−262	218−221	142−148	73−76	35−39	17−20
p_3	210−214	188−192	108−111	55−58	27−30	18−20

	ta_5	LR	BV	SV	BR	
p_1	28−31	0.55−0.56	3.15−3.31	4.32−4.61	1.80−2.00	
p_2	24 27	0.65−0.66	3.80−3.89	5.61−6.23	1.33−1.50	
p_3	24−27	0.57−0.58	3.82−4.08	5.27−5.77	1.40−1.60	

anterior and median precorneals 6, between median and posterior precorneals 3; Dc_2 7 μm long, Dc_3 22 μm long and Dc_4 18 μm long, distance between Dc_2 and Dc_3 15 μm long, between Dc_3 and Dc_4 13 μm long. Wing sheath (Fig. 3.21B) with single row of 7 pearls.

Abdomen (Fig. 3.21C). Sternite I medially with few points; shagreen of small points on tergites II, VIII and IX; tergites III−VII with anteriorly located shagreen of small points, and posteriorly located spinules; tergites IV−VII with longitudinal rows of spinules. Segment I with 3 D-setae, 1 V-seta and 1 L-seta, some D- and V-setae in segments II−VII lamelliform, segments II−VIII each with 4 lamelliform setae. Anal lobe (Fig. 3.21D) 103 μm long, 111 μm wide with 25−26 filaments in fringe covering the lobes completely and 3 lamelliform lateral anal macrosetae about as long as anal filaments, anterior one located at lateral filamentous setae 10−11 μm long; genital sac length 55 μm long; G/F 0.53; ALR 1.8.

Larva (Description from Hazra et al., 2003)

Total length 2.0 μm long.

Antenna (Fig. 3.22A). AR 0.90, length of antennal segments (I−IV): 107, 63, 52, 4 μm; segments II and III darkened; distance of ring organ from the base 59 μm long; blade 33 μm long.

Labrum. All S-setae simple; S I as in Fig. 3.22B. Premandible 24 μm long (Fig. 3.22C).

Mandible (Fig. 3.22D). 37 μm long with 1 apical and 4 inner teeth, apical tooth shorter than first inner one.

Mentum (Fig. 3.22E). 26 μm wide, triangular with a small tooth present in between the bases of median teeth.

Body. Each abdominal segment with 2 setae ventrally. Procercus 13 long with 4 apical setae, each 148 μm long. Basiventral spine (Fig. 3.22F) of posterior parapod dark brown, plumose at base, 44 μm long; anterior parapods 111 μm long and posterior parapods 129 μm long.

Distribution. OR: India (West Bengal).

3.17 *CORYNONEURA CHANDERTALI* SINGH & MAHESHWARI, 1987 (FIG. 3.23A−F)

Corynoneura chandertali Singh & Maheshwari, 1987: 13.

Diagnostic characters. Male antenna with 8 flagellomeres. AR 1.26. This species comes near to *Corynoneura carinata*, but differs in keel on gonostylus being absent, anal point being reduced and by the presence of

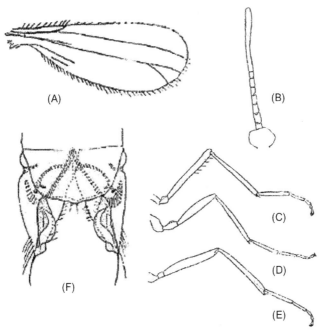

Figure 3.23 *Corynoneura chandertali* Singh & Maheshwari, 1987, Male imago (From Singh, S. & Maheshwari, G. (1987). Chironomidae (Diptera) of Chandertal Lake, Lahul Valley (Northwest Himalaya). *Annals of Entomology*, *5*(2), 11–20). (A) Wing. (B) Antenna. (C–E) Fore, mid and hind leg. (F) Hypopygium.

a strong seta on the apex of gonostylus; flagellum being seven segmented. It also differs in leg ratio, antennal ratio, venarum ratio and ratio of gono-coxite and gonostylus.

Male (Description from Singh & Maheshwari, 1987)

Total length 1.64 mm. Wing length 1.29 mm.

Coloration. Pedicel brown, flagellum yellow. Head yellow, eyes black, palpi light brown. Antepronotal lobe yellow, scutum dark brown, scutellum and postnotum light brown. Coxa and trochanter brown; femur, tibia and tarsal segment light yellow. Genitalia light brown.

Head. Ratio of maximum head width and width between eyes 2.16. Antenna (Fig. 3.23B) with 8 flagellomeres, ultimate flagellomere elongated, AR 1.26. Lengths of palpomeres (in μm): 5.2, 8.4, 12, 12.

Wing (Fig. 3.23A). Costa with 10 setae, R_{4+5} fused with costa and forms clavus. Costa ends about one-third of the length of wing. VR 2.60.

Legs (Fig. 3.23C–E). Fore trochanter with dorsal keel, fore femur with five setae on the distal half; fore tibia with two spurs, outer spur

Table 3.13 Lengths (in μm) and proportions of legs segments of male *Corynoneura chandertali* Singh & Maheshwari, 1987

	fe	ti	ta₁	ta₂	ta₃	ta₄	ta₅	LR
p₁	360	380	200	120	67	28	31	0.53
p₂	430	420	210	90	32	39	32	0.50
p₃	360	430	200	90	43	18	35	0.47

reaching up to the half of the inner one; middle tibial apex with two spurs, outer spur not reaching up to the half of inner one; fourth tarsal segment smaller than fifth and clavate. Apex of hind tibia enlarged, outer spur of hind tibia reaching up to 1/5 the inner spur. Lengths and proportions of legs as in Table 3.13.

Hypopygium (Fig. 3.23F). Setae present on the inner margin of gonocoxa. Gonocoxa 73 μm long. Gonostylus folded inwards, 44 μm long. Ratio of gonocoxa and gonostylus 1.90 μm.

Distribution. OR: India (Himachal Pradesh).

3.18 *CORYNONEURA COLLARIS* MAKARCHENKO & MAKARCHENKO, 2010 (FIGS. 3.24A−F AND 3.25A−E)

Corynoneura collaris Makarchenko & Makarchenko, 2010: 358; Fu & Sæther, 2012: 16.

Material examined. Canada: Northwest Territories, Horton River (67°41.2′N, 122°39.0′W), 6 males, Malaise trap, 19.vii.2000, D. C. Currie & B. V. Brown (CNC).

Diagnostic characters. The sternapodeme of male inverted U-shaped and the gonostylus bent apically, antenna has 9−10 flagellomeres and with long radial sensilla chaetica; superior volsella rounded, and inferior volsella developed and projecting in the median.

Male (*n* = 6)

Total length 1.01−1.13, 1.07 mm. Wing length 0.78−0.81, 0.80 mm. Total length/wing length 1.10−1.45, 1.29 mm. Wing length/profemur length 3.3 mm. Wing width/wing length 0.34−0.35 (3) mm.

Coloration. Apically entirely brown.

Head. Eyes bare. Antenna (Fig. 3.24C and D) with 9−10 flagellomeres, with more than 20 apical sensilla chaetica, 76−80 μm long and apically bent; ultimate flagellomere 101 μm long, apically expanded. AR 0.36. Tentorium (Fig. 3.24B) 115−121, 118 μm long, 12−18, 15 μm

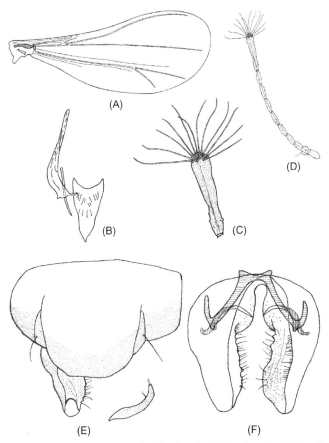

Figure 3.24 *Corynoneura collaris* Makarchenko & Makarchenko, 2010, Male imago (From Fu, Y., & Sæther, O. A. (2012). *Corynoneura* Winnertz and *Thienemanniella* Kieffer from the Nearctic region (Diptera: Chironomidae: Orthocladiinae). *Zootaxa, 3536*, 1—61. www.mapress.com/j/zt). (A) Wing. (B) Tentorium, stipes and cibarial pump. (C and D) Antenna. (E) Hypopygium, dorsal view. (F) Hypopygium, ventral view.

wide; stipes 44 μm long, 3 μm wide. Anterior margin of cibarial pump distinctly concave. Clypeus with 8—10, 9 setae. Lengths of palpomeres (in μm): 10—12; 12—14, 13; 21—22; 22—26, 24; 54—58, 56. Palpomere 5/3 ratio 2.5—2.6.

Thorax. Dorsocentrals 5, other setae hard to examine.

Wing (Fig. 3.24A). VR 3.1—3.3, 3.2. Cu/wing length 0.53—0.55, 0.54; C length 188—200, 193 μm long; Cu 440—450, 445 μm long; wing width/wing length: 0.33—0.35, 0.34; costa with 5—8, 7 setae.

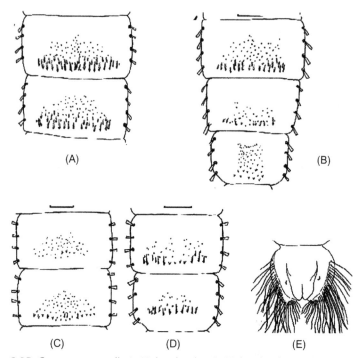

Figure 3.25 *Corynoneura collaris* Makarchenko & Makarchenko, 2010, Pupa imago (From Makarchenko, E. A., & Makarchenko, M. A. (2010). New data on the fauna and taxonomy of *Corynoneura* Winnertz (Diptera, Chironomidae, Orthocladiinae) for the Russian Far East and bordering territories. *Euroasian Entomological Journal, 9*(3), 353−370 + II). (A) Tergites IV−V. (B) Tergites VI−VIII. (C) Sternites V−VI. (D) Sternites VII−VIII. (E) Anal segment.

Legs. Fore trochanter with dorsal keel. Spur of fore tibia 16−20, 18 μm long; of mid tibia 8−10, 9 μm long; of hind tibia 11 μm and 26−28, 27 μm long. Width at apex of fore tibia 18−22, 20 μm; of mid tibia 17−19, 18 μm; of hind tibia (a) 32−36, 34 μm. Width of hind tibia 1/3 from apex (d) 19−25, 21 μm; elongation length (b) 30−37, 35 μm; length of maximum thickening (c_1) 63−69, 66 μm; total length of thickening (c_2) 99−104, 102 μm; a/d 1.3−1.8, 1.6; b/d 1.5−1.9, 1.6; c_1/d 2.8−3.2, 3.0; c_2/d 4.2−5.1, 4.8. Hind tibia expanded, with comb of 15−16 setae, 1 seta near spur strongly S-shaped. Lengths and proportions of legs as in Table 3.14.

Hypopygium (Fig. 3.24E and F). Laterosternite IX with 1 long seta. Superior volsella rounded and anteriomedially fused. Inferior volsella broad and fused with the inner margin of gonocoxite, with 3−5 strong

Table 3.14 Lengths (in μm) and proportions of legs segments of male *Corynoneura collaris* Makarchenko & Makarchenko, 2010 ($n = 6$) (based on Canadian material)

	fe	ti	ta_1	ta_2	ta_3	ta_4
p_1	240–244, 242	296	145	75	32	22
p_2	340–348, 344	308–328, 320	166–180, 172	81–85, 84	40–46, 44	20–24, 21
p_3	276–280, 278	300	162–168, 165	89–95, 92	38–42, 40	18–22, 20

	ta_5	LR	BV	SV	BR
p_1	24	0.49	4.5	3.7	2.0
p_2	32–36, 34	0.51–0.56, 0.54	4.4–5.0, 4.7	3.8–4.0, 3.9	1.9–2.2, 2.1
p_3	34–36, 35	0.54–0.56, 0.55	3.8–4.1, 4.0	3.4	2.3–2.5, 2.4

setae. Phallapodeme curved, 28—30, 29 μm long. Transverse sternapodeme 16—20, 18 μm. Attachment point of lateral sternapodeme with phallapodeme placed and directed caudally. Gonostylus curved to strongly curved, 29—39, 33 μm long; megaseta 6—7 μm long. Inner margin of gonocoxite with thin glandular setae. HR 2.2—2.9, 2.5; HV 2.6—3.9, 3.2.

Pupa (From Makarchenko & Makarchenko, 2010)

Total length 1.4 mm.

Cephalothorax. Median antepronotals 2. Wing sheath with 2 rows of pearls.

Abdomen (Fig. 3.25A—E). Shagreen and chaetotaxy as illustrated. Anal lobe 120—148, 130 μm long. Anal lobe fringe with 26—29 setae, 160 μm long. Anal macrosetae 200 μm long.

Remarks. The male and pupa were described by Makarchenko and Makarchenko (2010). More important additional characters of males are included in the present description.

Distribution. PA: Russia (Far East). **NE**: Canada.

3.19 *CORYNONEURA CONFIDENS* FU ET AL., 2009 (FIG. 3.26A—E)

Corynoneura confidens Fu et al., 2009: 6.

Material examined. Holotype male (BDN No. 13103), P.R. China: Sichuan Province, Kangding County, Wasigou, 30°03′N, 101°58′E, alt. 2560 m, light trap, 15.vi.1996, X. Wang.

Diagnostic characters. The species can be separated from other members of the genus by having an AR of 0.16, inferior volsella reduced, and phallapodeme not extending beyond posterior margin of tergite IX.

Male (*n* = 1)

Total length 1.16 mm. Wing length 0.70 mm. Total length/wing length 1.65. Wing length/length of profemur 3.4.

Coloration. Head dark brown. Antenna and palp yellow. Thorax brown. Legs yellowish brown. Tergites I—V yellow, tergites VI—IX brown. Wings light yellow to hyaline with pale yellow clava.

Head. Antenna (Fig. 3.26B) with 10 flagellomeres; ultimate flagellomere stout, 55 μm long, with short sensilla chaetica subapically, apex distinctly swollen, 80 μm long; AR 0.16. Tentorium (Fig. 3.26E) 115 μm long, 15 μm wide. Stipes 55 μm long. Cibarial pump slightly incurved.

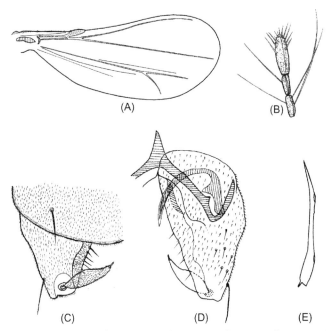

Figure 3.26 *Corynoneura confidens* Fu et al., 2009, Male imago (From Fu, Y., Sæther, O. A., & Wang, X. (2009). Corynoneura Winnertz from East Asia, with a systematic review of the genus (Diptera: Chironomidae: Orthocladiinae). *Zootaxa, 2287*, 1–44. www.mapress.com/j/zt). (A) Wing. (B) Antenna. (C) Hypopygium, dorsal view. (D) Hypopygium, ventral view. (E) Tentorium.

Palpomeres lengths (in μm): 13, 13, 20, 15, 50. Palpomere 2, 3 and 4 ellipsoid, 5 long and broad. Palpomere 5/3 ratio 2.5.

Thorax. Dorsocentrals more than 3, prealars 3, supraalar 1. Scutellum with 2 setae.

Wing (Fig. 3.26A). VR 3.2. C length 170 μm, C/wing length 0.24. Cu length 420 μm. Cu/wing length 0.6. Wing width/wing length 0.46. C with 7–8 setae.

Legs. Spur of fore tibia 15 μm long, spurs of mid tibia 8 μm long, of hind tibia 23 μm long. Width at apex of fore tibia 15 μm, of mid tibia 15 μm, of hind tibia 35 μm. Apex of hind tibia expanded, with comb of 12 setae, 1 seta near spur strongly S-shaped. Lengths and proportions of legs as in Table 3.15.

Hypopygium (Fig. 3.26C and D). Anal point absent. Tergite IX with posterior margin almost straight. Laterosternite IX with 1 long seta. Superior volsella anteriomedially fused. Inferior volsella inconspicuous. Sternapodeme inverted V-shaped; coxapodeme 23 μm long; attachment

Table 3.15 Lengths (in μm) and proportion of leg segments of male *Corynoneura confidens* Fu et al., 2009

	fe	ti	ta$_1$	ta$_2$	ta$_3$	ta$_4$	ta$_5$	LR	BV	SV	BR
p$_1$	205	250	130	70	35	20	30	0.52	3.8	3.5	2.0
p$_2$	295	275	155	73	38	18	28	0.56	4.7	3.7	1.8
p$_3$	238	260	143	78	33	15	25	0.55	4.3	3.5	2.3

Source: From Fu, Y., Sæther, O. A., & Wang, X. (2009). *Corynoneura* Winnertz from East Asia, with a systematic review of the genus (Diptera: Chironomidae: Orthocladiinae). *Zootaxa*, 2287, 1–44. www.mapress.com/j/zt.

point with phallapodeme placed in caudal third of lateral sternapodeme and directed caudally, phallapodeme 65 μm long, strongly curved with projection for joint with sternapodeme placed prelateral, not extending beyond posterior margin of tergite IX, with projection basally. Gonocoxite 63 μm long, with 1 seta apically. Gonostylus simple, 25 μm long; megaseta 3 μm long. HR 2.7, HV 4.6.

Distribution. OR: Oriental China (Sichuan).

3.20 *CORYNONEURA CORONATA* EDWARDS, 1924

Corynoneura coronata Edwards, 1924: 187; Schlee, 1968b: 24.

Diagnostic characters. Male antenna with 12 flagellomeres; AR 0.61.

Male (Description from Schlee, 1968b)

Head. Eyes bare. Antenna with 12 flagellomeres; AR 0.61.

Wing. Wing length 1 mm. Cu/wing length 0.63, C length/wing length 0.27, wing width/wing length 0.34.

Hypopygium. Basal joint. The posterior margin of tergite IX medially a little concave, each carrying 2 setae. Inferior volsella triangular. Gonostylus straightened and the same width with the exception of the distal end.

Distribution. PA: Bulgaria, Estonia, Finland, France, Germany, Great Britain, Ireland, Italy, Lebanon, Morocco, Netherlands, Norway, Romania, Russia (CET, NET), Spain, Sweden, Switzerland, Turkey.

3.21 *CORYNONEURA CRISTATA* FREEMAN, 1953 (FIG. 3.27A AND B)

Corynoneura cristata Freeman, 1953: 209.

Diagnostic characters. Male antenna with 10 flagellomeres. Inferior volsella developed and with a hooked apex. Transverse sternapodeme wide. Gonostylus apically curved (Similar to *Corynoneura dewulfi* Freeman

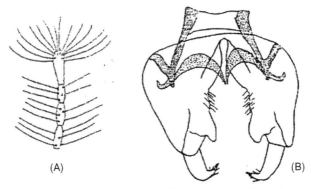

(A)   (B)

Figure 3.27 *Corynoneura cristata* Freeman, 1953; Male imago (From Freeman, P. (1953). Chironomidae from Western Cape Province — II . *Proceedings of the Royal Entomological Society of London (B), 22,* 201−253). (A) Apex of antenna. (B) Hypopygium: ventral view.

in size, color, wing venation and proportions of the legs, but differing in the structure of the apex of the antenna and in the hypopygium).

Male (Description from Freeman, 1953)

Head. Antenna with 10 flagellomeres (Fig. 3.27A); last segment slightly clubbed, about equal in length to the preceding two together, with an apical rosette of long hairs, the individual hairs being longer than the segment itself.

Hypopygium (Fig. 3.27B). Inferior volsella with a hooked apex. Sternapodeme curved into U-shaped, with wide transverse sternapodeme. Gonostylus straight basally and with apex sharply bent.

Distribution. AF: South Africa, Zimbabwe.

3.22 *CORYNONEURA CUSPIS* TOKUNAGA, 1936

Corynoneura cuspis Tokunaga, 1936: 48.

Diagnostic characters. Antenna with 11 flagellomeres, AR 0.6; ultimate segment pointed, bare at tip, with preapical pubescence; inferior volsella absent; gonostylus stout, with a large preapical projecting.

Male (From Tokunaga, 1936)

Total length 0.9−1.4 mm.

Coloration. Head dark brown. Thorax: prothorax dark brown; scutum dark brown, with vittae indistinct, shoulder parts brown; scutellum brown; postscutellum black, legs dark brown, abdomen dark brown.

Head. Antenna with 11 flagellomeres, AR 0.6; antenna pointed apically, without any setae at tip, with preapical pubescence; clypeus with about 10 setae.

Wing. Wing length 1.0 mm.

Hypopygium. Inferior volsella absent. Gonostylus stout, with a large preapical projecting.

Distribution. PA: Japan.

3.23 *CORYNONEURA CYLINDRICAUDA* FU ET AL., 2009 (FIG. 3.28A−E)

Corynoneura cylindricauda Fu et al., 2009: 8.

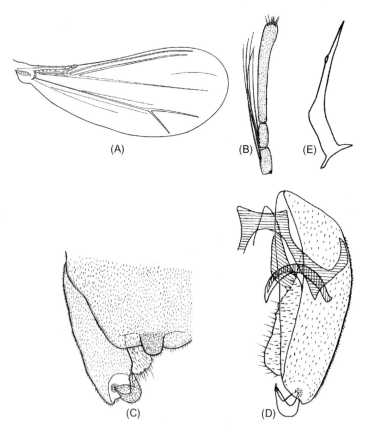

(A)   (B)   (E)

(C)   (D)

Figure 3.28 *Corynoneura cylindricauda* Fu et al., 2009, Male imago (From Fu, Y., Sæther, O. A., & Wang, X. (2009). *Corynoneura* Winnertz from East Asia, with a systematic review of the genus (Diptera: Chironomidae: Orthocladiinae). *Zootaxa, 2287*, 1−44. www.mapress.com/j/zt). (A) Wing. (B) Antenna. (C) Hypopygium, dorsal view. (D) Hypopygium, ventral view. (E) Tentorium.

Material examined. Holotype male (BDN No. 25092), P.R. China: Xinjiang Uygur Autonomous Region, Aletai City, Eerqisi river, 88°07′E, 47°55′N, alt. 400 m, sweep net, 3.xii.2002, X. Wang.

Diagnostic characters. The species is similar to *Corynoneura tokarape-quea* in having antenna with 11 flagellomeres and a similarly shaped sternapodeme, but can be separated by having a cylindrical anal point and long, large inferior volsella.

Male (n = 1)

Total length 1.32 mm. Wing length 0.84 mm. Total length/wing length 1.6. Wing length/length of profemur 3.2.

Coloration. Head brown. Antenna and palp yellow. Thorax dark brown. Legs yellow. Abdomen blackish brown. Wings light yellow to hyaline, with pale yellow clava.

Head. Antenna (Fig. 3.28B) with 11 flagellomeres; ultimate flagellomere 140 µm long, with rosette of apical sensilla chaetica, apically expanded, thickening 60 µm long; AR 0.42. Tentorium (Fig. 3.28E) 135 µm long, 15 µm wide. Stipes 63 µm long. Palpomeres length (in µm): 13, 15, 20, 25, 58. Palpomere 2, 3 and 4 ellipsoid, 5 long and slender. Palpomere 5/3 ratio 2.9.

Thorax. Dorsocentrals 5, prealars 4. Scutellum with 2 setae.

Wing (Fig. 3.28A). VR 3.2. C length 190 µm, C/wing length 0.23. Cu length 510 µm. Cu/wing length 0.61. Wing width/wing length 0.42. C with 7 setae.

Legs. Spur of fore tibia 20 µm and 13 µm long, spurs of mid tibia 13 µm and 8 µm long, of hind tibia 28 µm long. Width at apex of fore tibia 23 µm, of mid tibia 28 µm, of hind tibia 35 µm. Apex of hind tibia expanded, with comb of 17 setae, 1 seta near spur strongly S-shaped. Lengths and proportions of legs as in Table 3.16.

Hypopygium (Fig. 3.28C and D). Tergite IX with posterior margin straight, with many short setae. Anal point present, cylindrical, 10 µm

Table 3.16 Lengths (in µm) and proportion of leg segments of male *Corynoneura cylindricauda* Fu et al., 2009

	fe	ti	ta$_1$	ta$_2$	ta$_3$	ta$_4$	ta$_5$	LR	BV	SV	BR
p$_1$	265	300	165	100	58	28	33	0.55	3.4	3.4	1.7
p$_2$	375	310	210	93	45	20	33	0.68	4.7	3.3	2.7
p$_3$	310	315	185	100	43	20	33	0.59	4.2	3.4	2.0

Source: From Fu, Y., Sæther, O. A., & Wang, X. (2009). *Corynoneura* Winnertz from East Asia, with a systematic review of the genus (Diptera: Chironomidae: Orthocladiinae). *Zootaxa*, 2287, 1–44. www.mapress.com/j/zt.

long, 10 μm wide. Superior volsella well developed, inferior volsella large, rounded posteriorly, partly united with superior volsella. Sternapodeme inverted U-shaped; coxapodeme 33 μm long, attachment point with phallapodeme placed caudal and caudally directed; phallapodeme curved, not extending beyond posterior margin of tergite IX, 38 μm long. Gonocoxite 88 μm long. Gonostylus apically bent, 20 μm long; megaseta 5 μm long. HR 4.4, HV 6.6.

Distribution. PA: China (Xinjiang).

3.24 *CORYNONEURA DEWULFI* GOETGHEBUER, 1935 (FIG. 3.29A−F)

Corynoneura dewulfi Goetghebuer, 1935: 364; Freeman, 1956: 362; Lehmann, 1979: 37; Harrison, 1992: 153.

Corynoneura scotti Freeman, 1953: 209.

Diagnostic characters. The male imago is characterized by antenna with 9 flagellomeres, AR 0.3, phallapodeme protruding posteriorly alongside the inner edges of the gonocoxites. The female has Gonopophysis VIII undivided, apodeme lobe very large and strongly

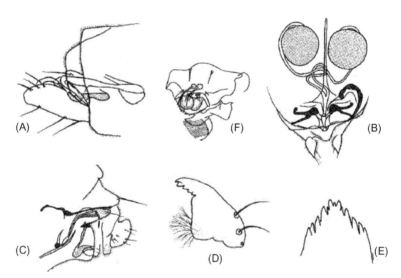

Figure 3.29 *Corynoneura dewulfi* Goetghebuer, 1935 (From Harrison, A. D. (1992). Chironomidae from Ethiopia, Part 2. Orthocladiinae with two new species and a key to *Thienemanniella* Kieffer (Insecta, Diptera). *Spixiana*, *15*, 149−195) Male imago. (A) Hypopygium. Female imago. (B and C) Genitalia. (D−F) Larva imago. (D) Mandible. (E) Mentum. (F) Labrum.

chitinized, gonocoxapodeme fairly straight; gonocoxite IX long and narrow with one seta; coxosternapodeme large, with chitinized lamellae protruding into the cavity. Labia fused, membranous centrally but strongly chitinized laterally. Seminal capsules spherical, mouth placed orally; spermathecal ducts large, joining to a common duct before discharging into the genital cavity. The larvae are separable by mentum triangular shaped, median tooth much smaller than first lateral tooth, lateral tooth smaller than rest.

Male

Freeman (1956) described the antennae, wings and hypopygium and showed how the three known African species can be differentiated on the structure of the last flagellomere and the apodemes.

Lehmann (1979) described the wings, the tip of the hind tibia with S-shaped apical setae and also the hypopygium in some detail.

The following details were given by Harrison (1992).

Male

Wing length 0.9 mm.

Head. Antenna with 9 flagellomeres, AR 0.3, apical one clubbed with terminal rosette of short setae as in Freeman (1956). Clypeus with 6 setae. Lengths of palpomeres (in µm): 13, 13, 16, 23, 46.

Thorax. Dorsocentrals 7, posterior prealars 2, scutellars 1 per side.

Wing. As illustrated by the above authors.

Legs. LR fore 0.50, mid 0.61, hind 0.60. Hind leg: tibia with conspicuous projection for comb with S-shaped seta terminally, inner side of tarsomere 1 with a row of 7 short robust setae. No sensilla chaetica.

Hypopygium (Fig. 3.29A). The ventral view is illustrated by Freeman and Lehmann. The Ethiopian specimens also had the long points of the phallapodeme protruding posteriorly alongside the inner edges of the gonocoxites. In addition, the Ethiopian specimens show that the enlarged phallapodeme and the lateral sternapodeme protrude just below laterosternite IX.

Female

Wing length 0.63−0.96 mm.

Head. Antenna with 5 flagellomeres, AR 0.34. Clypeus with 6 setae. Lengths of palpomeres (in µm): 14, 14, 16, 21, 37.

Thorax. Dorsocentrals 7, posterior prealars 2, scutellars 1 per side.

Wing. Similar to male. No setae on R veins, squama absent.

Legs. Similar to male, LR fore 0.48, mid 0.68, hind 0.61. No sensilla chaetica.

Genitalia (Fig. 3.29B and C). Gonopophysis VIII undivided, apodeme lobe very large and strongly chitinized, gonocoxapodeme fairly straight; tergite IX with caudal concavity (not in figures) and one seta per side, gonocoxite IX long and narrow with one seta; coxosternapodeme large, structure diffi- cult to discern, the two coxosternapodemes appear to lie in the domed roof of the genital cavity with chitinized lamellae protruding into the cavity, these lamellae appear to be joined, in some way, to the walls of the cavity. Segment X and postgenital plate well-developed, cerci small. Labia fused, membranous centrally but strongly chitinized laterally. Seminal capsules spherical, brown, no distinct neck, mouth placed orally; spermathecal ducts large, somewhat coiled, joining to a common duct before discharging into the genital cavity. The common duct appears to be surrounded by a sheath.

Larva

Head capsule. Length 240 μm. With surface sculpturing, similar to that of *C. scutellata* Winnertz (Cranston et al., 1983).

Coloration (preserved specimens). Head light, antenna segment 1 light, others dark. Body light brown, anterior and posterior claws and paraxial setae light.

Antenna. 1.02 times length of head capsule, 4 segments: 120, 48, 58, 3 μm; the ratio of ring organ with median spine to the base of segment I is 0.42; blade slightly curved, about 35 μm long.

Labrum. S-setae simple, one strongly developed pair appear to be ster- nite III; spines of peel epipharyngis small and shape obscure, first chaetulae lateralis obscure, shown in Fig. 3.29F with *dotted line*. Two outer chaetu- lae lateralis enlarged and flattened, anterior pair plumose, poster pair over- lapping and serrate. Premandible with large colorless and transparent brush and w small teeth.

Mandible (Fig. 3.29D). Apical tooth subequal to subapical tooth, seta subdentalis absent.

Mentum (Fig. 3.29E). Triangular shaped, median tooth much smaller than first lateral tooth, lateral tooth smaller than rest.

Maxilla. Palp normal with sensilla; setae maxillaris numerous and sim- ple; lacinial chaetae from serrate to almost plumose, but at least one spatu- late and simple; chaetulae of palpiger normal.

Remarks. The male gonostyli seem to be too weakly developed to act as proper claspers would appear that the enlarged and protruding phallapodemes and the lateral sternapodemes designed to grip within the genital cavity of the female, possibly assisted by the coxosternapodeme with their chitinized lamellae.

Ecology. Larvae in fast flowing rivers and streams.

Distribution. AF: Democratic Republic of Congo, Ethiopia, South Africa, Tanzania, Uganda, Zimbabwe.

3.25 *CORYNONEURA DIARA* ROBACK, 1957

Corynoneura diara Roback, 1957b; 10.

Diagnostic characters. Antenna with 11 flagellomeres; AR 0.55, last segment only slightly expanded apically, with apical rosette.

Male (Description from Roback, 1957b)

Total length 1.40 mm.

Coloration. Antenna light brown, antennal subapical dark brown; head dark brown; thorax dark brown; legs light brown; tergites brown.

Head. Eyes bare. Antenna with 11 flagellomeres; AR 0.55, last segment only slightly expanded apically, with apical rosette, as long as preceding four segments.

Thorax. Scutellum and postnotum dark brown.

Wing. Wing length 1.20 mm. Cu/wing length 0.67, C length/wing length 0.33.

Distribution. NE: USA (Utah).

3.26 *CORYNONEURA DIOGO* WIEDENBRUG ET AL., 2012 (FIGS. 3.30A−F, 3.31A−C, AND 3.32A−G)

Corynoneura diogo Wiedenbrug et al., 2012: 16.

Diagnostic characters. The male of *C. diogo* can be differentiated from others species with straight transversal sternapodeme and phallapodeme attached in the caudal apex of the sternapodeme by having the antennal apex with a distinct group of about 20 sensilla chaetica, slightly sinuous lateral sternapodeme and aedeagal lobe long and triangular. The group of sensilla at the antennal apex is also distinctive on the female, which also have a triangular apex, copulatory bursa in dorsal view with membranous oral extension and laterosternite IX without seta. The pupa has anal lobes laterally rounded, joined medially as an inverted "V," complete fringe of taeniate setae and shagreen on tergites IV−V of minute points, tergites and sternites points as long as wide. The larva has the head capsule integument sculptural mentum with two median teeth and the first pair of lateral teeth almost as strong as the median teeth.

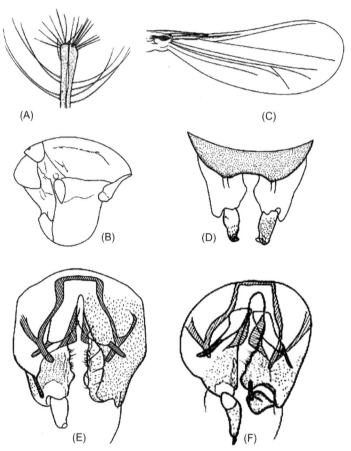

Figure 3.30 *Corynoneura diogo* Wiedenbrug et al., 2012; Male imago (From Wiedenbrug, S., Lamas, C. J. E., & Trivinho-Strixino, S. (2012). A review of the genus *Corynoneura* Winnertz (Diptera: Chironomidae) from the neotropical region. *Zootaxa*, *3574*, 1−61. www.mapress.com/j/zt). (A) Terminal flagellomere. (B) Thorax. (C) Wing. (D−F) Hypopygium. (D) Tergite IX and gonostylus included. (E and F) Tergite IX removed, sclerites hatched. (E) Ventral view, without gonostylus. (F) Dorsal view.

Male (Description from Wiedenbrug et al., 2012)

Total length 1.15 mm. Wing length 0.72 mm.

Coloration. Thorax brownish. Abdominal tergites I−IV whitish, other tergites and genitalia brownish. Legs whitish.

Head. AR 0.65−0.68. Antenna with 9 flagellomeres, apical flagellomere 170 μm. Flagellomeres, except first and last, with more than one row of setae each. Antennal apex with 20 sensilla (Fig. 3.30A). Eyes pubescent.

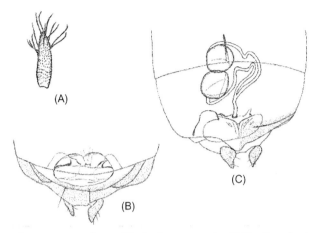

Figure 3.31 *Corynoneura diogo* Wiedenbrug et al., 2012; Female imago (From Wiedenbrug, S., Lamas, C. J. E., & Trivinho-Strixino, S. (2012). A review of the genus *Corynoneura* Winnertz (Diptera: Chironomidae) from the neotropical region. *Zootaxa, 3574*, 1–61. www.mapress.com/j/zt). (A) Terminal flagellomere. (B and C) Genitalia. (B) Dorsal view. (C) Ventral view.

Thorax. Antepronotal lobes dorsally tapering (Fig. 3.30B).

Wing. Clavus/wing length 0.20. Anal lobe absent (Fig. 3.30C).

Legs. Hind tibial scale 37−42 μm long, with one small S-seta.

Hypopygium (Fig. 3.30D−F). Tergite IX with 4 setae. Laterosternite IX without seta. Superior volsella absent. Inferior volsella slightly rounded. Aedeagal lobe long and triangular. Transverse sternapodeme straight, lateral sternapodeme sinuous, phallapodeme caudal attached. Phallapodeme elongated, with posterior margin sclerotized. Gonostylus thicker before apex.

Female (Description from Wiedenbrug et al., 2012)

Wing length 0.51 mm.

Coloration. Thorax brownish. Abdominal tergites III−IX with dark brown bands. Legs whitish.

Head. AR = 0.45. Antenna with 5 flagellomeres, apical flagellomere 45 μm long (Fig. 3.31A). Flagellomeres with more than one row of setae each. Antennal apex with 15 sensilla, triangular. Eyes pubescent.

Thorax. Antepronotal lobes dorsally tapering.

Wing. Clavus/wing length 0.40. Anal lobe absent.

Legs. Hind tibial scale 35 μm long, with one small S−seta.

Genitalia (Fig. 3.31B and C). Tergite IX with 2 setae. Laterosternite IX without seta. Two seminal capsules 32 μm long; one spermathecal duct with a loop, second straighter, both ducts join together shortly before seminal eminence, which has sclerotized outer borders. Notum 20 μm long.

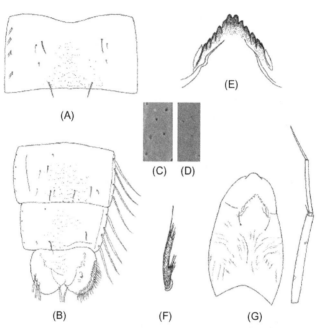

Figure 3.32 *Corynoneura diogo* Wiedenbrug et al., 2012; Immatures imago (From Wiedenbrug, S., Lamas, C. J. E., & Trivinho-Strixino, S. (2012). A review of the genus *Corynoneura* Winnertz (Diptera: Chironomidae) from the neotropical region. *Zootaxa, 3574*, 1–61. www.mapress.com/j/zt). (A–D) Pupa imago. (A) Tergite IV. (B) Tergites VII–IX and anal lobe (left is dorsal view of anal lobe, right is ventral view of anal lobe). (C) Detail of shagreen of tergite IV. (D) Detail of shagreen of sternite IV. (E–G) Larva imago. (E) Mentum. (F) Subbasal seta of posterior parapod. (G) Head, ventral view and separated antenna.

Membrane well sclerotized. Apodeme lobe median borders sclerotized. Coxosternapodeme curved, with one end at roof of copulatory bursa, copulatory bursa with dorsal-oral extension, oral median invaginated. Labia membranous, bare, apparently divided in two lobes. Gonocoxapodeme straight, gonapophyses median smoothly pointed. Cercus 25 μm long.

Pupa (Description from Wiedenbrug et al., 2012)

Total length 1.34 mm.

Coloration (exuviae). Cephalothorax light brown, abdomen transparent except brownish lateral margin and anal lobe.

Cephalothorax. Frontal apotome slightly granulated. Thorax suture orally with some spinules. Dc_3-seta wider than other three thin-taeniate setae. Dc_1 displaced ventrally. Wing sheaths with 1 to 2 rows of pearls.

Abdomen (Fig. 3.32A–D). Tergite and sternite I bare, tergites and sternites II–IX with homogeneous shagreen, shagreen points very small, not elongate. Conjunctives sternites IV / V — VII / VIII with small spinules. Segment I with 1, II with 3 L-setae and III — VIII with 4 long taeniate L-setae. Anal lobe rounded (Fig. 3.32B). Anal lobe with fringe almost complete, 3 taeniate macrosetae and inner setae taeniate.

Larva (Description from Wiedenbrug et al., 2012)

Head. Postmentum 175—187 μm long. Head capsule integument of exuviae, ventral with fine scratches (Fig. 3.32G). Mentum with two median teeth and five lateral teeth (Fig. 3.32E). Antenna 355 μm long, segments two and three darker (not drawn). First antennal segment subequal to postmentum length.

Abdomen. No modified ventral setae observed. Subbasal seta on posterior parapod serrated at one margin (Fig. 3.32F).

Remarks. The males of *C. diogo, Corynoneura salviniatilis* and *C. boraceia* are similar. All three species have the sensilla chaetica concentrated at the antennal apex, distinct straight transverse sternapodeme and more or less sinuous lateral sternapodeme. They are differentiated from each other by the shape of the aedeagal lobus and lateral sternapodeme. Larvae of *C. diogo* were found on *Eichhornia* sp. (Pontederiacea) from marginal lakes of the Mogi-Guaçú river at São Paulo State, Brazil.

Distribution. NT: Brazil (Luiz Antônio).

3.27 *CORYNONEURA DISINFLATA* FU & SÆTHER, 2012 (FIGS. 3.33A—F AND 3.34A—F)

Corynoneura n. sp. 2 Bolton, 2007: 31.

Corynoneura disinflata Fu & Sæther, 2012: 18.

Material examined. Holotype male, with associated larval and pupal exuviae, USA, Ohio, Delaware County, vernal stream, Seymore Woods S. N. P., 12.iv.1996, M. J. Bolton (ZMBN Type No. 456). Allotype female, with associated larval and pupal exuviae, as holotype (ZMBN). Paratypes: 1 male, 1 pupal and 1 larval exuviae as holotype except for 29. iv.1989; Ohio, Franklin County, Ohio, Sharon Woods Metro Park, woodland trickle, 3&5.iii.1986, 1 male, 1 pupal and 4 larval exuviae, M. J. Bolton (ZMBN, MJB).

Diagnostic characters. The male imago is characterized by having anterior margin of cibarial pump only slightly concave, hind tibia slightly expanded, superior volsella narrow; inferior volsella along the inner margin of gonocoxite and with many strong setae; transverse sternapodeme straight and weakly U-shaped, gonostylus expanded and

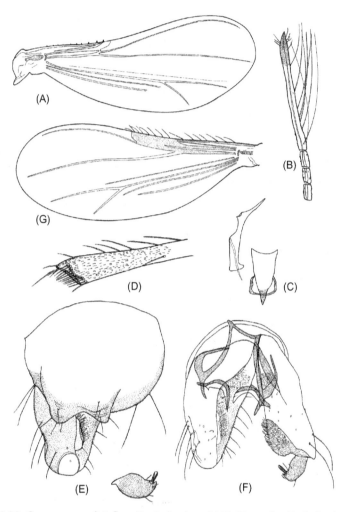

Figure 3.33 *Corynoneura disinflata* Fu & Sæther, 2012 (From Fu, Y., & Sæther, O. A. (2012). *Corynoneura* Winnertz and *Thienemanniella* Kieffer from the Nearctic region (Diptera: Chironomidae: Orthocladiinae). *Zootaxa, 3536,* 1−61. www.mapress.com/j/ zt). (A−F) Male imago. (A) Wing. (B) Antenna. (C) Tentorium, stipes and cibarial pump. (D) Hind tibial apex. (E) Hypopygium, dorsal view. (F) Hypopygium, ventral view. (G) Female imago: wing.

slightly curved; megaseta long. The female has 10 transparent fused lateral lamellae on inner side of coxosternapodeme and 17 lanceolate sensilla on labia; tergite IX medially concave, carrying 4 setae on each hump, gonocoxite with 1 long seta. The pupa has no taeniate L-setae on tergite I, tergite II with short L-setae, tergites III−VII with one

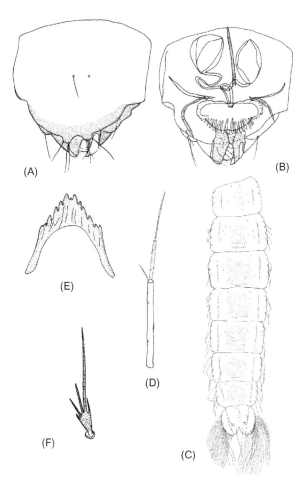

Figure 3.34 *Corynoneura disinflata* Fu & Sæther, 2012 (From Fu, Y., & Sæther, O. A. (2012). *Corynoneura* Winnertz and *Thienemanniella* Kieffer from the Nearctic region (Diptera: Chironomidae: Orthocladiinae). *Zootaxa, 3536*, 1−61. www.mapress.com/j/zt). (A and B) Female imago. (A) Genitalia, dorsal view. (B) Genitalia, ventral view. (C) Pupa: tergites II−IX. (D−F) Larva. (D) Antenna. (E) Mentum. (F) Subbasal seta of posterior parapod.

or two row thick shagreen posteriorly and 42 taeniae in fringe of anal lobe. The larvae are separable by mentum with three median teeth, the middle one very small, AR 0.89−0.91, and head capsule integument unsculptured.

Male (*n* = 2−3)

Total length 1.73−2.00, 1.87 mm. Wing length 1.02−1.10 mm. Total length/wing length 1.62−1.79. Wing length/profemur length 3.27−3.50.

Coloration. Head yellowish brown. Thorax yellowish brown, yellowish medially. Antenna, legs and abdomen yellowish.

Head. Antenna (Fig. 3.33B) with 12 flagellomeres; ultimate flagellomere 252−260, 255 μm long; AR 0.68−0.80, 0.72; antenna apically acute, with about 8−12 apical sensilla chaetica; longest antennal seta 391−408, 400 μm. Temporals absent. Tentorium, stipes and cibarial pump as in Fig. 3.33C; tentorium 137−147, 140 μm long, 15−18, 16 μm wide; stipes 62−83, 75 μm long, 3−4 μm wide. Anterior margin of cibarial pump slightly concave. Clypeus with 10−11 setae. Length of palpomeres (in μm): 18−22, 19; 18−23, 20; 21−36, 29; 30−40, 36; 62−68, 65. Palpomere 5/3 ratio 1.7−3.1, 2.4.

Thorax. Dorsocentrals 5−6, prealars 3. Supraalar 1. Scutellum with 2 setae. Anapleural suture 152−160, 156 μm long.

Wing (Fig. 3.33A). VR 2.0−2.2. Cu/wing length 0.56; C length 326−334 μm long; C length/wing length 0.30−0.32; Cu 570−620 μm long; wing width/ wing length 0.36−0.38. Costa with 5−7 setae.

Legs. Fore trochanter with dorsal keel. Spurs of fore tibia 26−30, 28 μm and 12 μm long, of mid tibia 12−14, 13 μm and 8−12, 10 μm long, of hind tibia 26−36, 33 μm, 15−17, 16 μm, and 14 μm long. Width at apex of fore tibia 21−25, 23 μm, of mid tibia 23−24 μm, of hind tibia (a) 35−38, 36 μm. Width of hind tibia 1/3 from apex (d) 21−28, 24 μm, elongation length (b) 21−22 μm, length of maximum thickening (c_1) 58−69, 65 μm, total length of thickening (c_2) 110 μm; a/d 1.4−1.7, 1.6; b/d 0.8−1.0, 0.9; c_1/d 2.1−3.3, 2.8; c_2/d 3.9−5.2, 4.7. Hind tibia (Fig. 3.33D) slightly expanded, with comb of 12 setae, without S-shaped seta. Lengths and proportions of legs as in Table 3.17.

Hypopygium (Fig. 3.33E and F). Tergite IX medially concave, with 8 long setae. Laterosternite IX with 2 long setae. Superior volsella long and narrow, anteriomedially separated. Inferior volsella with many strong marginal setae, placed caudally along the inner margin of gonocoxite. Phallapodeme curved, 48−51, 50 μm long, placed in caudal position of sternapodeme. Transverse sternapodeme 28−32, 31 μm wide. Gonocoxite 110−135, 124 μm long; gonostylus expanded, 35−41, 37 μm long; megaseta 10−17, 13 μm long. HR 3.1−3.9, 3.4; HV 3.2−5.1, 4.4.

Female ($n = 1$)

Total length 1.47 mm. Wing length 0.84 mm. Total length/wing length 1.76. Wing length/profemur length 3.95.

Coloration. Same as male.

Table 3.17 Lengths (in μm) and proportions of legs segments of male *Corynoneura disinflata* Fu & Sæther, 2012 (n = 3)

	fe	ti	ta_1	ta_2	ta_3	ta_4
p_1	312–340, 321	380–416, 395	208–216, 213	115–133, 124	61–71, 65	26–36, 29
p_2	444–456, 449	434–473, 454	240–261, 247	125–137, 132	61–71, 66	24–36, 29
p_3	383–416, 399	424–473, 446	256–269, 262	135–149, 142	57–71, 63	24–36, 28

	ta_5	LR	BV	SV	BR
p_1	30–51, 40	0.52–0.55, 0.54	3.3–3.8, 3.6	3.3–3.5, 3.4	1.5–2.0, 1.8
p_2	40–56, 45	0.51–0.57, 054	4.0–4.4, 4.2	3.5–3.9, 3.7	2.5
p_3	44–46, 45	0.55–0.61, 0.59	3.9–4.1, 4.0	3.1–3.4, 3.2	2.0–2.3, 2.2

Source: From Fu, Y., & Sæther, O. A. (2012). *Corynoneura* Winnertz and *Thienemanniella* Kieffer from the Nearctic region (Diptera: Chironomidae: Orthocladiinae). *Zootaxa*, 3536, 1–61. www.mapress.com/j/zt.

Table 3.18 Lengths (in μm) and proportions of legs segments of female *Corynoneura disinflata* Fu & Sæther, 2012 (n = 1)

	fe	ti	ta_1	ta_2	ta_3	ta_4	ta_5	LR	BV	SV	BR
p_1	212	264	139	67	38	20	34	0.53	3.9	3.4	1.9
p_2	272	308	156	75	42	20	36	0.51	4.3	3.7	2.0
p_3	264	300	168	91	38	20	38	0.56	3.9	3.4	2.1

Source: From Fu, Y., & Sæther, O. A. (2012). *Corynoneura* Winnertz and *Thienemanniella* Kieffer from the Nearctic region (Diptera: Chironomidae: Orthocladiinae). *Zootaxa*, 3536, 1–61. www.mapress.com/j/zt.

Head. AR 0.40. Length of flagellomeres (in μm): 36, 30, 30, 32, 53. Ultimate flagellomere with 5 apical sensilla chaetica. Tentorium 97 μm long, 10 μm wide. Clypeus with 9 setae. Palpomere lengths (in μm): 15, 15, 25, 25, 50. Palpomere 5/3 ratio 2.0.

Thorax. Dorsocentrals 5; prealars 2. Scutellum with 2 setae.

Wing (Fig. 3.33G). Wing broader than in male. VR 1.96, two anal veins present, Cu 448 μm long, Cu/wing length 0.54, C 416 μm long, C/wing length 0.50, wing width/wing length 0.43. Costa with 14 setae.

Legs. Fore trochanter with keel. Spurs of fore tibia 10 μm and 11 μm long, of mid tibia 14 μm and 8 μm long, of hind tibia 32 μm and 12 μm long. Width at apex of fore tibia 19 μm wide, of mid tibia 15 μm, of hind tibia (a) 30 μm. Width of hind tibia 1/3 from apex (d) 19 μm, elongation length (b) 17 μm, length of maximum thickening (c_1) 57 μm, total length of thickening (c_2) 104 μm; a/d 1.58, b/d 0.89, c_1/d 3.0, c_2/d 5.5. Apex of hind tibia slightly elongate, with comb of 10 setae, without S-shaped spur. Lengths and proportions of legs as in Table 3.18.

Abdomen. Number of setae on tergites II−VIII as: 1, 1, 1, 1, 1, 3, 2.

Genitalia (Fig. 3.34A and B). Tergite IX with 8 long caudal setae. Cercus 55 μm long, 15 μm wide. Notum length 79 μm. Gonocoxite with 1 long seta. Coxosternapodeme with 10 transparent fused lateral lamellae on inner side of coxosternapodeme and 17 lanceolate sensilla on labia. Seminal capsule 85 μm long; neck 20 μm long, 6 μm wide. PcS_2 4 μm long, PcS_3 6 μm long. PcS_{1-3} almost in a line; PcS_1 4 μm from PcS_2.

Pupa (n = 1)

Total length 2.46 mm. Exuviae yellowish, abdomen yellowish with brown belt on each side.

Cephalothorax. Frontal setae 14 μm long. Median antepronotals 7, 8 μm and 10 μm long. Lateral antepronotals 10 μm and 12 μm long, anterior precorneal seta (PcS_1) 6 μm, PcS_2 4 μm from PcS_3, PcS_1 11 μm from PcS_3, PcS_3 47 μm from thoracic horn. Anterior dorsocentral (Dc_1) 8 μm long, Dc_2

10 µm long, Dc_3 7 µm long, Dc_4 not observed. Dc_1 located 23 µm from Dc_2, Dc_2 located 55 µm from Dc_3. Wing sheath with 7 rows of pearls.

Abdomen (Fig. 3.34C). Shagreen and chaetotaxy as illustrated. No taeniate L-setae on tergite I ; tergite II with short L-setae. Anal lobe 180 µm long. Anal lobe fringe with 40−43 setae, 320 µm long. Three taeniate anal macrosetae almost transparent; exact length hard to examine; median setae shorter than anal macrosetae, about 160 µm long.

Larva ($n = 1-2$)

Coloration. Head yellowish. Antenná with basal segment almost transparent, other segments brown. Abdomen yellowish.

Head. Capsule length 252−288 µm, width 196−204 µm, head capsule integument unsculptured.

Postmentum 220−256 µm long. Sternite I simple, II large, rising from small tubercle, III not visible. Premandible 40 µm long, brush of premandible invisible. Mentum as in Fig. 3.34E. Mandibles 65 µm long. Antenna (Fig. 3.34D), AR 0.89−0.91. Length of flagellomeres I−IV (in µm): 184−202, 91−103, 105−121, 4−6. Basal segment width 17−18 µm, length of blade at apex of basal segment 28−32 µm.

Abdomen. Length of anal setae 260 µm. Procercus 10 µm long, 8 µm wide. Subbasal seta of posterior parapods split as Fig. 3.34F.

Distribution. NE: USA (Ohio State).

3.28 *CORYNONEURA DORICENI* MAKARCHENKO & MAKARCHENKO, 2006B (FIGS. 3.35A−G AND 3.36A−F)

Corynoneura doriceni Makarchenko & Makarchenko, 2006b: 152; Fu & Sæther, 2012: 22.

Corynoneura n. sp. 4 Bolton, 2007: 29.

Corynoneura sp. D Epler, 2001: 7.48.

Material examined. USA, Ohio, Portage County, Frame Lake, 2 males with associated larval and pupal exuviae, 2 females with associated larval and pupal exuviae, 21.vi.1987, M. J. Bolton (ZMBN, MJB).

Diagnostic characters. Based on the specimens from USA, the adult male is characterized by straight posterior margin of tergite IX; superior volsella rectangular and partly overlapping with inferior volsella; transverse sternapodeme thick and medially strongly concave. The adult female has AR 0.47−0.56; with 6 transparent developed serrated lateral lamellae on inner side of coxosternapodeme and 6 smaller fused lanceolate sensilla on the margin of labia. The pupa has no taeniate L-setae and shagreen on tergites I−II, all the tergites with weak shagreen and few minute

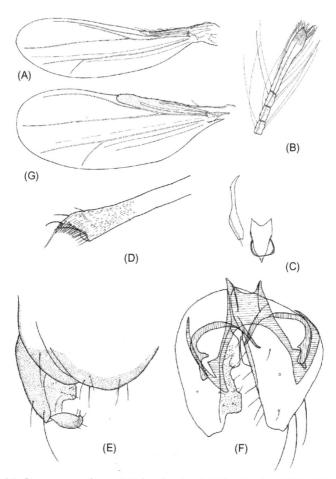

Figure 3.35 *Corynoneura doriceni* Makarchenko & Makarchenko, 2006b (From Fu, Y., & Sæther, O. A. (2012). *Corynoneura* Winnertz and *Thienemanniella* Kieffer from the Nearctic region (Diptera: Chironomidae: Orthocladiinae). *Zootaxa, 3536*, 1–61. www. mapress.com/j/zt). (A–F) Male imago. (A) Wing. (B) Antenna. (C) Tentorium, stipes and cibarial pump. (D) Hind tibial apex. (E) Hypopygium, dorsal view. (F) Hypopygium, ventral view. (G) Female imago: wing.

hooklets; and 29–31 taeniae in fringe of anal lobe. The larvae are separable by having mentum with three median teeth, the mentum has 6 lateral teeth, the lateral tooth adjacent to median tooth not small and the outer one very small, all antennal segments of the same color; length of antenna/length of head 2.35–2.41; and head capsule unsculptured.

Male (*n* = 1−2)

Total length 1.23–1.30 mm. Wing length 0.73–0.75 mm. Total length/wing length 1.64–1.80. Wing length/profemur length 2.9–3.0.

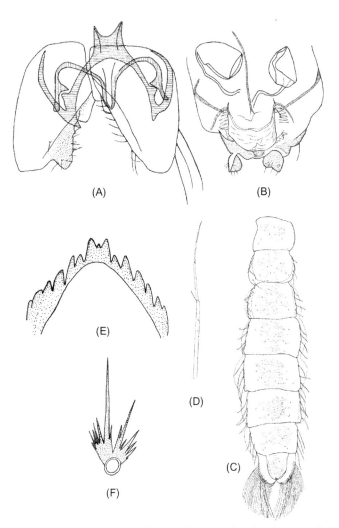

Figure 3.36 *Corynoneura doriceni* Makarchenko & Makarchenko, 2006b (From Fu, Y., & Sæther, O. A. (2012). *Corynoneura* Winnertz and *Thienemanniella* Kieffer from the Nearctic region (Diptera: Chironomidae: Orthocladiinae). *Zootaxa, 3536,* 1−61. www.mapress.com/j/zt). (A) Male imago: hypopygium, ventral view. (B) Female imago: genitalia, ventral view. (C) Pupa: tergites II−IX. D-F. Larva. (D) Antenna. (E) Mentum. (F) Subbasal seta of posterior parapod.

Coloration. Head brown. Antenna and legs yellowish. Thorax with upper antepronotum, upper scutum, scutellum, postnotum and preepisternum brown, medially yellowish. Abdomen: tergite VII medially pale yellow, marginally brown; other tergites brown.

Table 3.19 Lengths (in μm) and proportions of legs segments of male *Corynoneura doriceni* Makarchenko & Makarchenko, 2006b ($n = 2$)

	Fe	ti	ta_1	ta_2	ta_3	ta_4
p_1	248−252	288	154−164	89−93	38−44	18
p_2	340−344	304−312	184−188	89−91	40−44	16
p_3	280	288−294	166−172	99	38−40	11−18

	ta_5	LR	BV	SV	BR
p_1	32	0.57	3.8	3.3	1.0−1.5
p_2	30	0.59−0.62	4.6−4.9	3.4−3.6	2.3
p_3	21−28	0.58	4.0	3.3	2.1

Head. Antenna (Fig. 3.35B) with 12 flagellomeres, ultimate flagellomere 131 μm, AR 0.47−0.54; antenna apically expanded, with about 15−20 apical sensilla chactica, longest antennal seta 260−280 μm. Temporals absent. Tentorium, stipes and cibarial pump as in Fig. 3.35C, tentorium 121−131 μm long, 18−22 μm wide; stipes 65−81 μm long, 3−4 μm wide. Anterior margin of cibarial pump strongly concave. Clypeus with 8−11 setae. Lengths of palpomeres (in μm): 10−12, 17, 22−23, 30−32, 54−55. Palpomere 5/3 ratio 2.4−2.5.

Thorax. Dorsocentrals 4−5, prealars 2, supraalars 1. Scutellum with 2 setae. Anapleural suture 101−103 μm long.

Wing (Fig. 3.35A). VR 3.3. Cu/wing length 0.61−0.63, C length 196−212 μm long, Cu 460−468 μm long, C/wing length 0.26−0.29, wing width/wing length 0.34−0.35. Costa with 6−7 setae.

Legs. Fore trochanter with dorsal keel. Spurs of fore tibia 15−17 μm and 8 μm long, of mid tibia 14−15 μm and 6−7 μm long, of hind tibia 32 μm and 8 μm long. Width at apex of fore tibia 18−21 μm, of mid tibia 18−19 μm, of hind tibia (a) 30−35 μm. Width of hind tibia 1/3 from apex (d) 17−20 μm, elongation length (b) 21−23 μm, length of maximum thickening (c_1) 55−57 μm, total length of thickening (c_2) 97−100 μm, a/d 1.75−1.80, b/d 1.1−1.2, c_1/d 3.1−3.2, c_2/d 5.6−5.7. Hind tibia (Fig. 3.35D) expanded, with comb of 13 setae, 1 seta near spur hooked. Lengths and proportions of legs as in Table 3.19.

Hypopygium (Figs. 3.35E and F and 3.36A). Posterior margin of tergite IX almost straight, with 6 long setae. Laterosternite IX with 1 long seta. Superior volsella rectangle, anteriomedially separated. Inferior volsella well developed, partly overlapping with superior volsella. Phallapodeme strongly curved, 72−80 μm long, with lateral projection and joint with sternapodeme placed prelateral. Transverse sternapodeme thick and medially strongly concave, 17 μm wide. Large attachment point of lateral sternapodeme with phallapodeme placed caudal third and directed caudally. Gonostylus apical slightly curved, 22−30 μm long; megaseta 4 μm long. HR 2.9−4.1, HV 4.1−5.9.

Female ($n = 1-2$)

Total length 1.20 mm. Wing length 0.69−0.70 mm. Total length/ wing length 1.70. Wing length/profemur length 3.1−3.3.

Coloration. Head brown. Thorax yellowish brown, medially yellowish. Antenna and legs yellowish, abdomen yellowish brown.

Head. AR 0.47−0.56. Length of flagellomeres (in μm): 22−23, 25−28, 29−30, 30−32, 52−61. Ultimate flagellomere with 10−12 apical sensilla chaetica. Tentorium 104−105 μm long, 10−12 μm wide. Clypeus with 8−12 setae. Palpomere lengths (in μm): 14−15, 14−15, 17−23, 25−29, 41−50. Palpomere 5/3 ratio 2.2−2.4.

Thorax. Dorsocentrals 7, prealars 2. Scutellum with 2 setae.

Wing (Fig. 3.35G). Wing broader than in male. VR 2.4−2.6, two anal veins present, Cu 368−392 μm long, Cu/wing length 0.53−0.57, C 340 μm long; C/wing length 0.49, wing width/wing length 0.41−0.43. Costa with 10−11 setae.

Legs. Fore trochanter with keel. Spurs of fore tibia 11−14 μm and 10 μm long, of mid tibia 10−12 μm and 14−15 μm long, of hind tibia 33−35 μm and 10−11 μm long. Width at apex of fore tibia 18−21 μm wide, of mid tibia 19 μm, of hind tibia (a) 35−37 μm. Width of hind tibia 1/3 from apex (d) 18−19 μm, elongation length (b) 25−28 μm, length of maximum thickening (c_1) 62−81 μm, total length of thickening (c_2) 90−104 μm, a/d 1.9, b/d 1.4−1.5, c_1/d 3.3−4.5, c_2/d 5.0−5.5. Apex of hind tibia expanded, with comb of 14−16 setae, with one slightly hooked spur (same as in the male). Lengths and proportions of legs as in Table 3.20.

Abdomen. Number of setae on tergites II −VIII as: 1, 1, 1, 1, 1, 3−4, 2.

Genitalia (Fig. 3.36B). Tergite IX with 5−6 long caudal setae. Cercus 32−33 μm long, 32 μm wide. Notum length 106−120 μm. Gonocoxite with 1 long seta. Coxosternapodeme with 6 transparent well-developed serrated lateral lamellae on inner side and 6 smaller lanceolate sensilla on the margin of labia. Seminal capsule 72−81 μm long, neck 8−14 μm long, 6 μm wide.

Pupa ($n = 1-3$)

Total length 1.73−1.94, 1.89 mm. Exuviae yellowish including cephalothorax.

Table 3.20 Lengths (in μm) and proportions of legs segments of female *Corynoneura doriceni* Makarchenko & Makarchenko, 2006b ($n = 2$)

	fe	ti	ta_1	ta_2	ta_3	ta_4
p_1	208−224	260	125−128	65−71	36−38	20
p_2	280−304	288	156−160	77−79	38	18
p_3	220−256	264−272	144−148	81−83	36	16−18

	ta_5	LR	BV	SV	BR
p_1	26−30	0.48−0.49	3.9−4.0	3.7−3.8	1.0
p_2	26−30	0.54−0.56	4.4−4.7	3.6−3.7	1.5−1.8
p_3	32−34	0.54−0.55	3.8−4.0	3.4	1.7−2.0

Cephalothorax. Frontal setae 24−28, 26 μm long. Median antepronotals 14−16 μm and 16−18 μm long. Lateral antepronotals not visible. Anterior precorneal seta (PcS_1) 14 μm long, PcS_2 21 μm long, PcS_3 21 μm long. PcS_{1-3} almost in a line, PcS_1 30 μm from PcS_2, PcS_2 3 μm from PcS_3, PcS_1 33 μm from PcS_3, PcS_3 35 μm from thoracic horn. Anterior dorsocentral (Dc_1) not observed, Dc_2 14 μm long, Dc_3 10 μm long, Dc_4 20 μm long. Dc_2 located 67 μm from Dc_3, Dc_3 located 14 μm from Dc_4. Wing sheath with very week pearls.

Abdomen (Fig. 3.36C). Shagreen and chaetotaxy as illustrated. Caudal hooklets on tergal conjunctives minute, almost absent. No taeniate L-setae on tergites I−II. Anal lobe 127−133, 130 μm long. Anal lobe fringe with 29−31 setae, 240−260 μm long. Anal macrosetae 182−202 μm long, median setae 61−80 μm long.

Larva ($n = 2-3$)

Coloration. Head yellowish. Antenna with all segments yellowish. Abdomen yellowish.

Head. Capsule length 304−320 μm, width 192−200 μm, unsculptured. Postmentum 258−264 μm long. Sternite II obvious, arising from small tubercle, I and III not visible. Premandible 32−44 μm long. Mentum as in Fig. 3.36E. Mandibles 61−71 μm long. Antenna as in Fig. 3.36D, AR 0.94−0.98. Lengths of flagellomeres I −IV (in μm): 348−392, 168−172, 192−208, 3−4. Basal segment width 16−20 μm, length of blade at apex of basal segment 41−44 μm. Length of antenna/ length of head 2.35−2.41.

Abdomen. Length of anal setae 280−320 μm. Procercus 11 μm long, 8−11 μm wide. Subbasal seta of posterior parapods split as in Fig. 3.36F, 55−61 μm long. Posterior parapods 131−162 μm long.

Remarks. The male is nearly identical to the original description by Makarchenko and Makarchenko (2006b) except the slightly smaller size. The female, pupa and larva previously were unknown.

Distribution. PA: Russia (Far East). **NE**: USA.

3.29 *CORYNONEURA ECPHORA* FANG, WANG, FU, 2014 (FIG. 3.37A−F)

Corynoneura ecphora Fang et al., 2014: 2.

Material examined. Holotype male, P.R. China: Guangdong Province, Guangzhou City, Conghua County, Wenquan Town, 16. i.2011, H. Q. Tang (HBMY Type No. 0001). Paratypes: 1 male as holotype (HBMY No. 0002).

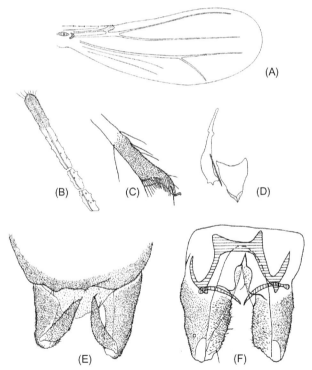

Figure 3.37 *Corynoneura ecphora* Fang et al., 2014, Male imago (From Fang, X. L., Wang, X., & Fu, Y. (2014). A new species of Corynoneura Winnertz from oriental China (Diptera, Chironomidae, Orthocladiinae), *Zootaxa, 3884*(6), 567−572. www. mapress.com/j/zt). (A) Wing. (B) Antenna. (C) Hind tibial apex. (D) Tentorium, stipes and cibarial pump. (E) Hypopygium, dorsal view. (F) Hypopygium, ventral view.

Diagnostic characters. The male imago is characterized by having an antenna with 11 flagellomeres, AR 0.38−0.39; superior volsella developed, like a collar with rounded corner; inferior volsella broad, with dented edge, placed caudally on gonocoxite; sternapodeme curved into U-shape, with developed oral projection, and lateral sternapodeme with caudal attachment point. Phallapodeme slightly curved, placed in caudal position of sternapodeme, gonostylus tapering.

Male ($n = 2$)

Total length 1.60−1.70 mm. Wing length 0.73−0.78 mm. Total length/wing length 2.14−2.16. Wing length/profemur length 3.3−3.4.

Coloration. Head yellow-brown, with dark-brown eyes. Thorax yellow-brown. Legs yellowish. Abdomen light brown.

Head. Antenna with 11 flagellomeres, AR 0.38−0.39, ultimate flagellomere 127−134 μm long, ultimate flagellomere slightly expanded apically, with about 7−8 apical sensilla chaetica (Fig. 3.37B). Temporals absent. Tentorium, stipes and cibarial pump as in Fig. 3.37D, tentorium 122−124 μm long; 16−19 μm wide; stipes 60−63 μm long, 3−5 μm wide. Anterior margin of cibarial pump (Fig. 3.37D) slightly concave. Clypeus with 9−10 setae. Length of palpomeres (in μm): 10−12, 14, 17−19, 22−25, 53−55. Palpomere 5/3 ratio 3.0−3.2.

Thorax. Antepronotals 5. Dorsocentrals 5. Scutellum with 2 setae.

Wing (Fig. 3.37A). VR 3.1−3.2. Cu/wing length 0.55−0.57; C 168−170 μm long;

Cu 375−380 μm long; wing width/wing length 0.40−0.41. Brachiolum with 1 seta, costa with 5−7 setae.

Legs. Fore trochanter with dorsal keel. Spur of fore tibia 22−24 μm long, of mid tibia 10 μm long, and spurs of hind tibia 30−32 μm long and 12 μm long. Width at apex of fore tibia 18−20 μm, of mid tibia 20 μm, of hind tibia (a) 34−36 μm. Width of hind tibia 1/3 from apex (d)18−20 μm, elongation length (b) 36−39 μm, length of maximum thickening (c_1) 60−62 μm, total length of thickening (c_2) 85−90 μm; a/d 2.0−2.1; b/d 2.1−2.3; c_1/d 3.3−3.5; c_2/d 4.7−5.1. Hind tibia expanded, with comb of 12−14 setae, 1 seta near spur strongly S-shaped (Fig. 3.37C). Lengths and proportions of legs as in Table 3.21.

Hypopygium (Fig. 3.37E and F). Tergite IX and laterosternite IX without long setae. Tergite IX medially slightly incurved, and pair of low caudal mounds carrying setae. Superior volsella anteriomedially fused and developed, like collar. Inferior volsella broad, with many glandular setae and dented margin, along the inner margin of gonocoxite and placed caudally. Phallapodeme slightly curved, 24−31 μm long, with projection for

Table 3.21 Lengths (in μm) and proportions of legs segments of male *Corynoneura ecphora* Fang et al., 2014 (*n* = 2)

	fe	ti	ta$_1$	ta$_2$	ta$_3$	ta$_4$
p$_1$	245−265	284−304	156−163	84−96	41−58	24
p$_2$	363−382	323−343	185−192	84−86	40−43	20−22
p$_3$	274−314	313−323	156−168	86−96	35−38	19−20

	ta$_5$	LR	BV	SV	BR	
p$_1$	34−36	0.54−0.55	3.4−3.8	3.4−3.5	2.3−2.5	
p$_2$	29−31	0.56−0.57	4.8−5.0	3.7−3.8	2.2−2.3	
p$_3$	29−34	0.50−0.52	4.2−4.5	3.8−3.9	2.5−2.8	

Source: From Fang, X. L., Wang, X., & Fu, Y. (2014). A new species of Corynoneura Winnertz from oriental China (Diptera, Chironomidae, Orthocladiinae). *Zootaxa*, 3884 (6): 567−572. www.mapress.com/j/zt.

joint with sternapodeme placed caudally. Transverse sternapodeme 31−36 μm wide. Small attachment point of lateral sternapodeme with phallapodeme placed and directed caudally. Gonostylus slightly curved tapering, 31−36 μm long, with 1 subapical setae; megaseta 4.8 μm long. HR 2.1−2.2; HV 5.2−5.4.

Remarks. The new species is similar to *Corynoneura sundukovi* Makarchenko *et* Makarchenko 2010, both species have the sternapodeme inverted U-shaped and oral projections developed, but can be separated by the antenna with 11 flagellomeres, AR 0.38−0.39 and superior volsella with big rounded corner in *C. ecphora* while *C. sundukovi* has antenna with 12 flagellomeres, AR 0.54−0.74 and superior volsella triangular.

Distribution. OR: China (Guangdong).

3.30 *CORYNONEURA EDWARDSI* BRUNDIN, 1949 (FIGS. 3.38A−C AND 3.39A−D)

Corynoneura edwardsi Brundin, 1949: 698; Hirvenoja & Hirvenoja, 1988: 217; Sasa, 1997: 49; Schlee, 1968b: 31; Fu et al., 2009: 9; Fu & Sæther, 2012: 27.

Material examined. P.R. China: Xinjiang Uyger Autonomous Region, Aketao County, Kelakuli Lake, 39°8′N, 75°56′E, alt. 4000 m, sweep net, 1 male, 19.viii.2002, H. Tang; Xinjiang Uyger Autonomous Region, Fukang County, Tianchi Lake, 44°8′N, 87°56′E, alt. 1980 m, sweep net, 1 male, 1.ix.2002, H. Tang. USA, Ohio, Summit County, Long Lake on *Nymphaea*, 1 male, 1 female, 2 larval and 2 pupal exuviae, 25.x.1986, M. J. Bolton (MJB).

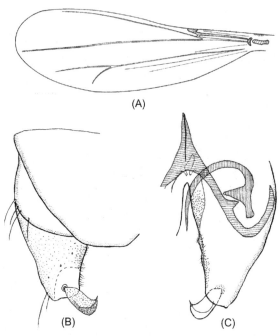

Figure 3.38 *Corynoneura edwardsi* Brundin, 1949; Male imago. (A) Wing. (B) Hypopygium, dorsal view. (C) Hypopygium, ventral view.

Diagnostic characters. The male are antenna with 10 flagellomeres, AR 0.94−1.0, 0.97; inferior volsellae absent, Sternapodeme inverted V-shaped. The female has AR 0.62, hind tibia without S-shaped spur, 4 setae on gonocoxite IX, 4 on tergite IX and coxosternapodeme with 8−9 lateral lamellae. The pupa is 1.6−1.8 mm long, has no taeniate L-setae on tergite I and only one on tergite II, and 46−49 taeniae in fringe of anal lobe; tergites III−VII with thick hooklets. The larvae are separable by AR 1.17−1.20, mentum with three median teeth, the central tooth smaller than the lateral median one; the first lateral teeth as large as adjacent tooth; and weakly sculptured head capsule.

Male (*n* = 2)

Total length 1.47−1.76, 1.60 mm. Wing length 1.15−1.25, 1.2 mm. Total length/wing length 1.28−1.4, 1.34. Wing length/length of profemur 3.6−3.7, 3.65.

Coloration. Head dark brown. Antenna and palpomere pale yellow-brown. Thorax dark brown. Abdominal segments brown. Legs yellow-brown. Wings transparent and yellowish, with pale yellow clava.

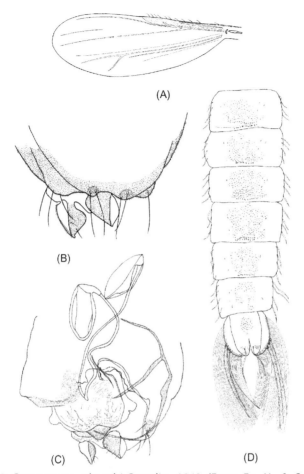

Figure 3.39 *Corynoneura edwardsi* Brundin, 1949 (From Fu, Y., & Sæther, O. A. (2012). *Corynoneura* Winnertz and *Thienemanniella* Kieffer from the Nearctic region (Diptera: Chironomidae: Orthocladiinae). *Zootaxa, 3536*, 1−61. www.mapress.com/j/ zt). (A−C) Female imago. (A) Wing. (B) Genitalia, dorsal view. (C) Genitalia, ventral view. (D) Pupa: tergites III−IX.

Head. Eyes bare, reniform. Antenna with 10 flagellomeres; ultimate flagellomere 290−330, 310 µm long, apex with short sensilla chaetica which extends a little way back from tip, apically acuate. AR 0.94−1.0, 0.97. Clypeus with 9 setae. Temporal setae lacking. Tentorium 140−170, 155 µm long, 20−23, 22.5 µm wide. Stipes: 60−65, 63 µm long. Palpomeres length (in µm): 18−20, 18; 18−20, 18; 25−35, 30; 35; 65. Palpomere 2 ellipsoid, 3 and 4 rectangular, 5 long and slender. Palpomere 5/3 ratio 2.6 (1).

Thorax. Antepronotum without lateral setae. Dorsocentrals 4(1), prealars 1, other setae cannot be seen.

Wing (Fig. 3.38A): VR 2.8. C length 310−330, 320 μm, C/wing length 0.26−0.27, 0.27. Cu length 510 μm. Cu/wing length 0.44. Wing width/wing length 0.34−0.35, 0.35. C with 8−10, 9 setae.

Legs. Fore trochanter with keel. Spur of front tibia 20−25, 23 μm and 13−15, 14 μm long, spurs of middle tibia 10−13, 12 μm and 10 μm long, of hind tibia 43−45, 44 μm long. Width at apex of fore tibia 25−33, 29 μm, of mid tibia 25 μm, of hind tibia 40−43, 41 μm. Tip of hind tibia expanded, with comb of 19−21, 20 setae and 1 seta near spur slightly hooked. Lengths and proportions of legs as in Table 3.22.

Hypopygium (Fig. 3.38B and C). Posterior margin weakly bilobed, and with many short setae, laterosternite IX with 3−4 long setae. Anal point absent. Inferior volsellae absent, sternapodeme inverted V-shaped, coxapodeme 37.5 μm long. Phallapodeme strongly curved not extending beyond tergite IX, 105−110, 107 μm long. Gonocoxite 100−115, 107.5 μm long with 2 long setae apically. Gonostylus curved apically, 33−50, 41.5 μm long. Megaseta 5−7.5, 6 μm long. HR 2.0−3.5, 2.8, HV 3.6−4.5, 4.0.

Female ($n = 1$)

Total length 2.10 mm. Wing length 1.04 mm. Total length/wing length 2.0. Wing length/profemur length 3.8.

Coloration. Head brown. Thorax brown, medially yellowish. Antenna and legs yellowish, abdomen yellowish brown.

Head. AR 0.62. Length of flagellomeres (in μm): 40, 35, 30, 33, 82. Ultimate flagellomere with many preapical sensilla chaetica. Clypeus with 10 setae. Palpomere lengths (in μm): 19, 22, 28, 39, 62. Palpomere 5/3 ratio 2.2.

Thorax. Dorsocentrals 5, prealars 2. Scutellum with 2 setae.

Wing (Fig. 3.39A). Wing broader than in male. VR 2.6, two anal veins present, Cu 600 μm long, Cu/wing length 0.58, C 489 μm long, C/wing length 0.47, wing width/wing length 0.39. Costa with 13 setae.

Legs. Fore trochanter with keel. Spurs of fore tibia 14 μm long, of mid tibia 11 μm and 10 μm long, of hind tibia 47 μm and 14 μm long. Width at apex of fore tibia 23 μm wide, of mid tibia 22 μm, of hind tibia (a) 39 μm. Width of hind tibia 1/3 from apex (d) 19 μm, elongation length (b) 35 μm, length of maximum thickening (c_1) 83 μm, total length of thickening (c_2) 124 μm, a/d 2.1, b/d 1.8, c_1/d 4.4, c_2/d 6.5. Apex of hind tibia expanded, with comb of 14 setae, without S-shaped spur. Lengths and proportions of legs as in Table 3.23.

Table 3.22 Lengths (in μm) and proportions of legs segments of male *Corynoneura edwardsi* Brundin, 1949 (based on Chinese specimens)

	fe	ti	ta$_1$	ta$_2$	ta$_3$	ta$_4$
p$_1$	310−350, 330	375−430, 403	220−240, 239	110−115, 113	63−68, 65	23−28, 25
p$_2$	430−510, 470	410−480, 445	230−260, 245	113−120, 116	55−65, 60	15−30, 23
p$_3$	375−420, 403	400−440, 420	230−265, 248	120−138, 129	50−60, 55	23−28, 25

	ta$_5$	LR	BV	SV	BR
p$_1$	43−45, 44	0.56−0.59, 0.58	3.7−4.1, 3.9	3.1−3.3, 3.2	1.4−1.6, 1.5
p$_2$	30−45, 38	0.54−0.56, 055	3.3−5.0, 4.2	3.7−3.9, 3.8	1.6−1.8, 1.7
p$_3$	40−45, 43	0.58−0.60, 0.59	4.2−4.3, 4.25	3.2−3.4, 3.3	1.4−1.5, 1.45

Table 3.23 Lengths (in μm) and proportions of legs segments of female *Corynoneura edwardsi* Brundin, 1949 ($n = 1$) (based on Nearctic specimen)

	fe	ti	ta$_1$	ta$_2$	ta$_3$	ta$_4$	ta$_5$	LR	BV	SV	BR
p$_1$	272	348	170	77	40	20	36	0.49	4.6	3.6	1.5
p$_2$	400	380	220	85	48	22	40	0.58	5.1	3.5	1.6
p$_3$	348	368	200	101	38	20	38	0.54	4.6	3.6	1.8

Abdomen. Number of setae on tergites II−VIII as: 2, 2, 2, 2, 2, 2, 2.

Genitalia (Fig. 3.39B and C). Tergite IX with 4 long caudal setae. Cercus 36 μm long, 30 μm wide. Notum length 101 μm. Gonocoxite with 4 long setae. Coxosternapodeme with 8−9 transparent fused lateral lamellae on inner side of coxosternapodeme. Seminal capsule 101 μm long; neck 12 μm long, 10 μm wide.

Pupa ($n = 1-2$)

Total length 2.31−2.40 mm. Exuviae yellowish brown, abdomen yellowish.

Cephalothorax. Frontal setae 20−25 μm long. Median antepronotals 20−22 μm and 28−30 μm long. Lateral antepronotals 15−18 μm and 18−20 μm long. Anterior precorneal setae (PcS$_1$) 20−26 μm long, PcS$_2$ 16−20 μm long, PcS$_3$ 16−20 μm long. PcS$_{1-3}$ almost in a line, PcS$_1$ 6−8 μm from PcS$_2$, PcS$_2$ 6−9 μm from PcS$_3$, PcS$_1$ 16−18 μm from PcS$_3$, PcS$_3$ 40−45 μm from thoracic horn. Dorsocentrals Dc$_{1-4}$ not clear. Wing sheath with 7 rows of pearls.

Abdomen (Fig. 3.39D). Shagreen and chaetotaxy as illustrated. No taeniate L-setae on tergite I. Anal lobe 91−94 μm long. Anal lobe fringe with 46−49 setae, 360−380 μm long. Three taeniate anal macrosetae 172−180 μm long; median setae shorter than anal macrosetae, about 101−120 μm long.

Larva ($n = 1-2$)

Coloration. Head brown. Antenna with basal segment yellowish, other segments yellowish brown. Abdomen yellowish.

Head. Capsule length 320−322 μm, width 220−224 μm, weakly sculptured. Postmentum 272−275 μm long. Sternite II obvious, rising from small tubercle, I and III not visible. Premandible 40−44 μm long. Mandible 61 μm long. AR 1.17−1.20. Length of flagellomeres I−IV (in μm): 168, 75, 89, 6. Basal segment width 19−21 μm, length of blade at apex of basal segment 28−30 μm. Length of antenna/length of head 0.98.

Abdomen. Length of anal setae 272 μm. Procercus 14 μm long, 17 μm wide. Subbasal seta of posterior parapods 48 μm long.

Remarks. This species is close to *C. gratias*, but can be separated easily by the sternapodeme, phallapodeme, gonostylus. This species is close to *C. arctica*, but can be separated by the posterior margin, apex of hind tibia, character of the hypopygium. This species is also close to *C. scutellata*, but can be separated by the inferior volsellae, apex of hind tibia. When we reviewed the literature regarding this species, we found all previous descriptions and figures too simple to recognize it, so we redescripted and illustrated the species based on Chinese specimens and made other relevant researchers convenient. Hirvenoja and Hirvenoja (1988) redescribed the male, pupa, and larva in detail.

Distribution. PA: Bulgaria, China (Xinjiang), Denmark, Finland, France, Germany, Great Britain, Ireland, Italy, Japan, Kaliningrad, Lithuania, Macedonia, Netherlands, Norway, Romania, Russia (CET, NET, Far East), Spain, Sweden, Switzerland.

3.31 *CORYNONEURA ELONGATA* FREEMAN, 1953 (FIG. 3.40A AND B)

Corynoneura elongata Freeman, 1953: 210.

Diagnostic characters. Male antenna with 10 flagellomeres. Inferior volsella absent. Sternapodeme inverted V-shaped (Similar to *C. dewulfi* Freeman and *C. cristata* Freeman in size, color, wing venation and leg proportions but differing in antennal and hypopygial structure).

Male (Description from Freeman, 1953)

Head. Antenna with 10 flagellomeres (Fig. 3.40A); apical segment rather longer than the preceding four segments together, narrow and not clubbed, with plume hairs for nearly three-quarters of its length, at the apex with short hairs not arranged in a rosette.

Hypopygium (Fig. 3.40B). Inferior volsella absent. Sternapodeme curved into V-shaped. Gonostylus apically curved.

Distribution. AF: South Africa, Zimbabwe.

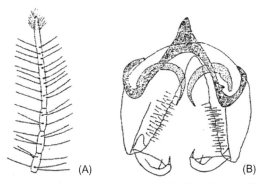

(A)   (B)

Figure 3.40 *Corynoneura elongata* Freeman, 1953; Male imago (From Freeman, P. (1953). Chironomidae from Western Cape Province — II. *Proceedings of the Royal Entomological Society of London (B), 22*, 201−253). (A) Apex of antenna. (B) Hypopygium: ventral view.

3.32 *CORYNONEURA ESPRAIADO* WIEDENBRUG ET AL., 2012 (FIGS. 3.41A−F, 3.42A−E, AND 3.43A−H)

Corynoneura espraiado Wiedenbrug et al., 2012: 21.

Diagnostic characters. The male of *Corynoneura espraiado* is very similar to *Corynoneura humbertoi*, both can be differentiated from other species by the following characters, antenna plumose with 10 flagellomeres, apex of hind tibia with a S-shaped seta, phallapodeme sclerotized on posterior margin and attached in the caudal apex of sternapodeme, aedeagal lobe triangular with a wide base, superior volsella low bearing short setae, inferior volsella present on the apical margin of gonocoxite. The coloration of the abdominal segments separate males of both species;

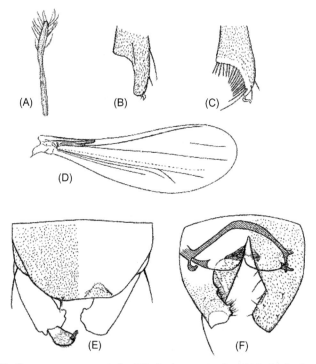

Figure 3.41 *Corynoneura espraiado* Wiedenbrug et al., 2012; Male imago (From Wiedenbrug, S., Lamas, C. J. E., & Trivinho-Strixino, S. (2012). A review of the genus *Corynoneura* Winnertz (Diptera: Chironomidae) from the neotropical region. *Zootaxa, 3574*, 1−61. www.mapress.com/j/zt). (A) Terminal flagellomeres. (B and C) Apex of hind tibia. (D) Wing. (E and F) Hypopygium. (E) Tergite IX; left is dorsal view and gonostylus, right is ventral view. (F) Tergite IX and gonostylus removed, sclerites hatched; left is dorsal view, right is ventral view.

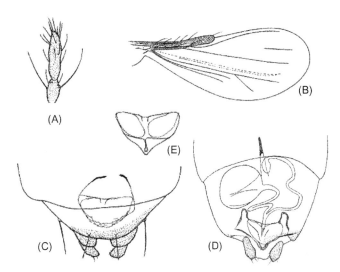

Figure 3.42 *Corynoneura espraiado* Wiedenbrug et al., 2012; Female imago (From Wiedenbrug, S., Lamas, C. J. E., & Trivinho-Strixino, S. (2012). A review of the genus *Corynoneura* Winnertz (Diptera: Chironomidae) from the neotropical region. *Zootaxa*, *3574*, 1—61. www.mapress.com/j/zt). (A) Terminal flagellomeres. (B) Wing. (C—E) Genitalia. (C) Dorsal view of genitalia. (D) Ventral view of genitalia. (E) Ventral view of labia.

C. espraiado tergites I—V are whitish and the rest of the segments are brownish; *C. humbertoi* tergites I—IV are whitish. The females of *C. humbertoi* are characterized by having one seminal capsule larger, about 57—60 μm, second Sca smaller and less sclerotized, coxosternapodeme orally curved with long oral dorsal rami, copulatory bursa not oral median invaginated. The pupa can be recognized by the sternite II with fine shagreen, tergites IV—VI posterior shagreen points larger, posterolateral points larger than posteromedian and anal lobes rounded separated from each other through a straight, perpendicular margin. The larva of *C. espraiado* has the first antennal segment longer than postmentum length, second antennal segment longer than half of the first and longer than the third, mentum with two median teeth, first lateral teeth adpressed to median teeth.

Male (Description from Wiedenbrug et al., 2012)

Total length 0.76—0.90 mm. Wing length 0.52—0.64 mm.

Coloration. Thorax brownish. Abdominal tergites I—V whitish, other tergites and genitalia brownish. Legs whitish.

Head. AR = 0.30—0.31. Antenna with 9 flagellomeres, apical flagellomere 118—143 μm (Fig. 3.41A). Flagellomeres with more than one row of setae each. Eyes pubescent.

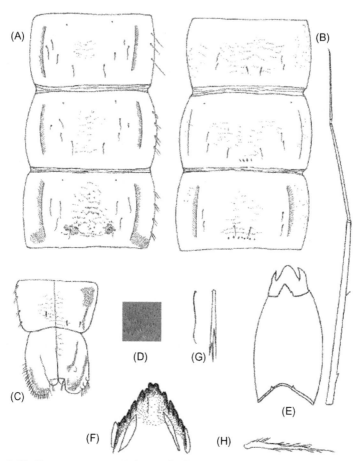

Figure 3.43 *Corynoneura espraiado* Wiedenbrug et al., 2012; Immatures imago (From Wiedenbrug, S., Lamas, C. J. E., & Trivinho-Strixino, S. (2012). A review of the genus *Corynoneura* Winnertz (Diptera: Chironomidae) from the neotropical region. *Zootaxa*, *3574*, 1–61. www.mapress.com/j/zt). (A–D) Pupa imago. (A) Tergites II–IV. (B) Sternites II–IV. (C) Segments VIII, IX and anal lobe; left is ventral view, right is dorsal view. (D) Detail of posterior shagreen of tergite IV. (E–H) Larva imago. (E) Head, ventral view and separated antenna. (F) Mentum. (G) Abdominal setae. (H) Subbasal seta of posterior parapods.

Thorax. Antepronotal lobes dorsally tapering.

Wing (Fig. 3.41D). Clavus/wing length 0.19–0.22. Anal lobe absent.

Legs. Hind tibial scale 22–35 μm long, with one S-seta (Fig. 3.41B and C).

Hypopygium (Fig. 3.41E and F). Tergite IX with 4 setae. Laterosternite IX with 1 seta. Superior volsella absent, anterior median margin of gonocoxite with short setae, aedeagal lobe triangular with a wide base, inferior volsella present apically on gonocoxite margin. Sternapodeme rounded, phallapodeme caudal attached. Phallapodeme with posterior margin sclerotized, slightly curved.

Female (Description from Wiedenbrug et al., 2012)

Total length 0.71 mm. Wing length 0.54—0.56 mm.

Coloration. Thorax light brown. Abdominal tergite I, whitish, II –IX brownish. Legs whitish.

Head. AR = 0.50—0.62. Antenna with 5 flagellomeres, apical flagellomere 55—60 μm long (Fig. 3.42A). Flagellomeres with more than one row of setae each. Eyes bare.

Thorax. Antepronotal lobes dorsally tapering.

Wing (Fig. 3.42B). Clavus/wing length 0.38—0.39. Anal lobe absent.

Legs. Hind tibial scale 30—45 μm long, with one small S-seta.

Genitalia (Fig. 3.42C—E). Tergite IX with 2—4 setae. Laterosternite IX with one seta. One large seminal capsules 57—60 μm long; second smaller less sclerotized difficult to see, one spermathecal duct with a loop, second straighter, both ducts join together shortly before seminal eminence, which has sclerotized outer borders. Notum 30 μm long. Membrane well sclerotized. Apodeme lobe with median border sclerotized. Coxosternapodeme, curved, with two rami, first ending at roof of copulatory bursa, second long, oral, dorsal. Labia membranous, bare, funnel shaped, apically building the accessory gonopore. Gonocoxapodeme straight, gonapophyses median smoothly pointed. Cercus 32—35 μm long.

Pupa (Description from Wiedenbrug et al., 2012)

Total length 1.30—1.51 mm.

Coloration (exuviae). Cephalothorax light brown, abdomen transparent except brownish muscle markings, lateral margin and anal lobe.

Cephalothorax. Frontal apotome rugose. Thorax suture smooth. All Dc-setae thin taeniate. Dc_3 longer. Dc_1 displaced ventrally. Wing sheaths with up to four rows of pearls sometimes apparently absent.

Abdomen (Fig. 3.43A—D). Tergite and sternite I bare, tergite and sternite II with few small shagreen points. Tergites III, VII—IX with fine, quite homogeneous shagreen. Tergites IV—VI with posterior shagreen points larger, posterolateral points larger than posteromedian. Conjunctives sternites III/IV—VII/VIII with small spinules. Segment I with 1, II with 3 L-setae and III—VIII with 4 long taeniate L-setae. Anal lobe rounded (Fig. 3.43C). Anal lobe with fringe not complete, 3 taeniate macrosetae and inner setae taeniate.

Larva (Description from Wiedenbrug et al., 2012)

Head (Fig. 3.43E). Postmentum 188−195 μm long. Head capsule integument smooth. Mentum with two median teeth, first lateral teeth small and adpressed to median, five additional lateral teeth (Fig. 3.43F). Antenna 663−690 μm long, segments two and three darker (not drawn). First segment longer than postmentum length.

Abdomen. Ventral setae modified, wider and apical split (Fig. 3.43G). Subbasal seta on posterior parapod serrated at both margins (Fig. 3.43H).

Remarks. Larvae of *C. espraiado* were collected in artificial substrate of stones and leaves at slow flowing, lowland streams at São Paulo State, Brazil.

Distribution. NT: Brazil (São Carlos).

3.33 *CORYNONEURA FERELOBATA* SUBLETTE & SASA, 1994 (FIGS. 3.44A−E, 3.45A−C, AND 3.46A−H)

Corynoneura ferelobatus Sublette & Sasa, 1994: 9; Fu et al., 2009: 10.

Corynoneura ferelobata Sublette, Sasa: Wiedenbrug et al., 2012: 25.

Material examined. P.R. China: Sichuan Province, Yajiang County, Sandaoqiao, 30°18′N, 101°E, alt. 2460 m, light trap, 1 male (BDN No. 11608), 9.vi.1996, X. Wang; Yunnan Province, Lijiang County, Heilongtan, 26°52′N, 100°15′E, alt. 2400 m, light trap, 1 male (BDN No. 10553), 28.v.1996, X. Wang; Yunnan Province, Hutiaoxia County, 30°N, 98°E, alt. 1700−5369 m, light trap, 2 males, 26.v.1996, X. Wang; Ningxia Autonomous Region, Liupanshan County, 35°14′N−39°14′N, 104°17′E−109°39′E, alt. 1100−1200 m, light trap, 1 male (BDN No. 1177), 7.viii.1987, X. Wang.

Diagnostic characters. This species is similar to *C. celtica* Edwards sharing an antenna with an AR of about 0.20−0.30, and an inverted V-shaped sternapodeme, but can be separated by the digitiform inferior volsella and the antenna with 9 flagellomeres. The female has two Sca of less than 50 μm long, copulatory bursa and labia posteromedially strongly invaginated, and coxosternapodeme curved with lamellae. The pupa has tergites with homogeneous fine shagreen, shagreen points longer than wider. Sternite I bare, II with very fine shagreen, without longer spinules. Segment IX and anal lobe together almost forming a circle, as long as wide. The larva of *C. ferelobata* has the dorsal integument granulated, first antennal segment subequal to postmentum length, mentum

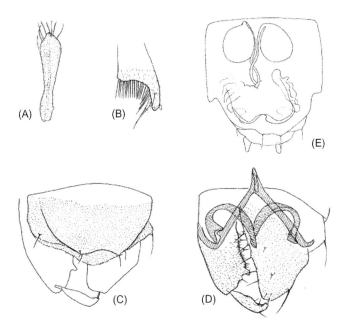

Figure 3.44 *Corynoneura ferelobata* Sublette & Sasa, 1994 (From Wiedenbrug, S., Lamas, C. J. E., & Trivinho-Strixino, S. (2012). A review of the genus *Corynoneura* Winnertz (Diptera: Chironomidae) from the neotropical region. *Zootaxa, 3574*, 1–61. www.mapress.com/j/zt). (A–D) Male imago. (A) Terminal flagellomere. (B) Apex of hind tibia. (C and D) Hypopygium. (C) Tergite IX. (D) Tergite IX removed, sclerites hatched; left is dorsal view, right is ventral view. (E) Female genitalia, dorsal view with view of labia and Sca. Specimens from Brazil, SP, Jundiaí, Serra do Japí.

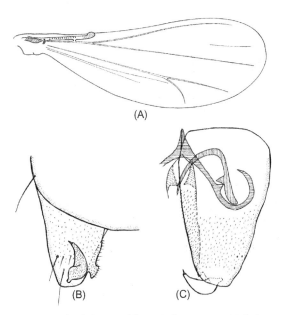

Figure 3.45 *Corynoneura ferelobata* Sublette & Sasa, 1994, Male imago. (A) Wing. (B) Hypopygium, dorsal view. (C) Hypopygium, ventral view.

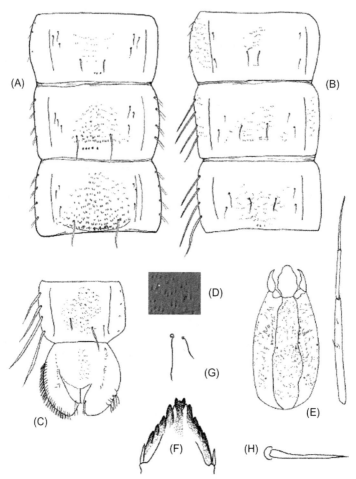

Figure 3.46 *Corynoneura ferelobata* Sublette & Sasa, 1994, Immatures imago (From Wiedenbrug, S., Lamas, C. J. E., & Trivinho-Strixino, S. (2012). A review of the genus *Corynoneura* Winnertz (Diptera: Chironomidae) from the neotropical region. *Zootaxa, 3574,* 1−61). (A−D) Pupa imago. (A) Tergites II−IV. (B) Sternites II−IV. (C) Segments VIII, IX and anal lobe; left is ventral view, right is dorsal view. (D) Detail of posterior shagreen of tergite IV. (E−H) Larva imago. (E) Head, dorsal view and separated antenna. (F) Mentum. (G) Abdominal seta. (H) Subbasal seta of posterior parapod.

with three median teeth, median tooth minute, seta on posterior parapod apparently simple.

Male (*n* = 5)

Total length 1.12−1.46, 1.34 mm. Wing length 0.7−0.85, 0.76 mm. Total length/wing length 1.7−2.1, 2.0. Wing length/length of profemur 2.9−3.7, 3.2.

Coloration. Head yellow-brown. Antenna and palpomere pale yellow. Thorax brown. Legs yellow-brown. Wings light yellow to hyaline, with pale yellow clava.

Head. Eyes bare, reniform. Antenna with 9 flagellomeres; ultimate flagellomere 70−95, 90 μm long, antenna apically slightly expanded, with rosette of apical sensilla chaetica (Fig. 3.44A), with maximum width 20 μm; AR 0.27−0.34, 0.3. Clypeus with 6−8, 7 setae. Temporal setae lacking. Tentorium 105−118, 112 μm long. Stipes: 35−40, 38 μm long. Palpomeres length (in μm): 10−13, 12; 10−15, 13; 15−18, 17; 18−23, 20; 48−50, 49. Palpomere 2, 3 and 4 ellipsoid, palpomere 5 long and slender. Palpomere 5/3 ratio 2.7−3.3, 3.0.

Thorax. Antepronotum without lateral setae. dorsocentrals 4−6, 5(3), prealars 2−5, 3(2). Scutellum with 1−2 setae.

Wing (Fig. 3.45A). VR 3.2−3.6, 3.4. C length 180−190, 186 μm, C/wing length 0.23−0.27, 0.25. Cu length 425−510, 460 μm. Cu/wing length 0.60−0.7, 0.66. Wing width/wing length 0.41−0.44, 0.41. C with 6−8, 7 setae.

Legs. Fore trochanter with keel. Spur of front tibia 18−20, 19 μm and 13−15, 14 μm long, spurs of middle tibia 7.5−10, 8 μm long, of hind tibia 25−30, 28 μm long. Width at apex of fore tibia 20−23, 21 μm, of mid tibia 15−23, 18 μm, of hind tibia 25−33, 18 μm. Tip of hind tibia expanded, with comb of 14−16, 15 setae and 1 seta near spur strong S-shaped (Fig. 3.44B). Lengths and proportions of legs as in Table 3.24.

Hypopygium (Figs. 3.44C and D and 3.45B and C). Posterior margin weakly bilobed, and with many short setae, laterosternite IX with 1 long seta. Anal point absent. Inferior volsellae digitiform, sternapodeme inverted V-shaped, coxapodeme 25−35, 30 μm long. Phallapodeme strongly curved extending beyond tergite IX, 50−70, 60 μm long, with basal projection. The enlarged phallapodeme and the lateral sternapodeme protrude just below laterosternite IX. Gonocoxite 58−78, 68 μm long with 2 setae apically.

Gonostylus curved apically, 23−28, 27 μm long. Megaseta 2.5−5, 3 μm long. HR 2.7−3.6, 3.1, HV 4.5−5.6, 5.3.

Female (Description from Wiedenbrug et al., 2012)

Thorax. 0.35−0.48 mm. Antepronotal lobes dorsally tapering.

Head. AR = 0.33−0.39. Antenna with 5 flagellomeres, apical flagellomere 30−45 μm long. Flagellomeres with more than one row of setae each. Eyes pubescent.

Legs. Hind tibial scale 35−37 μm long, with one small S-seta.

Table 3.24 Lengths (in μm) and proportions of legs segments of male *Corynoneura ferelobata* Sublette & Sasa, 1994 ($n = 5$) (based on Chinese specimens)

	fe	ti	ta_1	ta_2	ta_3	ta_4
p_1	200–240, 230	275–280, 277	120–133, 128	75–80, 77	48–50, 49	23–25, 24
p_2	300–310, 306	265–280, 273	150–170, 162	65–80, 74	35–43, 38	20
p_3	245–255, 250	265–290, 280	135–145, 140	75–78, 76	33–35, 34	15–20, 17

	ta_5	LR	BV	SV	BR	
p_1	30–33, 31	0.43–0.50, 0.46	3.5	3.7–4.3, 4.0	1.6–2.6, 2.1	
p_2	20–28, 24	0.57–0.62, 0.59	4.5–4.8, 4.6	3.4–3.8, 3.6	1.7–2.2, 1.9	
p_3	28–30, 29	0.48–0.55, 0.52	3.7–4.4, 4.0	3.6–3.9	1.4–2.0, 1.7	

Genitalia (Fig. 3.44E) Tergite IX with 4 setae. Laterosternite IX with one seta. Two seminal capsules 37−42 μm long; spermathecal ducts subequal, join together shortly before seminal eminence. Membrane well sclerotized. Coxosternapodeme curved with lamellae, copulatory bursa posteriorly strong invaginated, posterolateral folded. Labia bare membranous, posteriorly strongly invaginated. Notum 25 μm long. Apodeme lobe pointed. Gonocoxapodeme straight. Cercus 17−22 μm long.

Pupa (Description from Wiedenbrug et al., 2012)

Total length 1.24−1.41 mm.

Coloration (exuviae). Cephalothorax light gray, abdomen transparent except muscle markings, lateral margin and anal lobe.

Cephalothorax. Frontal apotome rugose. Thorax suture smooth slightly rugose, with some small spinules at the scutal tubercle region. All Dc-setae thin taeniate, except Dc_3 about 2 μm wide and 80 μm long. Dc_1 displaced ventrally. Wing sheaths with two to four rows of pearls.

Abdomen (Fig. 3.46A−D). Tergite and sternite I bare, tergites and sternites II −IX with homogeneous shagreen, points on sternites much smaller than points on tergites. Conjunctives tergites II /III − VI /VII and sternites III/ IV − VII/VIII with small spinules. Segment I with 1, II with 3 L-setae and III−VIII with 4 long taeniate L-setae. Anal lobe rounded (Fig. 3.46C). Anal lobe with fringe complete, 3 taeniate macrosetae and inner setae taeniate.

Larva (Description from Wiedenbrug et al., 2012)

Head. Postmentum 163 μm long. Head capsule integument dorsal granulated in reticulate pattern (Fig. 3.46E). Mentum with three median teeth, median tooth minute, first lateral teeth small, less sclerotized and adpressed to median, five additional lateral teeth (Fig. 3.46F). Antenna 348 μm long, segments two and three darker (not drawn). First segment longer than postmentum length.

Abdomen. Ventral setae not modified (Fig. 3.46G). Subbasal seta on posterior parapod not serrate, simple (Fig. 3.46H).

Remarks. Sublette and Sasa (1994) described the species based on specimens from Guatemala and according to Wiedenbrug et al. (2012). It also occurs in Brazil, Costa Rica. The Chinese specimens show some variations when compared with the original description. The Chinese specimens are paler, and the ultimate flagellomere is slightly curved. However, some variations also exist between specimens collected in southern and northern China.

Distribution. NT: Guatemala. **PA:** China (Ningxia). **OR:** China (Sichuan, Yunnan).

3.34 *CORYNONEURA FITTKAUI* SCHLEE, 1968 (FIG. 3.47A−F)

Corynoneura fittkaui Schlee, 1968b: 19; Langton & Visser, 2003: 324; Langton & Pinder, 2007: 92, Fu & Sæther, 2012: 29; Krasheninnikov & Makarchenko, 2009; Oliver et al., 1990: 22.

Diagnostic characters. Male antenna with 12−13 flagellomeres, when antenna with 13 flagellomeres, AR 0.31−0.44; when antenna with 12 flagellomeres, AR 0.43−0.55. Gonocoxite with basal wide groove-like cavity from where are the long bristles. Inferior volsella wide and stepwise outstanding.

Male (Description from Schlee, 1968b)

Head. Male antenna (Fig. 3.47A) with 12−13 flagellomeres, when antenna with 13 flagellomeres, AR 0.31−0.44; when antenna with 12 flagellomeres, AR 0.43−0.55.

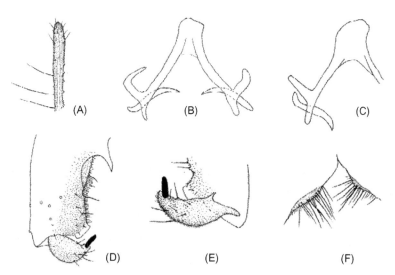

(A)   (B)   (C)

(D)   (E)   (F)

Figure 3.47 *Corynoneura fittkaui* Schlee, 1968 (From Makarchenko, E. A., & Makarchenko, M. A. (2010). New data on the fauna and taxonomy of *Corynoneura* Winnertz (Diptera, Chironomidae, Orthocladiinae) for the Russian Far East and bordering territories. *Euroasian Entomological Journal, 9*(3), 353−370 + II). (A−E) Male imago. (A) Distal part of terminal antennal flagellomere. (B and C) Sternapodema and phallapodemes. (D) Gonocoxite and gonostylus. (E) Gonostylus. (F) Inner basal angles of gonocoxite, from below.

Thorax. Suture 110 μm long.

Wing. Wing length 0.9−1.0 mm, VR 3.3. Cu/wing length 0.57−0.63, C length/wing length 0.28−0.30, wing width/wing length 0.35−0.39.

Legs. Fore trochanter with dorsal keel.

Hypopygium (Fig. 3.47B−F). The posterior margin of tergite IX circle-like. Laterosternite IX with 2 long seta. Basal joint. Gonocoxite with basal wide groove-like cavity from where are the long bristles. In the area of the cavity described above are numerous very long, straight, hyaline, cau-dally directed bristles. Inferior volsella wide and stepwise outstanding. Gonostylus long and slender, channel-shaped, slightly curved, standing at an acute angle to the longitudinal axis of the gonocoxite end part.

Distribution. NE: USA (North and South Carolina). **PA:** Austria, Denmark, Faroe Island, Finland, France, Germany, Great Britain, Italy, Luxembourg, Norway, Poland, Russia (NET, Far East), Spain.

3.35 *CORYNONEURA FLORIDAENSIS* FU & SÆTHER, 2012 (FIGS. 3.48A−G AND 3.49A−F)

Corynoneura sp. B Epler, 2001: 7.47.

Corynoneura celeripes (Winnertz): Simpson & Bode, 1980: 34.

Corynoneura "celeripes" (sensu Simpson & Bode, 1980): Bolton, 2007: 28.

Corynoneura floridaensis Fu & Sæther, 2012: 29.

Material examined. Holotype male, with associated larval and pupal exuviae, USA, Florida, Suwannee County, Suwannee River at State Road 51 near Luraville, 3.vi.1993, J. H. Epler, R. A. Mattson (ZMBN Type No. 457). Allotype female, with associated larval and pupal exuviae, USA, Florida, Hamilton County, Withlacoochee River above Suwannee River, 3.vi.1993, J. H. Epler, R. A. Mattson (ZMBN). Paratypes: 4 males, 4 females, 4 larval and 4 pupal exuviae as allotype; 3 males, 2 pupal exuviae, 1 male with pupal exuviae as allotype except for 11.xii.1995, 9. xii.1996, and 7.iii.2000; USA, Florida, Gilchrist County, Suwannee River near Pock Bluff at SR 340, 1 male with pupal exuviae, 6.ix.2001, J. H. Epler; 3 males, 3 pupal and 3 larval exuviae, USA, Ohio, Delaware County, Olentangy River Highbanks Park, 24.ix.1987, M. J. Bolton (ZMBN, MJB, JHE).

Diagnostic characters. The adult male is characterized by antenna with 11 flagellomeres, AR 0.30−0.48, 0.36; superior volsella well-developed, triangular; inferior volsella elongate rectangular with broadly

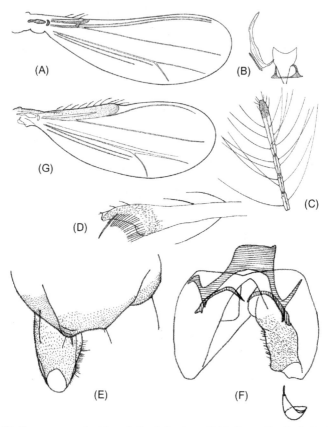

Figure 3.48 *Corynoneura floridaensis* Fu & Sæther, 2012 (From Fu, Y., & Sæther, O. A. (2012). *Corynoneura* Winnertz and *Thienemanniella* Kieffer from the Nearctic region (Diptera: Chironomidae: Orthocladiinae). *Zootaxa, 3536,* 1−61. www.mapress.com/j/zt). (A−F) Male imago. (A) Wing. (B) Tentorium, stipes and cibarial pump. (C) Antenna. (D) Hind tibial apex. (E) Hypopygium, dorsal view. (F) Hypopygium, ventral view. (G) Female imago: wing.

rounded distal angle and placed caudally; gonostylus curved. The adult female has AR 0.37, 2−3 transparent developed fused lateral lamellae on inner side of coxosternapodeme, and labia not well developed. The pupa has no taeniate L-setae on tergites I−II; no shagreen on tergite I, tergites II−IX with fine shagreen and only tergites IV−VII with several hooklets; and 26−34, 31 taeniae in fringe of anal lobe. The larvae are separable by mentum with two median teeth, the lateral tooth adjacent to median tooth very small, basal segment yellowish; second segment light brown, other segments dark brown; length of antenna/length of head 1.32−1.38, 1.35, and head capsule smooth, unsculptured.

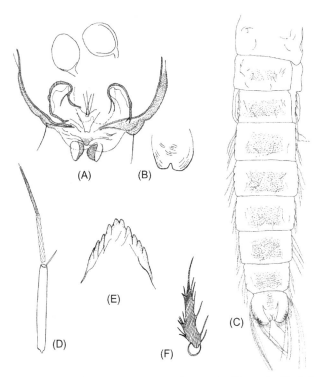

Figure 3.49 *Corynoneura floridaensis* Fu & Sæther, 2012 (From Fu, Y., & Sæther, O. A. (2012). *Corynoneura* Winnertz and *Thienemanniella* Kieffer from the Nearctic region (Diptera: Chironomidae: Orthocladiinae). *Zootaxa, 3536*, 1–61. www.mapress.com/j/zt). (A and B) Female imago. (A) Genitalia, ventral view. (B) Labia. (C) Pupa: tergites I –IX. (D–F) Larva. (D) Antenna. (E) Mentum. (F) Subbasal seta of posterior parapod.

Male ($n = 8-10$)

Total length 0.98–1.73, 1.32 mm. Wing length 0.57–0.65, 0.62 mm. Total length/ wing length 1.6–2.4, 1.8. Wing length/profemur length 2.8–3.0, 2.9.

Coloration. Head brown. Antenna and legs yellowish. Thorax and abdomen yellowish brown.

Head. Antenna (Fig. 3.48C) with 11 flagellomeres, ultimate flagellomere 75–133, 97 µm, AR 0.30–0.48, 0.36; ultimate flagellomere apically expanded, with about 8–12, 10 apical sensilla chaetica; longest antennal seta 220–240, 232 µm. Temporals absent. Tentorium, stipes and cibarial pump as in Fig. 3.48B. Tentorium 99–127, 109 µm long; 10–16, 12 µm wide. Stipes 50–67, 56 µm long, 3–4 µm wide. Clypeus with 7–10, 8 setae.

Length of palpomeres (in μm): 6−11, 9; 10−15, 11; 14−18, 16; 19−28, 23; 46−59, 48. Palpomere 5/3 ratio 2.6−3.5, 3.1.

Thorax. Dorsocentrals 4, prealars 2, supraalar 1. Scutellum with 2 setae. Anapleural suture 77−85, 83 μm long.

Wing (Fig. 3.48A). VR 2.6−3.2, 2.8. Cu/wing length 0.54−0.59, 0.57; C length 124−168, 139 μm long; Cu 372−368, 336 μm long; C/wing length 0.22−0.26, 0.23; wing width/wing length 0.42−0.44, 0.43. Costa with 4−6 setae.

Legs. Fore trochanter with dorsal keel. Spurs of fore tibia 15−22, 19 μm long; of mid tibia 11−15, 13 μm and 8−10, 9 μm long; of hind tibia 25−37, 31 μm and 7−14, 11 μm long. Width at apex of fore tibia 14−17, 16 μm; of mid tibia 14−15 μm; of hind tibia (a) 28−35, 31 μm. Width of hind tibia 1/3 from apex (d) 14−17, 15 μm; elongation length (b) 33−36, 35 μm; length of maximum thickening (c_1) 59−75, 67 μm; total length of thickening (c_2) 97−104, 100 μm; a/d 1.9−2.1, 2.0; b/d 2.1−2.4, 2.3; c_1/d 4.2−4.6, 4.4; c_2/d 5.9−6.9, 6.6. Hind tibia (Fig. 3.48D) expanded, with comb of 15−16 setae, 1 seta near spur strongly hooked. Lengths and proportions of legs as in Table 3.25.

Hypopygium (Fig. 3.48E and F). Posterior margin of tergite IX with 4−6 long setae. Laterosternite IX with 1−2 long setae. Superior volsella triangular, anteriomedially separated. Inferior volsella low and placed caudally. Phallapodeme slightly curved, 25−41, 32 μm long. Transverse sternapodeme broad; 17−28, 21 μm wide. Attachment point of lateral sternapodeme with phallapodeme placed and directed caudally. Gonostylus apical curved, 21−35, 26 μm long; megaseta 3−4 μm long. HR 2.0−2.9, 2.3; HV 3.3−4.9, 4.6.

Female ($n = 1$)

Total length 0.92 mm. Wing length 0.58 mm. Total length/wing length 1.59.

Coloration. Head brown. Thorax and abdomen brown. Antenna and legs yellowish.

Head. AR 0.37. Length of flagellomeres (in μm): 25, 26, 25, 23, 36. Ultimate flagellomere with 9 sensilla chaetica. Clypeus with 6 setae. Palpomere lengths (in μm): 7, 10, 15, 17, 43. Palpomere 5/3 ratio 2.9.

Thorax. Dorsocentrals 4, prealars 2. Scutellum with 2 setae.

Wing (Fig. 3.48G). VR 2.1, two anal veins present, Cu 320 μm long, Cu/wing length 0.55, C 261 μm long, C/wing length 0.45, wing width/ wing length 0.44. Costa with 11 setae.

Table 3.25 Lengths (in μm) and proportions of legs segments of male *Corynoneura floridaensis* Fu & Sæther, 2012 ($n = 8-10$)

	fe	ti	ta_1	ta_2	ta_3	ta_4
p_1	186–224, 205	208–252, 232	119–131, 125	63–71, 68	34–42, 39	18–20, 19
p_2	268–304, 288	228–272, 251	152–168, 161	61–73, 68	32–40, 36	14–20, 17
p_3	212–264, 233	208–248, 236	123–145, 136	65–77, 72	28–32, 30	14–18, 16
	ta_5	LR	BV	SV	BR	
p_1	26–28, 27	0.50–0.60, 0.54	3.4–4.4, 3.8	3.3–3.8, 3.5	2.4–3.0, 2.6	
p_2	20–26, 24	0.61–0.67, 064	4.4–5.5, 4.8	3.3–3.4	2.0–2.7, 2.2	
p_3	24–28, 26	0.55–0.59, 0.58	4.1–4.5, 4.3	3.3–3.6, 3.4	1.5–2.0, 1.8	

Source: From Fu, Y., & Sæther, O. A. (2012). *Corynoneura* Winnertz and *Thienemanniella* Kieffer from the Nearctic region (Diptera: Chironomidae: Orthocladiinae). *Zootaxa*, 3536, 1–61. www.mapress.com/j/zt.

Legs. Fore trochanter with keel. Spurs of mid tibia 10 μm and 8 μm long, of hind tibia 25 μm and 11 μm long. Width at apex of mid tibia 14 μm, of hind tibia (a) 23 μm. Width of hind tibia 1/3 from apex (d) 12 μm, elongation length (b) 33 μm, length of maximum thickening (c_1) 81 μm, total length of thickening (c_2) 121 μm, a/d 1.9, b/d 2.8, c_1/d 6.8, c_2/d 10.1. Apex of hind tibia expanded, with comb of 15 setae, with S-shaped spur. Lengths and proportions of legs as in Table 3.26.

Abdomen. Number of setae on tergites not measurable.

Genitalia (Fig. 3.49A and B). Tergite IX with 2 long caudal setae. Cercus 17 μm long, 10 μm wide. Notum 63 μm long. Gonocoxite with 1 long seta. Coxosternapodeme with 2−3 transparent fused lateral lamellae on inner side of coxosternapodeme. Seminal capsule 33 μm long; neck 7 μm long, 4 μm wide.

Pupa ($n = 4-5$)

Total length 1.61−1.70, 1.65 mm. Exuviae yellowish including cephalothorax.

Cephalothorax. Frontal setae 7−10, 8 μm long. Median antepronotals 7 μm and 10 μm long. Lateral antepronotals 8 μm long. Anterior precorneal seta (PcS_1) 8−15, 10 μm long; PcS_2 8−14, 10 μm long; PcS_3 8−14, 11 μm long. PcS_{1-3} in a line; PcS_1 6 μm from PcS_2; PcS_2 4−6, 5 μm from PcS_3; PcS_1 11−14, 12 μm from PcS_3; PcS_3 28−45, 35 μm from thoracic horn. Anterior dorsocentral (Dc_1) 4 μm long; Dc_2 6 μm long; Dc_3 4 μm long; Dc_4 not observed. Dc_1 located 10 μm from Dc_2, Dc_2 located 44 μm from Dc_3. Wing sheath with 3−4 rows of pearls.

Abdomen (Fig. 3.49C). Shagreen and chaetotaxy as illustrated. No taeniate L-setae on tergites I−II. Anal lobe 91−139, 117 μm long. Anal lobe fringe with 26−34, 31 setae; 326−400, 362 μm long; 3 taeniate anal macrosetae about 284−360, 328 μm long; median setae about 61−81, 72 μm long.

Larva ($n = 4-6$)

Table 3.26 Lengths (in μm) and proportions of legs segments of female *Corynoneura floridaensis* Fu & Sæther, 2012 ($n = 1$)

	fe	ti	ta₁	ta₂	ta₃	ta₄	ta₅	LR	BV	SV	BR
P_1	−	−	−	−	−	−	−	−	−	−	−
P_2	244	216	135	61	30	16	22	0.63	4.6	3.4	2.0
P_3	212	216	103	61	28	12	28	0.48	4.1	4.2	1.9

Source: From Fu, Y., & Sæther, O. A. (2012). *Corynoneura* Winnertz and *Thienemanniella* Kieffer from the Nearctic region (Diptera: Chironomidae: Orthocladiinae). *Zootaxa*, 3536, 1−61. www.mapress.com/j/zt.

Coloration. Head yellowish brown. Antenna: basal segment yellowish; second segment light brown, other segments dark brown. Abdomen yellowish.

Head. Capsule length 188−208, 198 μm, width 107−141, 126 μm; smooth, unsculptured. Postmentum 156−174, 164 μm long. Sternite II obvious, rising from small tubercle, I and III invisible. Premandible 21−29, 25 μm long. Mentum as Fig. 3.49E. Mandible 40−48, 45 μm long. Antenna (Fig. 3.49D), AR 0.72−0.87, 0.81. Length of flagellomeres I−IV (in μm): 106−129, 118; 65−73, 69; 68−76, 72; 6. Basal segment width 15−17, 16 μm; length of blade at apex of basal segment 30−33, 32 μm. Length of antenna/length of head 1.32−1.38, 1.35.

Abdomen. Length of anal setae 160−192, 175 μm. Procercus 7−8 μm long, 7 μm wide. Subbasal seta of posterior parapods split as in Fig. 3.49F, 34−40, 36 μm long.

Distribution. NE: USA (Florida).

3.36 *CORYNONEURA FORTISPICULA* WIEDENBRUG & TRIVINHO-STRIXINO, 2011 (FIGS. 3.50A−F, 3.51A−D, AND 3.52A−M)

Corynoneura-group spec. 5: Wiedenbrug, 2000: 102.

Corynoneura fortispicula Wiedenbrug & Trivinho-Strixino, 2011: 2; Wiedenbrug et al., 2012: 29.

Diagnostic characters. The adult males are separable from other species except *Corynoneura hirvenojai* by the antenna with 7 flagellomeres, 6 distinct separated and the last composed of 5 fused flagellomeres, antenna sparsely plumose with one row of seta on each flagellomere, AR 1.0−1.3; eyes pubescent; abdominal tergites I−IV whitish, V−IX brown; attachment of the phallapodeme caudal on the lateral sternapodeme, phallapodeme with posterior margin sclerotized, rounded and curved anteriorly; gonostylus slender, tapering apically. Adult females are separable from other species by the funnel shaped labia, seminal capsules size subequal, abdominal segments IV−IX brownish, III median whitish, lateral brownish. The pupae are distinguished from other species by the shagreen of tergites IV and V with posterior row of strong spines longer than wide (Fig. 3.52F), pearl row usually present, 3 short lateral setae on segments IV−VII, if four the last minute, anal lobe rectangular, anal lobe fringe restricted to posterior margin. The larvae can be recognized by the head integument without sculptures, mentum with 3 median teeth and 6 lateral teeth, antenna longer than postmentum, AR 0.76−0.88, second

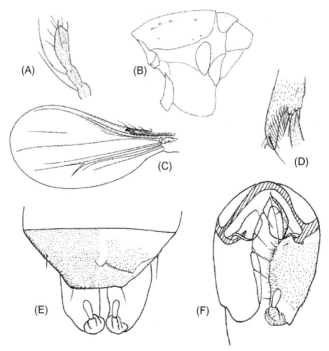

Figure 3.50 *Corynoneura fortispicula* Wiedenbrug & Trivinho-Strixino, 2011, Male imago (From Wiedenbrug, S., & Trivinho-Strixino, S. (2011). New species of the genus *Corynoneura* Winnertz (Diptera, Chironomidae) from Brazil. *Zootaxa, 2822*, 1—40. www.mapress.com/j/zt). (A) Terminal flagellomeres. (B) Thorax. (C) Wing. (D) Apex of hind tibia. (E) Tergite IX; left is dorsal view, right is ventral view. (F) Tergite IX removed, sclerites hatched; right is dorsal view, left is ventral view.

antennal segment subequal or shorter than the third, basal seta on posterior parapods split from the base.

Male (Description from Wiedenbrug & Trivinho-Strixino, 2011)

Total length 0.75 mm. Wing length 0.47 mm. Total length/wing length 1.58. Wing length/ length of profemur 3.17.

Coloration. Thorax brownish; legs whitish, brownish ring on apex of femur and proximal tibia; tergites I—IV whitish, V—IX brownish.

Head. AR = 1.05—1.34. Antenna with 7 flagellomeres, flagellomeres 1—6 distinctly separated from each other, ultimate flagellomere 105—160 μm long, composed of 5 distinct but fused flagellomeres (Fig. 3.50A), two sensilla chaetica at the apex of the last flagellomere. Flagellomeres, except first and last, with one row of seta each, fused flagellomeres also with one row of setae each. Temporal setae absent. Clypeus 30—32 μm long, with 6—7

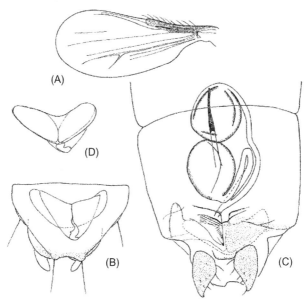

Figure 3.51 *Corynoneura fortispicula* Wiedenbrug & Trivinho-Strixino, 2011, Female imago (From Wiedenbrug, S., & Trivinho-Strixino, S. (2011). New species of the genus *Corynoneura* Winnertz (Diptera, Chironomidae) from Brazil. *Zootaxa, 2822*, 1−40. www.mapress.com/j/zt). (A) Wing. (B) Dorsal view of labia. (C) Ventral view of genitalia; left is gonapophysis removed. (D) Ventral view of labia.

setae. Tentorium 93 μm long; 5 μm wide. Palpomere lengths (in μm): 7, 10, 12, 17, 35−42. Third palpomere with one sensilla clavata. Eyes pubescent.

Thorax (Fig. 3.50B). Antepronotum with 1 lateral seta. Dorsocentrals 3, prealars 2. Scutellars 2. Antepronotal lobes dorsally reduced, tapering.

Wing (Fig. 3.50C). VR 2.93−4.64. Clavus 25−35 μm wide, ending 112−125, 115 μm from arculus. Clavus/wing length 0.24−0.25. Anal lobe absent.

Legs. Spur of front tibia 20−22 μm long; spur of middle tibia 5 μm and 5−12 μm long; spurs of hind tibia 30 μm long, second spur S-shaped. Width of apex of front tibia 12−17 μm, of middle tibia 12−15 μm, of hind tibia 30−32 μm (Fig. 3.50D). Front tibial scale 7−10 μm long; hind tibial scale 37−42 μm long. Leg measurements (in μm) and ratios as in Table 3.27.

Abdomen. Tergite VI 0−1 seta, VII 3 setae, VIII 1 seta.

Hypopygium (Fig. 3.50E and F). Tergite IX with 2 setae. Laterosternite IX with 0−1 seta. Superior volsella absent. Inferior

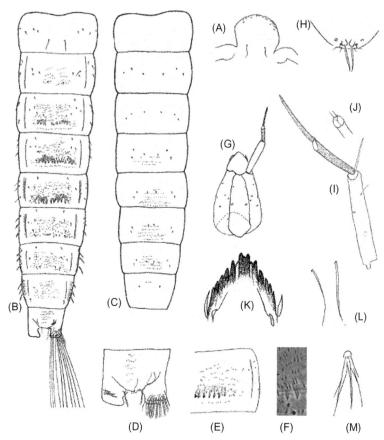

Figure 3.52 *Corynoneura fortispicula* Wiedenbrug & Trivinho-Strixino, 2011, Immatures imago (From Wiedenbrug, S., & Trivinho-Strixino, S. (2011). New species of the genus *Corynoneura* Winnertz (Diptera, Chironomidae) from Brazil. *Zootaxa, 2822,* 1−40. www.mapress.com/j/zt). (A−F) Pupa imago. (A) Frontal apotome. (B) Abdomen, dorsal view. (C) Sternites I −VIII. (D) Detail of anal lobe; left is dorsal view without fringe, right is ventral view. (E) Detail of tergite IV. (F) Detail of posterior shagreen of tergite IV. (G−M) Larva imago. (G) Dorsal view of the head and antenna. (H) Dorsal view of labrum. (I) Antenna. (J) Detail of lauterborn organs. (K) Mentum. (L) Abdominal seta. (M) Subbasal seta of posterior parapod.

volsella smooth with longer microtrichia directed to median. Phallapodeme 22−25 μm long, rounded and sclerotized on posterior margin; transverse sternapodeme 12−15 μm long. Gonocoxite 60−67 μm long. Gonostylus tapering apically, 17 μm long; megaseta 5−10 μm long. HR 3.86.

Female (Description from Wiedenbrug & Trivinho-Strixino, 2011)

Table 3.27 Lengths (in μm) and proportions of legs segments of male *Corynoneura fortispicula* Wiedenbrug & Trivinho-Strixino, 2011

	fe	ti	ta$_1$	ta$_2$	ta$_3$	ta$_4$
p$_1$	150−178	182−220	75−90	50−63	30−35	15−20
p$_2$	217−248	195−215	102−125	52−65	27−33	15
p$_3$	162−200	170−200	92−103	55−65	20−25	12−15

	ta$_5$	LR	BV	SV	BR
p$_1$	22−25	0.40−0.41	3.42−3.47	4.41−4.43	1.50−2.00
p$_2$	22	0.52−0.58	4.35−4.38	3.70−4.03	1.75−2.50
p$_3$	22−25	0.51−0.54	3.78−3.94	3.59−3.90	1.75−2.50

Source: From Wiedenbrug, S., & Trivinho-Strixino, S. (2011). New species of the genus Corynoneura Winnertz (Diptera, Chironomidae) from Brazil. *Zootaxa*, 2822: 1−40. www.mapress.com/j/zt.

Total length 0.91−0.93 mm. Wing length 0.63 mm. Total length/ wing length 1.45. Wing length/length of profemur 3.98.

Coloration. Thorax brownish; legs whitish with brown ring posterior on femur and anterior on tibia; tergites I−II whitish, III whitish on median region brownish lateral, IV−IX brownish.

Head. AR 0.37−0.50. Antenna with 5 flagellomeres, apical flagellomere 37−50 μm long; two sensilla chaetica on flagellomeres 3−5, first and second flagellomeres not possible to observe. Temporal setae absent. Clypeus 25 μm long, with 8 setae. Tentorium 60−65 μm long, 2−4 μm wide. Palpomere lengths (in μm): 7−10, 10, 15, 17, 35−40. One sensilla clavata on third palpomere. Eyes pubescent.

Thorax. Antepronotals 2 median, 1 lateral. Dorsocentrals 2, prealars 2. Scutellars 2. Antepronotal lobes dorsally tapering.

Wing (Fig. 3.51A). VR 2.64. Clavus 37 μm wide, ending 247 μm from arculus. Clavus/wing length 0.40.

Legs. Spur of front tibia 12 μm long; spur of middle tibia 10 μm; spurs of hind tibia 30−32 μm long, second spur small S-shaped. Front tibial scale 10 μm long, hind tibial scale 32−40 μm long. Width of apex of front tibia 12−15 μm, of middle tibia 12−15 μm, of hind tibia 30−32 μm. Leg measurements (in μm) and ratios as in Table 3.28.

Abdomen. One seta on tergite VI, 3 setae on tergite VII, 1 seta on tergite VIII.

Genitalia (Fig. 3.51B−D). Gonocoxite IX with 1 seta. Tergite IX with 2 setae. Two seminal capsules respectively 37 μm and 42 μm long; one spermathecal duct with a loop, second straighter, both ducts join together shortly before seminal eminence, which has sclerotized outer borders. Notum 30−35 μm long. Membrane well sclerotized. Apodeme lobe well

Table 3.28 Lengths (in μm) and proportions of legs segments of female *Corynoneura fortispicula* Wiedenbrug & Trivinho-Strixino, 2011

	fe	ti	ta_1	ta_2	ta_3	ta_4
p_1	147−157	177−210	75−77	47−52	27	15−17
p_2	210−225	195−237	107−120	50−52	25−30	15−17
p_3	150−205	175−200	77−102	47−60	22−25	15

	ta_5	LR	BV	SV	BR
p_1	20−25	0.37−0.42	3.55−3.77	4.33−4.74	2.00
p_2	20−22	0.50−0.55	4.55−4.85	3.76−3.85	1.75−2.33
p_3	22	0.44−0.51	3.74−4.14	3.95−4.19	2.00

Source: From Wiedenbrug, S., & Trivinho-Strixino, S. (2011). New species of the genus Corynoneura Winnertz (Diptera, Chironomidae) from Brazil. *Zootaxa*, 2822: 1−40. www.mapress.com/j/zt.

sclerotized, apically pointed. Coxosternapodeme, with one tiny lateral lamellae, strongly curved, coxosternapodeme with one end at roof of copulatory bursa, last semicircle-shaped. Labia membranous, bare, separated, dorsal funnel shaped, apically building the accessory gonopore, ventral divided in two lobes. Gonocoxapodeme straight, gonapophyses median smoothly pointed. Postgenital plate squared with few microtrichia. Cercus 27−30 μm long.

Pupa (Description from Wiedenbrug & Trivinho-Strixino, 2011) Total length 1.32 mm.

Cephalothorax. Frontal apotome granulated (Fig. 3.52A). Frontal setae 45−57 μm long. Median antepronotals 32−60 μm. Distance between Dc_1 and Dc_2 57−70 μm; Dc_2 and Dc_3 22−27 μm; Dc_3 and Dc_4 10−20 μm. Dc_1 displaced ventrally. Wing sheaths with one row of pearls, sometimes difficult to see.

Abdomen (Fig. 3.52B−F). Tergite I , sternites I − II and usually III without shagreen. Tergite II with few shagreen points. Tergites III − VI with fine shagreen grading to a row of stronger spines posteriorly, particularly strong on tergites IV and V (Fig. 3.52E and F). Tergites VII − IX with shagreen with spinules of the same size. Sternites III − IV and VIII with fine shagreen posteriorly, V − VII with fine shagreen. Sternite VIII on female with few shagreen points. Conjunctive tergite II /III with 0 − 24 spinules; tergite III/IV with 18 − 30 hooklets; tergite IV/ V with 15 − 31 hooklets; tergite V /VI with 0 − 5 spinules; tergite VI/VII with 0 − 2 spinules; sternite IV/ V with 13 − 22 spinules; sternite V /VI with 13 − 27 spinules; sternite VI/VII with 13 − 26 spinules, sternite VII/VIII with 6 − 20 spinules. Segment I with 4 D-setae, 1 L-seta, without V -seta; segments

II−IV 4 D-setae, 3 L-setae and 3 V-setae; segments V−VII 4 D-setae, 3V-setae and 3 thin taeniate L-setae, on segments VI−VII sometimes one additional minute L$_4$-setae; segment VIII with 1 D-seta, 1 V-seta and 3 short taeniate L-setae and sometimes one additional minute L$_4$-setae. O-setae visible from tergites II−VII. Anal lobe rectangular 82−102 μm long (Fig. 3.52D). Anal lobe fringe posteriorly with 8−11 taeniate setae, 200−450 μm long, one median and 1−3 lateral thin and shorter fringe setae; 3 macrosetae taeniate; inner setae taeniate.

Larva (Description from Wiedenbrug & Trivinho-Strixino, 2011)

Head (Fig. 3.52G). Head capsule integument smooth. Frontal apotome 137−170 μm long; head width 100−117 μm; postmentum 110−127 μm; postmentum/head width 1.08−1.11. Sternite I simple (Fig. 3.52H). Premandible with the outer lateral lamellae with brush. Mentum with three median teeth, intermediate teeth minute and six adjacent teeth (Fig. 3.52K). Distance between setae submenti 35−40 μm. Mandible length 37 μm. Antennae (Fig. 3.52I and J): AR 0.76−1.00. Length of segment I 60−75 μm, II 30−45 μm, III 27−47 μm, IV 2−5 μm; part of the third antennal segment not sclerotized; basal segment width 12 μm; antennal blade 27−37 μm long; ring organ at 10−20 μm from the base of first antennal segment. Antennal segments two and three brown.

Abdomen. Ventral setae modified, slightly wider and longer than dorsal setae (Fig. 3.52L). Subbasal seta on posterior parapod split from base (Fig. 3.52M).

Remarks. The larvae from *C. fortispicula* were collected from litter standing near the water surface from small mountain streams of the Atlantic Forest. This species was found in Minas Gerais, São Paulo and Rio Grande do Sul States. According to Wiedenbrug et al. (2012), the species was also collected in small streams of a semideciduous forest in Mato Grosso do Sul and Goiás States, Brazil.

Distribution. NT: Brazil.

3.37 *CORYNONEURA FRANCISCOI* WIEDENBRUG ET AL., 2012 (FIGS. 3.53A−E AND 3.54A−I)

Corynoneura franciscoi Wiedenbrug et al., 2012: 29.

Diagnostic characters. The male of *Corynoneura franciscoi* can be differentiated from other species with straight transverse sternapodeme and with phallapodeme attachment placed in the caudal apex of the sternapodeme by following characters, antenna with 11 flagellomeres, apex of

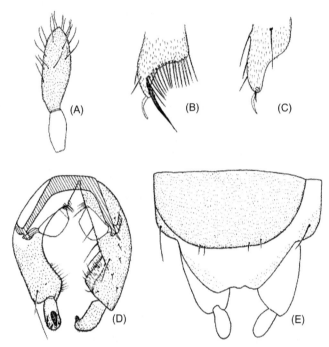

Figure 3.53 *Corynoneura franciscoi* Wiedenbrug et al., 2012; Male imago (From Wiedenbrug, S., Lamas, C. J. E., & Trivinho-Strixino, S. (2012). A review of the genus *Corynoneura* Winnertz (Diptera: Chironomidae) from the neotropical region. *Zootaxa, 3574*, 1–61. www.mapress.com/j/zt). (A) Terminal flagellomeres. (B and C) Apex of hind tibia. (D and E) Hypopygium. (D) Tergite IX removed, sclerites hatched; left is dorsal view, right is ventral view. (E) Tergite IX.

hind tibia with a curved shaped seta, anterior margin of gonocoxite with several longer setae directed to median, phallapodeme posterior margin rounded and gonostylus without crista dorsalis. The pupa has a scutal tubercle, abdomen with 3 short lateral setae and anal lobe rectangular, without macrosetae and median setae. The larva of *C. franciscoi* is characterized by the antennae shorter than postmentum length, mentum with three subequal median teeth, spine on posterior pseudopods long (longer than 100 μm) simple or slightly split at the base.

Male (Description from Wiedenbrug et al., 2012)

Total length 1.25 mm.

Head. AR = 0.24. Antenna with 11 flagellomeres, apical flagellomere 65 μm (Fig. 3.53A). Flagellomeres with more than one row of setae each. Eyes pubescent.

Thorax. Antepronotal lobes dorsally tapering.

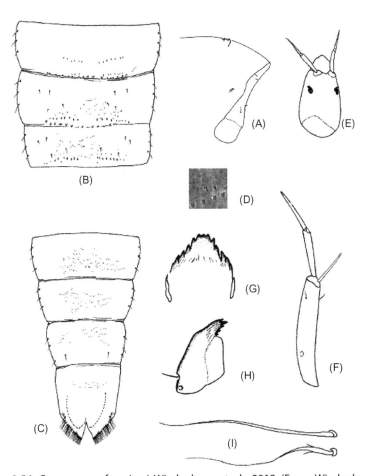

Figure 3.54 *Corynoneura franciscoi* Wiedenbrug et al., 2012 (From Wiedenbrug, S., Lamas, C. J. E., & Trivinho-Strixino, S. (2012). A review of the genus *Corynoneura* Winnertz (Diptera: Chironomidae) from the neotropical region. *Zootaxa, 3574,* 1–61. www.mapress.com/j/zt). (A–H) Immatures imago. (A–D) Pupa. (A) Thorax, anterior part. (B) Tergites III–V. (C) Tergites VI–IX and anal lobe. (D) Detail of posterior shagreen of tergite IV. (E–I) Larva. (E) Head, dorsal view. (F) Antenna. (G) Mentum. (H) Mandible. (I) Subbasal seta of posterior parapod.

Legs. Hind tibial scale with curved seta (Fig. 3.53B and C).

Hypopygium (Fig. 3.53D and E). Tergite IX with 4 setae. Laterosternite IX with 1 seta. Gonocoxite anterior median margin with longer setae. Superior volsella absent. Inferior volsella low apically on gonocoxite margin. Aedeagal lobe apparently absent. Transverse sternapodeme straight, phallapodeme caudal attached. Phallapodeme posterior margin rounded. Gonostylus without crista dorsalis.

Pupa (Description from Wiedenbrug et al., 2012)

Total length 1.30 mm.

Cephalothorax. Thorax slightly rugose, scutal tubercle present, with wide and short spinules (Fig. 3.54A). Wing sheaths apparently without row of pearls.

Abdomen (Fig. 3.54B—D). Sternites and tergites I and II not observed. Tergite III with posterior row of spinules median interrupted, if fine shagreen present not observed. Tergites IV and V with fine shagreen and posterior row of slightly larger spinules, median interrupted. Tergites VI—IX with fine homogeneous shagreen. Conjunctives tergite III/IV and tergite IV/V with hooklets. Segments III—VIII with 3 short L-setae. Anal lobe rectangular (Fig. 3.54C). Anal lobe with fringe restricted to posterior margin, macrosetae and inner setae absent. Corner of anal lobe darker sclerotized with small spine.

Larva (Description from Wiedenbrug et al., 2012)

Head (Fig. 3.54E). Postmentum 155—165 μm long. Head capsule integument smooth. Mentum with three median teeth, intern median tooth slightly smaller than extern median teeth, six additional lateral teeth (Fig. 3.54G). Antenna 128—135 μm long, segments two and three brown, not drawn (Fig. 3.54F).

Abdomen. Body setae simple, strong and long, about half as long as segment width. Subbasal seta on posterior parapod long apparently simple, but in some cases slightly split at the base (Fig. 3.54I).

Remarks. All larvae were sampled from larvae of *Corydalus* spp. (Megaloptera) in small streams in São Paulo State, Brazil. The pupal cases were found attached to the body of *Corydalus*, as well as fourth and third instar larvae. However more information of the benthos is necessary to evaluate whether the life cycle of *C. franciscoi* is restricted to *Corydalus* spp. or whether the species can also be found in the benthos.

Additional material from larvae of *Corynoneura* found on larva of *Argia* (Odonata) from Córrego Gallarada, Campos do Jordão, SP, leg Peruquetti 12.xi.1999, was examined by Wiedenbrug et al. (2012). This however has proven to be a different species, since the larvae had a smaller antenna (postmentum length 188 μm, antenna 115 μm long, antenna I 73 μm, antenna II 27 μm, antenna III 12 μm, antenna IV 3 μm and seta on posterior parapod simple, 97 μm) and although the associated pupa also showed no macrosetae, the scutal tubercle was composed of pointed spinules differently than by *C. franciscoi*. The pharate female pupa indicated a presence of two Sca, one slightly smaller than the other and labia funnel shaped.

Other similar pupal exuviae with rectangular anal lobe, developed scutal tubercle, short L-setae and without macrosetae were also recorded by Wiedenbrug (2000; *Corynoneura* group spec. 17) from mountain streams in Rio Grande do Sul State and were also collected at streams from Minas Gerais State (Toca da Raposa, Vila de Maringá, Bocaína de Minas) by the first author.

Distribution. NT: Brazil (Pedregulho, São Carlos).

3.38 *CORYNONEURA FUJIUNDECIMA* SASA,1985 (FIG. 3.55A−C)

Corynoneura fujiundecima Sasa, 1985: 130; Fu et al., 2009: 10.

Material examined. Male holotype (No. 90: 81), Japan, Honshu, Yamanashi Prefecture, Lake Yamanaka, sweep net, 13.v.1983, M. Sasa.

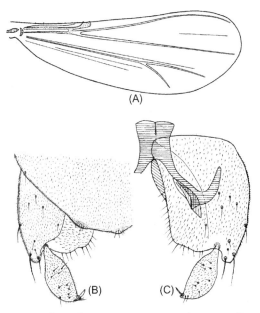

Figure 3.55 *Corynoneura fujiundecima* Sasa, 1985; Male imago (From Fu, Y., Sæther, O. A., & Wang, X. (2009). *Corynoneura* Winnertz from East Asia, with a systematic review of the genus (Diptera: Chironomidae: Orthocladiinae). *Zootaxa, 2287,* 1−44. www.mapress.com/j/zt). (A) Wing. (B) Hypopygium, dorsal view. (C) Hypopygium, ventral view.

Table 3.29 Lengths (in μm) and proportions of legs segments of male *Corynoneura fujiundecima* Sasa, 1985

	fe	ti	ta$_1$	ta$_2$	ta$_3$	ta$_4$	ta$_5$	LR	BV	SV	BR
p$_1$	325	395	215	120	63	28	40	0.54	3.7	3.3	1.8
p$_2$	470	420	245	115	58	28	43	0.58	4.7	3.6	1.5
p$_3$	385	415	250	135	80	28	43	0.60	3.7	3.2	1.6

Diagnostic characters. The species can be separated from other members of the genus by having an antenna with 12 flagellomeres, AR 0.62−0.70; superior volsella not clear, nearly fused with the gonocoxite, and inferior volsella broad. Posterior margin of tergite IX almost straight, sternapodeme inverted U-shaped.

Additional description and corrections. Scutellum and postnotum brown, abdominal tergites yellowish brown while the original description Sasa (1985) mentions scutellum and postnotum entirely black, abdominal tergites I−V yellowish brown, VI to hypopygium dark brown. "R$_1$, R$_{2+3}$ and R$_{4+5}$ fused and shortened" should be corrected to "R$_1$, R$_{2+3}$ and costa fused."

Some additional characters should be added: total length 1.86−2.13 mm. Wing length 1.14−1.24 mm. Antenna 12 flagellomeres, AR 0.62−0.70. VR 2.7. Wing width/wing length 0.35. Clypeus with 10 setae, Tentorium 138 μm long 23 μm wide. VR 2.7. C length 350 μm, C/wing length 0.31. Cu length 700 μm. Cu/wing length 0.62. Wing width/wing length 0.35. C with 9 setae. Leg measurements (in μm) and ratios as in Table 3.29. Posterior margin of tergite IX almost straight, with several short setae. Sternapodeme inverted U-shaped, coxapodeme 25 μm long, attachment point with phallapodeme placed caudal and caudally directed; phallapodeme short, straight and broad, not extending beyond posterior margin of tergite IX, 50 μm long. Gonocoxite 98 μm long, with 6 long setae apically. Gonostylus 50 μm long; megaseta 10 μm long. HR 1.95, HV 3.25. The wing and hypopygium of holotype have been redrawn (Fig. 3.55A−C).

Distribution. PA: Japan, Russia (Far East).

3.39 *CORYNONEURA GRATIAS* SCHLEE, 1968 (FIG. 3.56A AND B)

Corynoneura gratias Schlee, 1968b; 26. Hirvenoja & Hirvenoja, 1988: 218; Fu et al., 2009: 12.

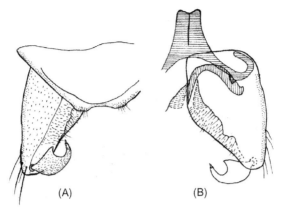

Figure 3.56 *Corynoneura gratias*, Male imago. (A) Hypopygium, dorsal view. (B) Hypopygium, ventral view.

Material examined. P.R. China: Liaoning Province, Panjin County, Guchengzi, 40°41′N−41°27′N, 121°34′E, alt. 50 m, 1 male (BDN No. 03374), 13.iv.1992, J. Wang.

Diagnostic characters. Wing length 1.0 mm. Antenna with 10 flagellomeres; apex with short sensilla chaetica subapically, apically acute, AR 0.92. Hind tibia without S-shaped seta. Posterior margin of tergite IX bilobed, laterosternite IX with 1 seta. Inferior volsella absent, sternapodeme inverted U-shaped, attachment point with phallapodeme placed caudal and ventrally directed; phallapodeme strongly curved.

Male (*n* = 1)

Wing length 1.0 mm. Wing length/length of profemur 3.45.

Coloration. Head brown. Thorax yellow-brown. Abdominal segments yellow. Legs yellow. Wings transparent and yellowish, with pale yellow clava.

Head. Eyes bare, reniform. Antenna with 10 flagellomeres; ultimate flagellomere 265 µm long, apex with short sensilla chaetica which extends a little way back from tip, apically acuate. AR 0.92. Clypeus with 9 setae. Temporal setae lacking. Tentorium 140 µm long, 20 µm wide. Stipes: 40 µm long. Palpomeres length (in µm): 13, 15, 20, 30, 55. Palpomere 2 and 3 ellipsoid, 4 rectangular, 5 long and slender. Palpomere 5/3 ratio 2.75.

Thorax. Dorsocentrals 8, cannot see any other setae.

Wing. Kinkled, cannot see the vein clearly.

Legs. Fore trochanter with keel. Spur of front tibia 27.5 µm long, spurs of middle tibia 15 µm long, of hind tibia 42.5 µm and 10 µm long. Width

Table 3.30 Lengths (in μm) and proportions of legs segments of male *Corynoneura gratias* (based on Chinese specimen)

	fe	ti	ta$_1$	ta$_2$	ta$_3$	ta$_4$	ta$_5$	LR	BV	SV	BR
p$_1$	290	350	208	100	63	28	28	0.59	3.6	3.1	1.7
p$_2$	400	385	230	98	55	30	43	0.58	4.5	3.5	2.0
p$_3$	325	365	210	108	50	25	45	0.58	4.0	3.3	2.2

at apex of fore tibia 27.5 μm, of mid tibia 27.5 μm, of hind tibia 50 μm. Tip of hind tibia expanded, with comb of 16 setae. Lengths and proportions of legs as in Table 3.30.

Hypopygium (Fig. 3.56A). Posterior margin bilobed, and with many short setae, laterosternite IX with 1 long seta. Anal point absent. Inferior volsellae absent, sternapodeme inverted U-shaped, coxapodeme 30 μm long, with caudal projection basally phallapodeme strongly curved, 65 μm long. Gonocoxite 90 μm long with 4 setae apically. Gonostylus curved apically, 45 μm long, median protuberant. Megaseta 7.5 μm long. HR 2.0.

Remarks. Schlee (1968b) described the species based on specimen from Stuttgart, Germany. *C. scutellata* Winnertz listed by Wang (2000) from Liaoning Province was wrongly identified and should be corrected to *C. gratias* Schlee.

Distribution. PA: China (Liaoning), Corsica, Czech Republic, Finland, France, Germany, Great Britain, Ireland, Norway, Russia (NET, Far East), Slovakia, Spain, Sweden.

3.40 *CORYNONEURA GUANACASTE* WIEDENBRUG ET AL., 2012 (FIG. 3.57A−F)

Corynoneura guanacaste Wiedenbrug et al., 2012: 32.

Diagnostic characters. The adult males are separable from other species with the attachment of the phallapodeme caudal on the sternapodeme, and S-seta on the apex of hind tibia by the superior volsella rounded bearing long seta directed to posteromedian, inferior volsella large and rectangular and aedeagal lobe pointed.

Male (Description from Wiedenbrug et al., 2012)

Total length 1.36 mm. Wing length 0.59 mm.

Coloration. Thorax brownish. Abdominal tergites I−II whitish, other tergites and genitalia brownish. Legs whitish.

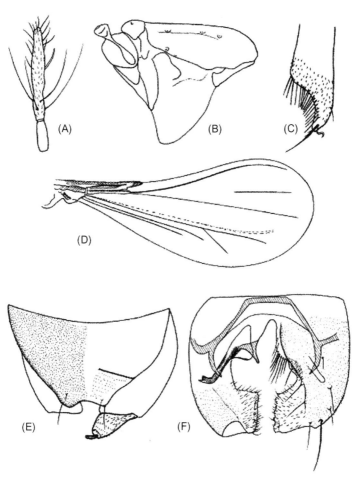

Figure 3.57 *Corynoneura guanacaste* Wiedenbrug et al., 2012; Male imago (From Wiedenbrug, S., Lamas, C. J. E., & Trivinho-Strixino, S. (2012). A review of the genus *Corynoneura* Winnertz (Diptera: Chironomidae) from the neotropical region. *Zootaxa, 3574*, 1−61. www.mapress.com/j/zt). (A) Terminal flagellomeres. (B) Thorax. (C) Apex of hind tibia. (D) Wing. (E and F) Hypopygium. (E) Tergite IX; left is dorsal view, right is ventral view of tergite IX and gonostylus. (F) Tergite IX and gonostylus removed, sclerites hatched; right is ventral view, left is dorsal view.

Head. AR = 0.72. Antenna with 9 flagellomeres, apical flagellomere 220 µm. Flagellomeres with more than one row of setae each. Antennal apex as in Fig. 3.57A. Eyes pubescent.

Thorax. Antepronotal lobes dorsally tapering (Fig. 3.57B).

Wing. Clavus/wing length 0.24. Anal lobe absent (Fig. 3.57D).

Legs. Hind tibial scale 37−42 µm long, with one S-seta (Fig. 3.57C).

Hypopygium (Fig. 3.57E and F). Tergite IX with 2 setae. Laterosternite IX apparently without seta. Superior volsella rounded bearing long setae medially directed. Inferior volsella large and rectangular. Aedeagal lobe pointed. Transverse sternapodeme straight, phallapodeme caudal attached. Phallapodeme spatula-shaped. Gonostylus with a low crista dorsalis.

Remarks. *Corynoneura guanacaste* hypopygium has a pointed aedeagal lobe and a distinct inferior volsella similarly to *Corynoneura septadentata* Wiedenbrug *et* Trivinho-Strixino, however the long setae directed to median on the superior volsella and the size of the inferior volsella separate this two species. The species is only known from the type locality, Guanacaste Province in Costa Rica.

Distribution. NT: Costa Rica.

3.41 *CORYNONEURA GYNOCERA* TUISKUNEN, 1983 (FIGS. 3.58A−D AND 3.59A AND B)

Corynoneura gynocera Tuiskunen, 1983: 100.

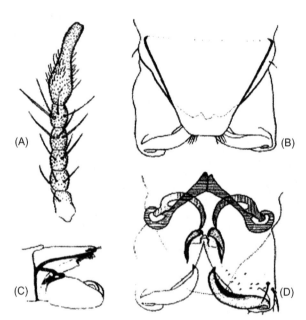

Figure 3.58 *Corynoneura gynocera* Tuiskunen, 1983; Male imago (From Tuiskunen, J. (1983). A description of *Corynoneura gynocera* sp. n. (Diptera, Chironomidae) from Finland. *Annales Entomologica Fennici, 49,* 100−102). (A) Antenna. (B−D) Hypopygium. (B) Dorsal view. (C) Lateral view. (D) Ventral view.

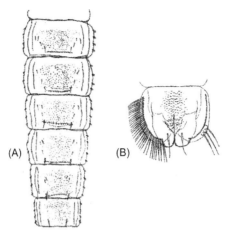

Figure 3.59 *Corynoneura gynocera* Tuiskunen, 1983; Pupa imago (From Tuiskunen, J. (1983). A description of *Corynoneura gynocera* sp. n. (Diptera, Chironomidae) from Finland. *Annales Entomologica Fennici, 49*, 100−102). (A) Tergites III−VIII. (B) Anal lobe.

Holotype location. Finland.

Diagnostic characters. Male antenna with 6 flagellomeres; AR 0.58−0.72. Sternapodeme inverted V-shaped. Tergite IX narrower and almost totally covering the hypopygium. Phallapodeme strongly curved with projection for joint with sternapodeme placed prelateral.

Male (Description from Tuiskunen, 1983)

Total length 1.4−1.7 mm.

Coloration (after 13 years' preservation in ethanol). Ground color of thorax light brown, scutal stripes darker. Wings very slightly brownish. Halteres pale. Abdomen annulate because of the light front and hind margins of the brown tergites.

Head. Eyes without dorsal projection. Frontal tubercles absent. Antenna with 6 flagellomeres (Fig. 3.6A); AR 0.58−0.72. Clypeus with 8−11 setae.

Thorax. Dorsocentrals 4−7, acrostichals absent; scutellars 2; prealars 2. Antepronotum with 0−2 setae in its ventral part. The oral part of anapleural suture reduced, the rest sometimes hard to see, only a narrow line.

Wing. Wing length 0.8−1.0 mm. Tip of wing reaches the posterior margin of the sixth tergite.

Legs. LR_1 0.47−0.53; LR_2 0.47−0.52; LR_3 0.47−0.53. Hind tibia with a well-developed process.

Hypopygium (Fig. 3.58B−D). Tergite IX characteristic, very long, but narrower than usual in *Corynoneura* spp., almost totally covering the hypopygium. The posterior margin of tergite IX straight. Laterosternite IX with 1 long seta. Sternapodeme inverted V-shaped; attachment point with phallapodeme placed in caudal third of lateral sternapodeme and directed caudally, phallapodeme strongly curved with projection for joint with sternapodeme placed prelateral, not extending beyond posterior margin of tergite IX. Gonocoxites without lobes in the distal half, basally with lobes densely covered with short hairs, and a hook in the oral part. Gonostylus tapering towards the tip, densely covered with short hairs on the inner surface.

Pupa (Description from Tuiskunen, 1983)

Total length 1.8−2.2 mm.

Coloration. Thorax light brownish, abdominal segments transparent, laterally slightly brownish.

Thorax. Thoracic horn absent. Antennal sheaths short and similar in the two sexes.

Wing. Wing sheaths with 3−5 apical rows of pearls. The sheath reaches the end of the second abdominal segment. Some granulation along the oral part of the thoracic suture.

Abdomen. Abdominal tergites I and II with weak spinules. No rows of stronger spines present. Segment II with three lateral setae. Abdominal segments 3−7 each with 4 thick lateral filaments and 4 tergal setae on each side. Spinules more numerous and stronger than on tergites I and II Each tergite with an anal row of 8−15 stronger spines (Fig. 3.59A), except tergite 8, which lacks such a row of spines, having only 2 dorsal setae on each side. Anal lobe fringed with 35−42 setae; 3 larger marginal setae and one more central seta on each side (Fig. 3.59B). Genital sheaths of the male exuviae conspicuous, brownish.

Distribution. PA: Finland.

3.42 *CORYNONEURA HERMANNI* WIEDENBRUG & TRIVINHO-STRIXINO, 2011 (FIGS. 3.60A−D, 3.61A−D, AND 3.62A−L)

Corynoneura-group spec. 4: Wiedenbrug, 2000: 101.

Corynoneura hermanni Wiedenbrug & Trivinho-Strixino, 2011: 7.

Diagnostic characters. The adult males are separable from other species by the antenna with 10 flagellomeres, AR 0.27; eyes pubescent; hypopygium with superior volsella bearing long setae directed to median,

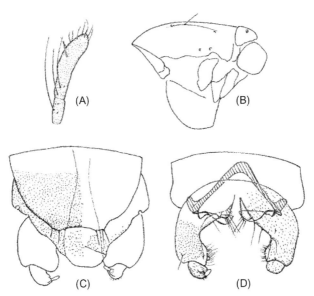

Figure 3.60 *Corynoneura hermanni* Wiedenbrug & Trivinho-Strixino, 2011, Male imago (From Wiedenbrug, S., & Trivinho-Strixino, S. (2011). New species of the genus *Corynoneura* Winnertz (Diptera, Chironomidae) from Brazil. *Zootaxa, 2822,* 1—40. www.mapress.com/j/zt). (A) Terminal flagellomeres. (B) Thorax. (C) Tergite IX; left is dorsal view, right is ventral view. (D) Tergite IX removed, sclerites hatched; left is dorsal view, right is ventral view.

attachment of the phallapodeme caudally on the lateral sternapodeme, phallapodeme posterior margin sclerotized, rounded, curved to anterior, crista dorsalis absent on gonostylus. Adult females are separable from other species except *Corynoneura mediaspicula* by the labia being funnel shaped, two seminal capsules subequal in size and abdominal segments III—IX brownish. The pupae are distinguished from other species by the median field of shagreen on tergites III and IV not separated from posterior field. Spines of posterior row narrow, with straight lateral margin. Anal lobe not joined medially as an inverted "V" but separated through a straight margin, posterior margin rounded with fringe. The larvae can be recognized by the head integument without sculptures, the mentum with 2 median and 6 lateral teeth, antenna longer than frontoclypeal apotome, head elongate (postmentum length/head width 1.35), AR 0.88, modified abdominal setae apically split, basal seta on posterior parapods with one slightly pectinate side.

   Male (Description from Wiedenbrug & Trivinho-Strixino, 2011)
   Thorax length 0.35 mm.

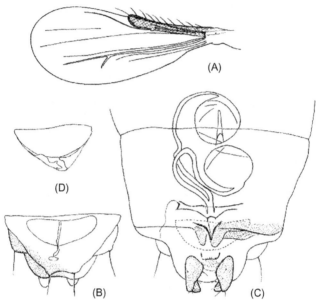

Figure 3.61 *Corynoneura hermanni* Wiedenbrug & Trivinho-Strixino, 2011, Female imago (From Wiedenbrug, S., & Trivinho-Strixino, S. (2011). New species of the genus *Corynoneura* Winnertz (Diptera, Chironomidae) from Brazil. *Zootaxa, 2822,* 1–40. www.mapress.com/j/zt). (A) Wing. (B) Dorsal with view of labia. (C) Ventral view of genitalia; left is gonapophysis removed. (D) Ventral view of labia.

Coloration. Thorax brownish; legs whitish with brownish ring posterior on femur and anterior on tibia; tergites I–IV whitish, V–IX brownish in anterior half, rest whitish.

Head. AR = 0.27. Antenna with 10 flagellomeres, ultimate flagellomere 55 μm long, at least one sensillum chaeticum on last flagellomere (Fig. 3.60A). Flagellomeres, except the first and last, with more than one rows of seta. Temporal setae absent. Clypeus 37 μm long, with 8 setae. Tentorium 77 μm long. Palpomere lengths (in μm): 7, 10, 15, 17, 32. Sensilla clavata not possible to see. Eyes pubescent.

Thorax (Fig. 3.60B). Antepronotum apparently without seta. Dorsocentrals 4, prealars 2. Scutellars 2. Antepronotal lobes dorsally reduced.

Wing. Not measurable.

Legs. Spur of front tibia 10 μm long, scale 5 μm long. Width of apex of front tibia 12 μm. Leg measurement (in μm) and ratios as in Table 3.31.

Abdomen setation. Three setae on tergite VII, 1 seta on tergite VIII.

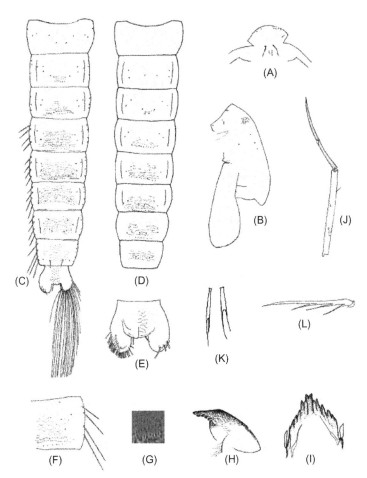

Figure 3.62 *Corynoneura hermanni* Wiedenbrug & Trivinho-Strixino, 2011 Immatures imago (From Wiedenbrug, S., & Trivinho-Strixino, S. (2011). New species of the genus *Corynoneura* Winnertz (Diptera, Chironomidae) from Brazil. *Zootaxa, 2822*, 1−40. www.mapress.com/j/zt). (A−G) Pupa imago. (A) Frontal apotome. (B) Thorax, lateral view. (C) Tergites I−IX and anal lobe, right is ventral, left is dorsal without fringe. (D) Sternites I−VIII. (E) Detail of anal lobe; right is dorsal view without fringe, left is ventral view. (F) Detail of tergite IV. (G) Detail of posterior shagreen of tergite IV. (H−L) Larva imago. (H) Mandible. (I) Mentum. (J) Antenna. (K) Abdominal setae. (L) Subbasal seta of posterior parapod.

Hypopygium (Fig. 3.60C and D). Tergite IX with 4 small setae with wide posterior hyaline projection bearing microtrichia. Laterosternite IX with 1 seta. Superior volsella rounded with long seta directed medially. Inferior volsella prominent with longer microtrichia medially directed.

Table 3.31 Lengths (in μm) and proportions of legs segments of male *Corynoneura hermanni* Wiedenbrug & Trivinho-Strixino, 2011

	fe	ti	ta$_1$	ta$_2$	ta$_3$	ta$_4$	ta$_5$	LR	BV	SV	BR
p$_1$	—	177	85	50	27	15	22	0.48	—	—	4.00

Source: From Wiedenbrug, S., & Trivinho-Strixino, S. (2011). New species of the genus Corynoneura Winnertz (Diptera, Chironomidae) from Brazil. *Zootaxa*, 2822: 1–40. www.mapress.com/j/zt.

Table 3.32 Lengths (in μm) and proportions of legs segments of female *Corynoneura hermanni* Wiedenbrug & Trivinho-Strixino, 2011

	fe	ti	ta$_1$	ta$_2$	ta$_3$	ta$_4$	ta$_5$	LR	BV	SV	BR
p$_1$	172	200	92	50	27	15	25	0.46	3.96	4.03	1.75
p$_2$	235	220	127	52	27	15	22	0.58	4.96	3.57	2.25
p$_3$	190	205	107	60	22	12	22	0.52	4.28	3.67	3.00

Source: From Wiedenbrug, S., & Trivinho-Strixino, S. (2011). New species of the genus Corynoneura Winnertz (Diptera, Chironomidae) from Brazil. *Zootaxa*, 2822: 1–40. www.mapress.com/j/zt.

Phallapodeme 17 μm long, with posterior margin rounded and sclerotized; transverse sternapodeme rounded. Gonocoxite 37 μm long. Gonostylus 12 μm long, apically tapering; megaseta 5 μm long. HR 3.

Female (Description from Wiedenbrug & Trivinho-Strixino, 2011)

Thorax length 0.70 mm. Wing length 0.55 mm. Wing length/length of profemur 3.18.

Coloration. Thorax light brown, with brown vittae, postnotum and preepisternum; legs whitish with a brown ring anterior on tibia; tergites I–II whitish, III–IX brownish, with posterior whitish band.

Head. AR 0.3. Antenna with 5 flagellomeres, apical flagellomere 30 μm long; two sensilla chaetica on flagellomeres 5, at least one on 3–4, first and second flagellomeres not possible to observe. Temporal setae absent. Clypeus 45 μm long, with 8 setae. Tentorium 75 μm long, 5 μm wide. Palpomeres length (in μm): 10, 12, 15, 20, 45. One sensilla clavata on third palpomere. Eyes pubescent.

Thorax. Antepronotal not possible to observe. Dorsocentrals 4, prealars 2. Scutellars 2. Antepronotal lobes dorsally tapering.

Wing (Fig. 3.61A). VR 2.60. Clavus 35 μm wide, ending 245 μm from arculus. Clavus/wing length 0.45.

Legs. Spur of front tibia 10 μm long; spur of middle tibia 7 μm; spurs of hind tibia 32 μm long, second spur small S-shaped. Front tibial scale 10 μm long, mid tibial scale 12 μm long, hind tibial scale 35 μm long. Width of apex of front tibia 15 μm, of middle tibia 15 μm, of hind tibia 25 μm. Leg measurements (in μm) and ratios as in Table 3.32.

Abdomen. One seta on tergite VI, 3 setae on tergite VII, 1 seta on tergite VIII.

Genitalia (Fig. 3.61B−D). Gonocoxite IX with 1 seta. Tergite IX with 2 setae. Two seminal capsules respectively 30 μm and 32 μm long, one spermathecal duct with a loop, second straighter both ducts join together a short distance before seminal eminence, which has sclerotized outer borders. Notum 50 μm long. Membrane well sclerotized. Apodeme lobe anterior margin curved to posterior, apically pointed. Coxosternapodeme strongly curved, apparently without lateral lamellae, copulatory bursa semicircle-shaped. Labia membranous, bare, dorsal funnel shaped, apically building the accessory gonopore, ventral shorter. Gonocoxapodeme straight, gonapophyses triangular medially. Postgenital plate with microtrichia. Cercus 30 μm long.

Pupa (Description from Wiedenbrug & Trivinho-Strixino, 2011) Total length 1.21−1.41 mm.

Cephalothorax. Frontal apotome smooth or with few granulations (Fig. 3.62A). Distance between Dc_1 and Dc_2 60−65 μm; Dc_2 and Dc_3 35−45 μm; Dc_3 and Dc_4 20−25 μm. Dc_1 displaced ventrally. Wing sheath with one, sometimes two, row of pearls (Fig. 3.62B).

Abdomen (Fig. 3.62C−G). Tergite I and sternites I−II without shagreen. Tergites II−III with few shagreen spinules posteriorly. Tergites VI−IX with fine shagreen, shagreen points slightly increasing in size posteriorly on segments III−V (Fig. 3.62F and G). Sternites III−IV with fine shagreen posteriorly, IV−VIII with fine shagreen. Conjunctive tergite II/III with 7 spinules; tergite III/IV with 19 hooklets; tergite IV/V with 13 hooklets; tergite V/VI with 9 spinules; sternite IV/V with 12 spinules; sternite V/VI with 13 spinules; sternite VI/VII with 11 spinules, sternite VII/VIII with 9 spinules. segment I with 4 D-setae, 1 L-seta, without V-seta; segments II−III 3 D-setae, 3 L-setae and 3 V-seta; segments IV−VII 3−4 D-setae, 3V-setae and 3 short taeniate L-setae; segment VIII with 2 D-setae, 1 V-seta and 4 taeniate L-setae. Anal lobe distally rounded 87−95 μm long (Fig. 3.62E), posterior with 21−24 taeniate setae, 130 μm long; 3 macrosetae taeniate; inner setae taeniate.

Larva (Description from Wiedenbrug & Trivinho-Strixino, 2011)

Head. Head capsule integument smooth. Frontal apotome 190−195 μm long; head width 120 μm; postmentum 155−162 μm long; postmentum/head width 1.35. Sternite I simple. Premandible not possible to see. Mentum with two median teeth and six adjacent teeth (Fig. 3.62I). Distance between setae submenti 32−35 μm. Mandible (Fig. 3.62H) length 37−40 μm with outer sclerotized protuberance.

Antennae (Fig. 3.62J): AR 0.88. Length of segment I 125 μm, II 60 μm, III 77 μm, IV 5 μm; basal segment width 12 μm; antennal blade 27 μm long; ring organ at 35 μm from the base of first antennal segment. Lauterborn organs inserted on second antennal segment not overreaching distal border of segment. Antennal segments two and three brown.

Abdomen. Ventral setae modified, slightly wider and slit at extremity (Fig. 3.62K). Subbasal seta on posterior parapod as Fig. 3.62L.

Remarks. The larvae from *Corynoneura hermanni* were collected from litter standing near the water surface from small mountain streams. This species was found in São Paulo and Rio Grande do Sul States, Brazil.

Distribution. NT: Brazil.

3.43 *CORYNONEURA HIRVENOJAI* SUBLETTE & SASA, 1994 (FIG. 3.63A−C)

Corynoneura hirvenojai Sublette & Sasa, 1994: 12.

Diagnostic characters. This species resembles *Corynoneura lacustris* (Edwards) in tibial features and genitalia. It may be differentiated from *C. lacustris* Edwards by having only 7−8 flagellomeres and a differently shaped inferior volsella on the gonocoxite.

Male (Description from Sublette & Sasa, 1994)

Coloration. Head and thorax largely dark with narrow yellow ground color visible on pleura. Halteres pale. Fore and middle femora weakly infuscate, the color apparently being formed by irregular, close set rows of microtrichiae; bases of all tibiae narrowly dark fasciate. Abdominal terga I−IV pale; V−IV mostly dark; genitalia dark but usually slightly paler than ninth tergum.

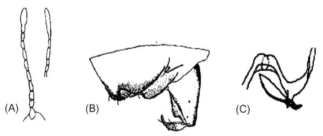

(A)   (B)   (C)

Figure 3.63 *Corynoneura hirvenojai* Sublette & Sasa, 1994, Male imago (From Sublette, J. E., & Sasa, M. (1994). Chironomidae collected in Onchocerciasis endemic areas of Guatemala (Insecta, Diptera). *Spixiana Supplement, 20*, 1−60). (A) Antenna. (B) Hypopygium, dorsal view. (C) Phallapodeme and sternapodeme.

Head. Eyes reniform, moderately exerted on head; glabrous. Antenna with 7—8 flagellomeres (Fig. 3.63A); antennal ratio 0.62—1.42; antennal segmentation variable with 3 or 4 fused flagellomeres visible in addition to the elongate terminal one. Clypeus at the base about the same width as the antennal pedicel; with 6—8 setae. Temporal setae lacking. Lengths of palpomeres (in μm): —, 16, 22, 44.

Thorax. Antepronotum wide in the basal third, abruptly tapered in the middle third, and weakly tapered in the apical third. Setae: lateral antepronotals 1; dorsocentrals 3, in 1 row; acrostichals lacking; prealars 2; scutellars 2.

Wing. Membrane with very fine microtrichia visible at $500 \times$. Costa extended 0.23—0.26 of the wing length. Wing length 0.55—0.60 mm. Venarum ratio 2.71.

Legs. Claws weakly bifurcate at tips. Spur of fore tibia 20 μm; Spur of mid tibia 8 μm. Hind tibial apex similar to that illustrated by Hirvenoja and Hirvenoja (1988, Fig. 6) and Schlee (1968b, Fig. 113). Spur of hind tibia 38 μm, with weak side teeth reaching 0.47 of the spur length; hind tibial comb of 15 setae. Leg ratios: LR of fore leg 0.37—0.45; P II 0.53—0.58; PIII 0.56—0.62.

Hypopygium (Fig. 3.63B and C). The posterior margin of tergite IX medially concave, with 4 short setae. Inferior volsella triangular. Sternapodeme curved into U-shaped, gonostylus apically curved.

Distribution. NT: Guatemala.

3.44 *CORYNONEURA HORTONENSIS* FU & SÆTHER, 2012 (FIG. 3.64A—F)

Corynoneura hortonensis Fu & Sæther, 2012: 33.

Material examined. Holotype male, Canada: Northwest Territories, Horton River (69°21′N, 126°49.8′W), 4.vii.2000, Bohdan Bilyj (CNC). Paratypes: 4 males, 1 female, as holotype (ZMBN).

Diagnostic characters. The male is similar to *Corynoneura kedrovaya* Makarchenko and Makarchenko in the shape of the inferior volsella and gonostylus, but the two species can easily be separated as the tergite IX of *C. kedrovaya* has 4—6 long setae, the phallapodeme is strongly curved and the joint with the sternapodeme is placed prelaterally, while in *Corynoneura hortonensis* tergite IX of the male is without long setae, the phallapodeme is slightly curved and placed in a caudal position of the sternapodeme. The male is also similar to *C. sundukovi* Makarchenko and Makarchenko in the shape of the inferior volsella and sternapodeme, but

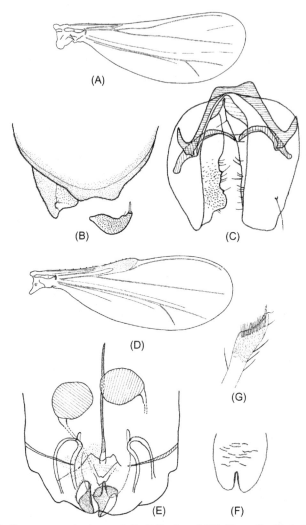

Figure 3.64 *Corynoneura hortonensis* Fu & Sæther, 2012 (From Fu, Y., & Sæther, O. A. (2012). *Corynoneura* Winnertz and *Thienemanniella* Kieffer from the Nearctic region (Diptera: Chironomidae: Orthocladiinae). *Zootaxa, 3536*, 1−61. www.mapress.com/j/zt). (A−C) Male imago. (A) Wing. (B) Hypopygium, dorsal view. (C) Hypopygium, ventral view. (D−F) Female imago. (D) Wing. (E) Genitalia, ventral view. (F) Labia. (G) Hind tibial apex.

the two species can easily be separated as the superior volsellae of *C. sundukovi* large with almost right-angled corner, while in *C. hortonensis* superior volsellae small. Genitalia of female very simple, without any long setae, coxosternapodeme with only 1 lateral lamella.

Table 3.33 Lengths (in μm) and proportions of legs segments of male *Corynoneura hortonensis* Fu & Sæther, 2012 (*n* = 3−5)

	fe	ti	ta$_1$	ta$_2$	ta$_3$	ta$_4$
p$_1$	236	276−284	139	83	46	18
p$_2$	320−340	280−304	170	73	42	20
p$_3$	252−260	268−280	143−152	81−89	32−34	16−28

	ta$_5$	LR	BV	SV	BR	
p$_1$	30	0.49	3.7	3.7	1.8	
p$_2$	28	0.61	4.7	3.5	2.2	
p$_3$	24−28	0.53−0.54	3.9−4.3	3.6	2.0−2.7	

Source: From Fu, Y., & Sæther, O. A. (2012). *Corynoneura* Winnertz and *Thienemanniella* Kieffer from the Nearctic region (Diptera: Chironomidae: Orthocladiinae). *Zootaxa*, 3536, 1−61. www.mapress.com/j/zt.

Male (*n* = 3−5)

Total length 0.96−0.98, 0.97 mm. Wing length 0.66−0.74, 0.70 mm. Total length/wing length 1.3−1.5, 1.4. Wing length /profemur length 2.8.

Coloration. Head brown; thorax and legs brown; abdominal tergites I−VI yellow, VII−IX brown.

Head. Eyes bare. Antennae in all specimens broken. Tentorium 110−120, 114 μm long, 11−15, 13 μm wide; stipes 41 μm long, 3 μm wide. Clypeus with 5−7, 6 setae. Lengths of palpomeres (in μm): 6−8, 7; 10−12, 11; 12−19, 17; 21−24, 22; 51−55, 53. Palpomere 5/3 ratio 2.1−2.9, 2.4.

Thorax. Antepronotals absent. Dorsocentrals 6−7, prealars 2. Scutellum with 2 setae.

Wing (Fig. 3.64A). VR 3.6−3.9, 3.7. Cu/wing length 0.56−0.60, 0.58; C length 326−358, 341 μm long; Cu 420−432, 426 μm long; wing width/wing length 0.44−0.46, 0.45; costa with 6−8, 7 setae.

Legs. Fore trochanter with dorsal keel. Spur of fore tibia 16−20, 18 μm long; of mid tibia 10−13, 11 μm long; of hind tibia 11 μm and 25−28, 26 μm long. Width at apex of fore tibia 18−22, 20 μm; of mid tibia 15−17, 16 μm; of hind tibia (*a*) 32−36, 33 μm. Width of hind tibia 1/3 from apex (*d*) 15−18, 17 μm; elongation length (*b*) 33−37, 35 μm; length of maximum thickening (c_1) 55−63, 59 μm; total length of thickening (c_2) 97−104, 99 μm; *a/d* 1.9−2.1, 2.0; *b/d* 1.8−2.3, 2.1; c_1/d 3.6−4.2, 3.8; c_2/d 5.7−6.4, 6.1. Hind tibia expanded, with comb of 12−18, 14 setae, 1 seta near spur strongly S-shaped (Fig. 3.64G). Lengths of leg segments and their proportions as in Table 3.33.

Hypopygium (Fig. 3.64B and C). Tergite IX medially concave with posterior humps. Laterosternite IX without a long seta. Superior volsella small and anteriomedially fused. Inferior volsella slightly rounded and fused with the inner margin of gonocoxite, with 1−3 strong setae. Phallapodeme curved, 30−35, 32 µm long. Transverse sternapodeme 18−22, 20 µm. Megaseta 4−6 µm long. HR 1.9−2.3, 2.1; HV 3.1−3.3, 3.2.

Female ($n = 1$)

Total length 0.89 mm. Wing length 0.71 mm. Total length/wing length 1.30.

Coloration. Head brown; thorax and legs yellowish brown; abdomen as in the male but paler.

Head. Eyes bare. Antenna broken. Tentorium 69 µm long, 7 µm wide. Palpomere lengths (in µm): 15, 14, 18, 25, 37. Palpomere 5/3 ratio 2.1. Palpomere 2 small, palpomeres 3, 4 and 5 elliptical. Clypeus with 6 setae.

Thorax. Antepronotals 0, dorsocentrals 7, prealars 2. Scutellum with 2 setae.

Wing (Fig. 3.64D). VR 2.75. Wing broader than in male. C 340 µm long; Cu 440 µm long. C/wing length 0.49, Cu/wing length 0.60. Costa with 13 setae.

Legs. Fore trochanter with distinct dorsal keel. Fore leg broken except femur; spur of mid tibia 10 µm long, of hind tibia 30 µm long. Width at apex of mid tibia 21 µm, of hind tibia (a) 30 µm. Width of hind tibia 1/3 from apex (d) 18 µm, elongation length (b) 33 µm, length of maximum thickening (c_1) 62 µm, total length of thickening (c_2) 102 µm; a/d 1.7, b/d 1.8, c_1/d 3.4, c_2/d 5.7. Hind tibia expanded, with comb of 12 setae, 1 seta near spur strongly S-shaped (Fig. 3.64G). Lengths of leg segments and their proportions as in Table 3.34.

Abdomen. Setae not observable.

Genitalia (Fig. 3.64E and F). Without any long setae. Cercus 30 µm long, 21 µm wide. Notum length 110 µm long. Coxosternapodeme with

Table 3.34 Lengths (in µm) and proportions of legs segments of female *Corynoneura hortonensis* Fu & Sæther, 2012 ($n = 1$)

	fe	ti	ta_1	ta_2	ta_3	ta_4	ta_5	LR	BV	SV	BR
p_1	200	—	—	—	—	—	—	—	—	—	—
p_2	252	248	—	—	—	—	—	—	—	—	—
p_3	248	260	141	79	36	20	32	0.54	3.9	3.6	2.5

Source: From Fu, Y., & Sæther, O. A. (2012). *Corynoneura* Winnertz and *Thienemanniella* Kieffer from the Nearctic region (Diptera: Chironomidae: Orthocladiinae). *Zootaxa*, 3536, 1−61. www.mapress.com/j/zt.

only 1 lateral lamella. Seminal capsules 44 μm long, 30 μm wide; neck length 10 μm long, width at apex 6 μm.

Remarks. The type material is not in a good condition, with the antennae and most of the legs broken. However, most of observable characters of the male separate the species from all other members of the genus.

Distribution. NE: Canada.

3.45 *CORYNONEURA HUMBERTOI* WIEDENBRUG ET AL., 2012 (FIGS. 3.65A−D, 3.66A−E, AND 3.67A−G)

Corynoneura humbertoi Wiedenbrug et al., 2012: 34.

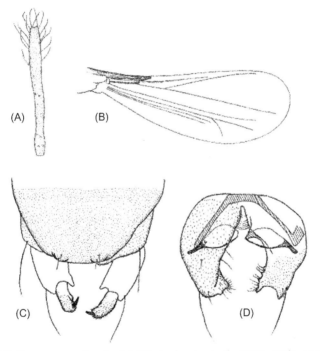

Figure 3.65 *Corynoneura humbertoi* Wiedenbrug et al., 2012; Male imago (From Wiedenbrug, S., Lamas, C. J. E., & Trivinho-Strixino, S. (2012). A review of the genus *Corynoneura* Winnertz (Diptera: Chironomidae) from the neotropical region. *Zootaxa, 3574*, 1−61. www.mapress.com/j/zt). (A) Terminal flagellomere. (B) Wing. (C and D) Hypopygium. (C) Tergite IX and gonostylus. (D) Tergite IX and gonostylus removed, sclerites hatched; right is dorsal view, left is ventral view.

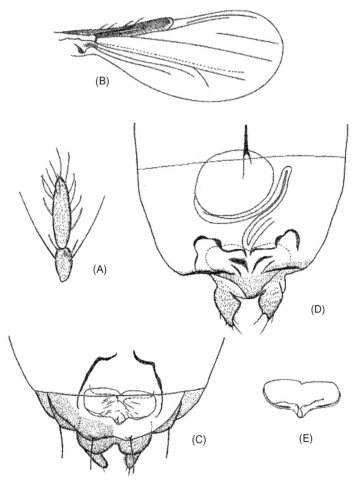

Figure 3.66 *Corynoneura humbertoi* Wiedenbrug et al., 2012 (From Wiedenbrug, S., Lamas, C. J. E., & Trivinho-Strixino, S. (2012). A review of the genus *Corynoneura* Winnertz (Diptera: Chironomidae) from the neotropical region. *Zootaxa, 3574*, 1–61). (A–D) Female imago. (A) Terminal flagellomere. (B) Wing. (C–E) Genitalia. (C) Dorsal view with view of labia. (D) Ventral view of genitalia. (E) Ventral view of labia.

Diagnostic characters. See male diagnostic characters of *C. espraiado.* The females are characterized by having one Sca larger, about 52–95 μm, second Sca less sclerotized, Csa orally curved with long oral dorsal rami and copulatory bursa oral median invaginated. The pupa cannot be differentiated from *C. hermanni* and *Corynoneura longiantenna*, it has rounded anal lobe separated through a straight perpendicular margin, shagreen on

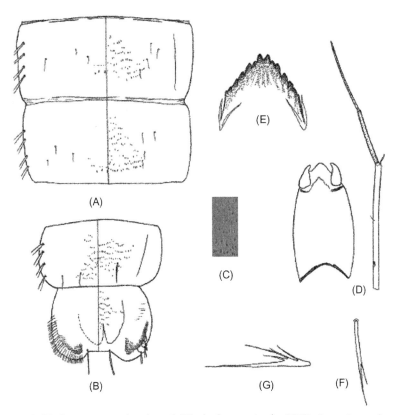

Figure 3.67 *Corynoneura humbertoi* Wiedenbrug et al., 2012; Immatures imago (From Wiedenbrug, S., Lamas, C. J. E., & Trivinho-Strixino, S. (2012). A review of the genus *Corynoneura* Winnertz (Diptera: Chironomidae) from the neotropical region. *Zootaxa, 3574,* 1–61). (A–C) Pupa imago. (A) Segments III–IV; left is ventral view, right is dorsal view. (B) Segments VIII, IX and anal lobe; left is ventral view, right is dorsal view. (C) Detail of posterior shagreen of tergite IV. (D–G) Larva imago. (D) Head, ventral view and separated antenna. (E) Mentum. (F) Abdominal setae. (G) Subbasal seta of posterior parapod.

tergites fine and homogeneous. The larva of *C. humbertoi* have the first antennal segment longer than postmentum length, second antennal segment shorter than the third, mentum with two median teeth, first lateral teeth slightly larger than the second.

Male (Description from Wiedenbrug et al., 2012)

Total length 1.06 mm. Wing length 0.57–0.70 mm.

Coloration. Thorax brownish. Abdominal tergites I–IV whitish, other tergites and genitalia brownish. Legs whitish.

Head. AR = 0.49−0.64. Antenna with 9−10 flagellomeres, apical flagellomere 120−153 μm (Fig. 3.65A). Flagellomeres with more than one row of setae each. Eyes pubescent.

Thorax. Antepronotal lobes dorsally tapering.

Wing. Clavus/wing length 0.22−0.24. Anal lobe absent (Fig. 3.65B).

Legs. Hind tibial scale 30−35 μm long, with one S-seta.

Hypopygium (Fig. 3.65C and D). Tergite IX with 4 setae. Laterosternite IX with 1 seta. Superior volsella low with longer setae, aedeagal lobe triangular with a wide base, inferior volsella present apically on gonocoxite margin. Transverse sternapodeme straight, phallapodeme caudal attached. Phallapodeme with posterior margin sclerotized slightly curved.

Female (Description from Wiedenbrug et al., 2012)

Total length 1.2 mm. Wing length 0.51−0.62 mm.

Coloration. Thorax light brown. Abdominal tergites I−III, whitish, IV−IX brownish. Legs whitish.

Head. AR = 0.51. Antenna with 5 flagellomeres, apical flagellomere 52−57 μm long (Fig. 3.66A). Flagellomeres with more than one row of setae each. Eyes pubescent.

Thorax. Antepronotal lobes dorsally tapering.

Wing. Clavus/wing length 0.42. Anal lobe absent (Fig. 3.66B).

Legs. Hind tibial scale 35−37 μm long, with one small S-setae.

Genitalia (Fig. 3.66C−E). Tergite IX with 2 setae. Laterosternite IX with 1−2 setae. One large seminal capsules 52−95 μm long; second difficult to see, one spermathecal duct with a loop, both spermathecal ducts join together shortly before seminal eminence. Notum 27 μm long. Apodeme lobe with median border sclerotized. Coxosternapodeme, curved, with two rami, first ending at roof of copulatory bursa, second long oral dorsal. Copulatory bursa oral median invaginated. Labia membranous, bare, funnel shaped, apically building the accessory gonopore. Gonocoxapodeme straight, gonapophyses median smoothly pointed. Cercus 35 μm long.

Pupa (Description from Wiedenbrug et al., 2012)

Total length 1.36 mm.

Coloration (exuviae). Cephalothorax light brown, abdomen transparent except light brown muscle markings, lateral margin and anal lobe.

Cephalothorax. Frontal apotome rugose. Thorax suture slightly rugose at scutal tubercle region. All Dc-setae thin taeniate, except Dc_3 wider. Dc_1 displaced ventrally. Rows of pearls at wing sheaths apparently absent.

Abdomen (Fig. 3.67A−C). Tergite and sternite I bare, tergites III−IX with fine and homogeneous shagreen. Sternites with very fine shagreen. Conjunctives sternites III/IV−VII/VIII with small spinules. Segment I with 1, II with 3 L-setae and III−VIII with 4 long taeniate L-setae. Anal lobe rounded (Fig. 3.67B). Anal lobe with fringe not complete, 3 taeniate macrosetae and inner setae taeniate.

Larva (Description from Wiedenbrug et al., 2012)

Head. Postmentum 180−200 μm long. Head capsule integument smooth. Mentum with two median teeth, first lateral teeth not adpressed to median, larger than second lateral tooth (Fig. 3.67E). Antenna 575−618 μm long, segments two and three darker as the first segment. First segment much longer than postmentum length (Fig. 3.67D).

Abdomen. Ventral setae modified, slightly wider and apical slit (Fig. 3.67F). Subbasal seta on posterior parapod with longer serrated rami, split at the base (Fig. 3.67G).

Remarks. According to H. F. Mendes, larvae of *C. humbertoi* were collected from leaves at the bottom of a lake in São Paulo State, additional material was found on *Eichhornia* sp. (Pontederiacea) in a river in Mato Grosso do Sul State, Brazil.

Distribution. NT: Brazil (Ribeirão Preto).

3.46 *CORYNONEURA INAWAPEQUEA* SASA ET AL., 1999 (FIG. 3.68A−C)

Corynoneura inawapequea Sasa et al., 1999: 21; Fu et al., 2009: 12.

Material examined. JAPAN: Honshu, Fukushima Prefecture, Lake Inawashiro, light trap, male (holotype No. 389: 96), 20.viii.1999, K. Kitami.

Diagnostic characters. The species can be separated from other members of the genus by having an antenna with 10 flagellomeres; AR 0.66; superior volsella almost fused with gonocoxite, inferior volsella digitiform; sternapodeme inverted V- to U-shaped; gonostylus with triangular crista dorsalis.

Additional description and corrections. R_1, R_{2+3} and costa fused to a short and thick vein with VR 3.2, while the original description Sasa, Kitami, and Suzuki (1999) mentions R_1 and R_{4+5} fused to a short and thick vein and VR 2.08. Apex of hind tibia with comb of 17 setae while the original description mentions apex of hind tibia with a comb of 12 setae. Gonocoxite with digitiform inferior volsella while the original description mentions gonocoxite without inner lobe, but with a small rounded and hyaline process.

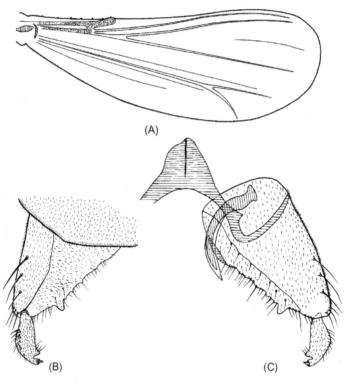

Figure 3.68 *Corynoneura inawapequea* Sasa et al., 1999, Male imago (From Fu, Y., Sæther, O. A., & Wang, X. (2009). *Corynoneura* Winnertz from East Asia, with a systematic review of the genus (Diptera: Chironomidae: Orthocladiinae). *Zootaxa, 2287,* 1–44). (A) Wing. (B) Hypopygium, dorsal view. (C) Hypopygium, ventral view.

Some additional characters should be added: clypeus with 12 setae. Tentorium 118 μm long, 10 μm wide. VR 3.2. C length 180 μm, C/wing length 0.27. Cu length 420 μm. Cu/wing length 0.64. Wing width/wing length 0.38. C with 12 setae. Lengths of leg segments and their proportions as in Table 3.35. Posterior margin of tergite IX almost straight. Sternapodeme inverted V- to U-shaped, coxapodeme 38 μm long, attachment point with phallapodeme placed in caudal third of lateral sternapodeme and directed caudally; phallapodeme weakly curved, extending beyond posterior margin of tergite IX, 65 μm long. Gonocoxite 85 μm long, with 5 long setae apically. Gonostylus 33 μm long, megaseta 3 μm long. HR 2.6, HV 3.7. The wing and hypopygium of holotype are redrawn (Fig. 3.68A–C).

Distribution. PA: Japan.

Table 3.35 Lengths (in μm) and proportions of legs segments of male *Corynoneura inawapequea* Sasa et al., 1999

	fe	ti	ta$_1$	ta$_2$	ta$_3$	ta$_4$	ta$_5$	LR	BV	SV	BR
p$_1$	215	260	155	85	40	20	33	0.60	3.5	3.1	1.7
p$_2$	300	280	168	73	40	20	35	0.60	4.5	3.5	1.6
p$_3$	250	270	160	80	33	18	33	0.59	4.1	3.3	1.8

3.47 *CORYNONEURA INCIDERA* HAZRA ET AL., 2003 (FIGS. 3.69A–D AND 3.70A–E)

Corynoneura incidera Hazra et al., 2003: 73.

Diagnostic characters. The adult male is characterized by AR 0.36; ultimate flagellomere clubbed and shorter than two preceding together with few plume hairs basally and an apical rosette of long hairs each being longer than the segment itself. Inferior volsella absent. The pupa has tergite I bare, tergite II with posteriorly located shagreen of few small points and single row of 7 lancolate spinules, tergite III with posteriorly situated shagreen of 13−14 small point and single row of 9 lanceolate spines, tergites IV−VIII with numerous small points. Anal lobe with 13−14 filaments in fringe restricted to the posterior half of the lobes.

Male (Description from Hazra et al., 2003)

Total length 1.05 mm. Wing length 0.35 mm. Total length/wing length 3.00. Wing length/profemur length 2.36.

Head. AR 0.36, length of flagellomeres (I−VII): 22, 18, 19, 29, 33, 30, 55; ultimate flagellomere clubbed and shorter than two preceding together (Fig. 3.69A) with few plume hairs basally and an apical rosette of long hairs each being longer than the segment itself.

Thorax. Dorsocentrals 4, prealars 3. Scutellum with 2 setae.

Wing. Venation not distinctly marked.

Legs. Fore trochanter as in Fig. 3.69B. Spur of fore tibia 18 μm long, of mid tibia 11 μm long, of hind tibia 26 μm long. Width of the apex of fore tibia 12 μm, mid tibia 13 μm and of hind tibia 18 μm. Hind tibia with comb of 14−15 setae and hind apex as in Fig. 3.69C. Lengths and proportions of legs as in Table 3.36.

Abdomen. Tergites II−VIII with 1 lateral seta on each side and 1 median seta.

Hypopygium (Fig. 3.69D). Tergite IX covering much of the gonocoxites with straight posterior margin. Gonocoxite 41 μm long with 2 setae

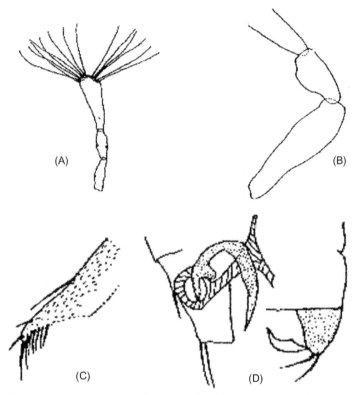

Figure 3.69 *Corynoneura incidera* Hazra et al., 2003, Male imago (From Hazra, N., Nath, S., & Chaudhuri, P. K. (2003). The genus *Corynoneura* Winnertz (Diptera. Chironomidae) from the Darjeeling-Sikkim Himalayas of India, with description of three new species. *Entomologist's Monthly Magazine, 139,* 69–82). (A) The last three flagellomeres of antenna. (B) Fore trochanter. (C) Hind tibial comb. (D) Hypopygium.

apical without inferior volsella; gonostylus simple 22 μm long; sternapodeme inverted V-shaped, phallapodeme strongly curved extending beyond tergite IX; HR 1.86; HV 4.77.

Pupa (Description from Hazra et al., 2003)

Total length of exuviae 1.2 mm.

Cephalothorax. Frontal apotome (Fig. 3.70A) with frontal setae 11 μm long. Median antepronotals 11 μm long each; distance between anterior and median precorneals 2, between median and posterior precorneals 3; near precorneals, thorax with patch of scales (Fig. 3.70B), distance between Dc_2 and Dc_3 24 μm long, between Dc_3 and Dc_4 11 μm long. Wing sheath (Fig. 3.70C) apically with double rows of 12 pearls.

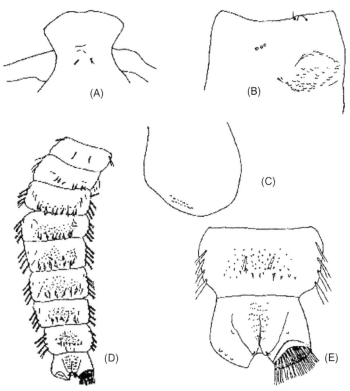

Figure 3.70 *Corynoneura incidera* Hazra et al., 2003, Pupa imago (From Hazra, N., Nath, S., & Chaudhuri, P. K. (2003). The genus *Corynoneura* Winnertz (Diptera. Chironomidae) from the Darjeeling-Sikkim Himalayas of India, with description of three new species. *Entomologist's Monthly Magazine, 139*, 69−82). (A) Frontal apotome. (B) Apex of thorax lateral view. (C) Tip of wing sheath. (D) Tergites. (E) Tergite VIII, anal lobe and genital sac.

Table 3.36 Lengths (in μm) and proportions of legs segments of male *Corynoneura incidera* Hazra, Nath *et* Chaudhuri, 2003

	fe	ti	ta$_1$	ta$_2$	ta$_3$	ta$_4$	ta$_5$	LR	BV	SV	BR
p$_1$	148	177	81	52	—	—	—	0.45	—	—	1.81
p$_2$	203	170	111	—	—	—	—	0.65	—	—	1.66
p$_3$	148	152	81	48	22	15	22	0.53	3.55	5.06	1.71

Abdomen (Fig. 3.70D). Tergite I bare, tergite II with posteriorly located shagreen of few small points and single row of 7 lanceolate spinules, tergite III with posteriorly situated shagreen of 13−14 small points and single row of 9 lanceolate spines, tergites IV−VIII with numerous small points, tergites IV−V with 8−9, tergites VI−VII with 5−6 lanceolate spines and tergite VIII with 5−6 spinules arranged posteriorly in a single row, tergite IX with only numerous small points. Segment I with 1 L-seta, segment II with 3 L-setae, segments III−VIII with 4 L-setae, some D-setae, and V-setae are lamelliform, 3 lamelliform lateral anal macrosetae on each side, anterior one located at lateral filamentous seta 3, and 2 lamelliform median macrosetae near the apex of genital sac. Anal lobe (Fig. 3.70E) 77 μm long, 107 μm wide with 13−14 filaments in fringe restricted to the posterior half of the lobes; genital sac 55 μm long; G/F 0.71; ALR 1.45.

Distribution. OR: India (Sikkim).

3.48 *CORYNONEURA INEFLIGIATA* FU ET AL., 2009 (FIG. 3.71A−E)

Corynoneura yoshimurai Wang, 2000: 635 (not Tokunaga, 1936); Fu et al., 2009: 13.

Material examined. Holotype male (BDN No. 03375), P.R. China: Liaoning Province, Kuandian County, 40°45′N, 124°46′E, alt. 400−1336 m, light trap, 22.ix.1992, X. Wang.

Diagnostic characters. Similar to *Corynoneura yoshimurai* in coloration and in having antenna with 10 flagellomeres, but can be separated by the weak anal lobe, phallapodeme with small inner projection, and the broad inferior volsella with dented edge.

Male (*n* = 1)

Total length 0.15 mm. Wing length 0.88 mm. Total length/wing length 1.65. Wing length/length of profemur 3.45.

Coloration. Head dark brown. Antenna and palpomere yellow. Thorax brown. Abdominal segments yellow to brown. Legs yellowish brown. Wings light yellow to hyaline.

Head. Antenna (Fig. 3.71B) with 10 flagellomeres; ultimate flagellomere 130 μm long; apex with rosette of short sensilla chaetica; apically expanded, thickening 70 μm long, 30 μm wide; AR 0.43. Clypeus with 6 setae. Tentorium (Fig. 3.71E) 125 μm long, 25 μm wide. Stipes 50 μm long. Palpomere lengths (in μm): 15, 13, 18, 25, 50. Palpomere 2 and 3 ellipsoid, 4 rectangular, 5 long and slender. Palpomere 5/3 ratio 2.86.

Thorax. Damaged.

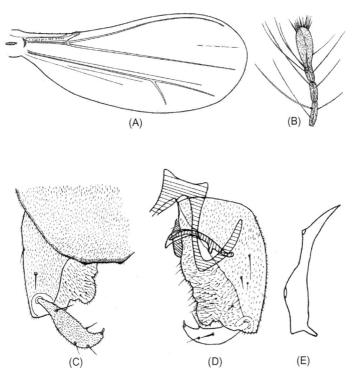

(A)

(B)

(C)   (D)   (E)

Figure 3.71 *Corynoneura inefligiata* Fu et al., 2009, Male imago (From Fu, Y., Sæther, O. A., & Wang, X. (2009). *Corynoneura* Winnertz from East Asia, with a systematic review of the genus (Diptera: Chironomidae: Orthocladiinae). *Zootaxa, 2287,* 1–44). (A) Wing. (B) Antenna. (C) Hypopygium, dorsal view. (D) Hypopygium, ventral view. (E) Tentorium.

Wing (Fig. 3.71A). Anal lobe weak. R_{4+5} clearly separate from M_1 in full length. VR 3.4. C length 220 μm. Cu/wing length 0.6. Wing width/ wing length 0.44. C with 7 setae.

Legs. Spur of fore tibia 27 μm long, spurs of mid tibia 10 μm and 13 μm long, of hind tibia 30 μm long. Width at apex of fore tibia 20 μm, of mid tibia 20 μm, of hind tibia 38 μm. Apex of hind tibia expanded, with comb of 12 setae and one hook-like seta near spur. Lengths and proportions of legs as in Table 3.37.

Hypopygium (Fig. 3.71C and D). Posterior margin of tergite IX straight. Anal point absent. Inferior volsella large and broad with dented edge. Sternapodeme inverted U-shaped, attachment point with phallapodeme placed caudal and ventrally directed; phallapodeme weakly curved, not extending beyond posterior margin of tergite IX, with small inner

Table 3.37 Lengths (in μm) and proportion of leg segments of male *Corynoneura inefligiata* Fu et al., 2009

	fe	ti	ta$_1$	ta$_2$	ta$_3$	ta$_4$	ta$_5$	LR	BV	SV	BR
p$_1$	255	325	158	93	60	30	35	0.49	3.0	3.7	2.0
p$_2$	370	340	195	93	50	23	35	0.57	4.5	3.6	1.6
p$_3$	300	360	193	110	45	25	35	0.54	3.8	3.4	1.5

projection. Gonocoxite 70 μm long. Gonostylus slightly curved apically, 35 μm long; megaseta 3 μm long. HR 2.0, HV 4.1.

Distribution. PA: China (Liaoning).

3.49 *CORYNONEURA ISIGAHEIA* SASA & SUZUKI, 2000

Corynoneura isigaheia Sasa & Suzuki, 2000a: 10.

Material examined. Japan: Mount Omotodake, sweep net, male holotype (No. 385: 32), 1. vii.1999, M. Sasa.

Diagnostic characters. The species can be separated from other members of the genus by antenna with 6 flagellomeres, AR 0.26, wing width/wing length 0.73, sternapodeme inverted V-shaped.

Male (Description from Sasa & Suzuki, 2000a)

Total length 0.82 mm. Wing length 0.40 mm.

Coloration. Scutal stripes and postnotum brown, other scutal potions and scutellum pale, abdominal tergites I−IV pale, V to hypopygium brown, leg yellow.

Head. Eyes bare, antenna with 6 flagellomeres, ultimate flagellomere very short and oval, AR 0.26, last segment with 16 long apical setae. Palp very short, segments II−IV very short and oval.

Thorax. Prealars 2. Scutellum with 2 setae.

Wing. Wing width/wing length 0.73.

Legs. Tip of fore tibia with a long terminal spur, tip of mid tibia with 2 terminal spurs, tip of hind tibia strongly expanded and with a long spur, and a comb composed of 14 long spines, short spur absent. LR$_1$ 0.49, LR$_2$ 0.61, LR$_3$ 0.48; BR$_1$ 2.3, BR$_2$ 1.8, BR$_3$ 2.2.

Hypopygium. Inferior volsella undeveloped, placed caudally. Phallapodeme strongly curved, sternapodeme inverted V-shaped, small attachment point with phallapodeme placed in caudal third of lateral sternapodeme and directed caudally; phallapodeme well developed, strongly curved with projection for joint with sternapodeme placed prelateral. Gonostylus apical curved.

Distribution. PA: Japan.

3.50 *CORYNONEURA KADALINKA* MAKARCHENKO & MAKARCHENKO, 2010 (FIG. 3.72A AND B)

Corynoneura kadalinka Makarchenko & Makarchenko, 2010: 360.

Diagnostic characters. Male: Antenna with 10 flagellomeres; apical 1/4 of terminal flagellomere with sensitive hairs, but apex is bare. Hind tibia with slightly curve seta in apex; $a/d = 2,0$; $b/d = 2,0$. Inferior volsella is absent. Superior volsella like large spine with sharp apex. Sternapodeme inverted V-shaped. Phallapodemes are high and connect with sternapodeme by lock.

Male (Description from Makarchenko & Makarchenko, 2010)

Total length 1.3 mm. Total length/wing length 1.86.

Head. Antenna (Fig. 3.72A) with 10 flagellomeres, AR 0.77. Clypeus with 9 setae. Lengths of palpomeres (in μm): —, 12, 16, 24, 44.

Thorax. Pronotum dark yellow. Dorsocentrals 5, prealars 2. Scutellum with 2 setae.

Wing. Wing length 0.70 mm.

Legs. Fore trochanter with dorsal keel. Spurs of fore tibia 24 μm long, of mid tibia 8 μm long, of hind tibia 20 μm long. Hind tibia expanded, with comb of 17−18 setae. Lengths and proportions of legs as in Table 3.38.

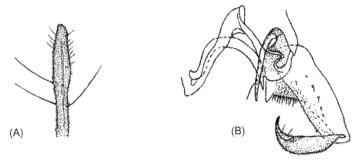

(A)   (B)

Figure 3.72 *Corynoneura kadalinka* Makarchenko & Makarchenko, 2010, Male imago (From Makarchenko & Makarchenko, 2010). (A) Distal part of terminal antennal flagellomere. (B) Total view of hypopygium, from above (tergite IX deleted).

Table 3.38 Lengths (in μm) and proportions of legs segments of male *Corynoneura kadalinka* Makarchenko & Makarchenko, 2010

	fe	ti	ta$_1$	ta$_2$	ta$_3$	ta$_4$	ta$_5$	LR	BV	SV	BR
p$_1$	280	288	164	92	52	20	36	0.57	3.50	3.27	1.43
p$_2$	340	320	176	84	44	20	36	0.55	4.54	3.75	2.3
p$_3$	280	296	172	100	40	20	36	0.58	3.82	3.35	1.5

Hypopygium (Fig. 3.72B). The posterior margin of tergite IX medially concave, with 4−6 short setae. Laterosternite IX with 2 short seta. Gonocoxite 100 μm, superior volsella triangular, with many bristles. Inferior volsella absent. Sternapodeme curved into V-shaped. Gonostylus narrow, slightly curved, 60 μm long.

Distribution. PA: Russia (Far East).

3.51 *CORYNONEURA KEDROVAYA* MAKARCHENKO & MAKARCHENKO, 2006B (FIG. 3.73A AND B)

Corynoneura kedrovaya Makarchenko & Makarchenko, 2006b: 152.

Diagnostic characters. Total length 0.7−0.9 mm. Antenna with 8−9 flagellomeres. AR 0.42−0.43. Gonocoxite with short rectangular or slightly roundish inferior volsella. Transverse sternapodeme is situated upward, with saddle-shaped apex. Phallapodemes is also upward and curved into a sickle, with slender apices, in basal half with short inner projection.

Male (Description from Makarchenko & Makarchenko, 2006b)

Total length 0.7−0.9 mm. Total length/wing length 1.25−1.57.

Coloration. Thorax and abdomen brown.

Head. Eyes bare. Antenna with 8 flagellomeres, AR 0.42−0.43. Lengths of palpomeres (in μm): —, 12, 16, 20, 44−48.

Thorax. Dorsocentrals 4, prealars 2.

Wing. Wing length 0.54−0.59 mm.

Legs. Fore trochanter with dorsal keel. Spurs of fore tibia 18 μm long, of mid tibia 12 μm long, of hind tibia 28 μm and 18 μm long. Hind tibia with comb of 15 setae. Lengths and proportions of legs as in Table 3.39.

Hypopygium (Fig. 3.73A and B). Tergite IX with 4−6 long setae. Inferior volsella small rectangular, located in caudal of gonocoxite.

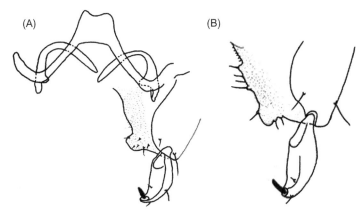

(A)   (B)

Figure 3.73 *Corynoneura kedrovaya* Makarchenko & Makarchenko, 2006b, Male imago (From Makarchenko & Makarchenko, 2006b). (A) Part of hypopygium (tergite IX removed). (B) Part of gonocoxite and gonostylus.

Table 3.39 Lengths (in μm) and proportions of legs segments of male *Corynoneura kedrovaya* Makarchenko & Makarchenko, 2006b

	fe	ti	ta$_1$	ta$_2$	ta$_3$	ta$_4$
p$_1$	192−204	236−248	116−128	64	40−44	20
p$_2$	260−280	236−240	140−144	60−64	32−36	16
p$_3$	204−220	212−226	120−128	64	24	16

	ta$_5$	LR	BV	SV	BR	
p$_1$	26	0.49−0.52	3.58−3.72	3.53−3.63	2.0	
p$_2$	24−28	0.59−0.60	4.68−4.74	3.54−3.61	2.5	
p$_3$	26−28	0.57	4.06−4.42	3.47−3.48	1.7−2.0	

Gonostylus apically curved, narrowed and elongated, transverse sternapodeme curved into U-shaped.

Distribution. NE: Canada (Northwest Territories). **PA:** Russia (Far East).

3.52 *CORYNONEURA KIBUNELATA* SASA, 1989 (FIGS. 3.74A−C AND 3.75A−I)

Corynoneura kibunelata Sasa, 1989: 61; Yamamoto, 2004: 18; Fu et al., 2009: 15; Makarchenko & Makarchenko, 2010: 362.

Corynoneura tenuistyla Tokunaga, 1936; Makarchenko & Makarchenko, 2006a: 491, Makarchenko & Makarchenko, 2006b: 157.

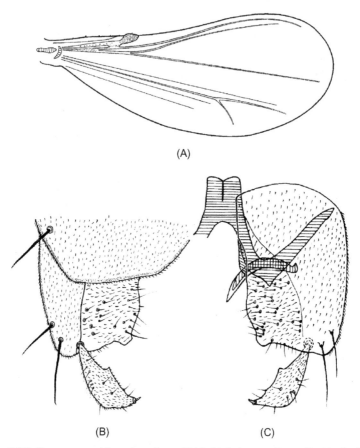

(A)

(B)   (C)

Figure 3.74 *Corynoneura kibunelata* Sasa, 1989; Male imago (From Fu, Y., Sæther, O. A., & Wang, X. (2009). *Corynoneura* Winnertz from East Asia, with a systematic review of the genus (Diptera: Chironomidae: Orthocladiinae). *Zootaxa, 2287*, 1–44). (A) Wing. (B) Hypopygium, dorsal view. (C) Hypopygium, ventral view.

Material examined. Japan: Honshu, Kyoto Prefecture, Kibune River, sweep net, male holotype (No. 166: 61), 11.x.1988, M. Sasa.

Diagnostic characters. The species can be separated from other members of the genus by having an antenna with 10 flagellomeres; AR 0.31; superior volsella (not clear in the available material) probably fused with the gonocoxite; inferior volsella broad and with a small prominent part; sternapodeme inverted U-shaped; gonostylus medially expanded.

Additional description and corrections. The holotype had 10 flagellomeres and an AR of 0.31, while the original description by Sasa (1989) mentions antenna with 9 flagellomeres and an AR of 0.43–0.56.

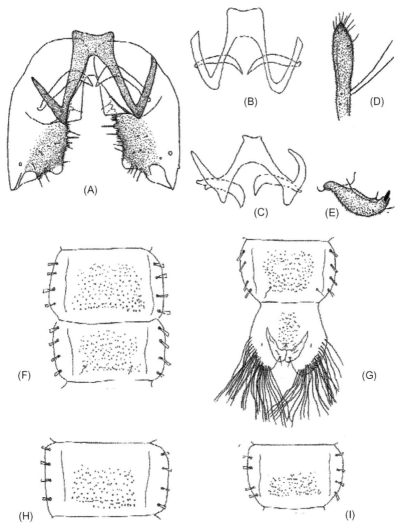

Figure 3.75 *Corynoneura kibunelata* Sasa, 1989; Male imago (From Makarchenko, E. A., & Makarchenko, M. A. (2010). New data on the fauna and taxonomy of *Corynoneura* Winnertz (Diptera, Chironomidae, Orthocladiinae) for the Russian Far East and bordering territories. *Euroasian Entomological Journal, 9*(3), 353−370 + II). (A) Total view of hypopygium, from above (tergite IX deleted). (B and C) Sternapodema and phallapodemes. (D) Distal part of terminal antennal flagellomere. (E) Gonostylus. (F−I) Pupa imago. F. Tergites VI−VII. (G) Tergite VIII and anal segment. (H) Sternite VII. (I) Sternite VIII.

Table 3.40 Lengths (in μm) and proportions of legs segments of male *Corynoneura kibunelata* Sasa, 1989 (based on holotype)

	fe	ti	ta$_1$	ta$_2$	ta$_3$	ta$_4$	ta$_5$	LR	BV	SV	BR
p$_1$	240	280	150	78	50	23	30	0.54	3.1	2.7	2.3
p$_2$	345	280	175	103	43	23	33	0.63	4.0	3.6	2.3
p$_3$	260	275	163	88	38	23	33	0.59	3.9	3.3	2.1

Table 3.41 Lengths (in μm) and proportions of legs segments of male *Corynoneura kibunelata* Sasa, 1989

	fe	ti	ta$_1$	ta$_2$	ta$_3$	ta$_4$
p$_1$	216−260	268−320	128−152	64−88	44−52	20−24
p$_2$	340−360	312−328	188−196	88−92	48	20−24
p$_3$	292	300−304	176−188	104	40−44	20

	ta$_5$	LR	BV	SV	BR
p$_1$	28−36	0.47−0.48	3.71−3.87	3.78−3.94	1.8−2.3
p$_2$	32	0.60	4.37−4.60	3.47−3.51	3.3
p$_3$	32−36	0.59−0.62	3.84−3.92	3.17−3.36	3.0

Source: From Makarchenko, E. A., & Makarchenko, M. A. (2010). New data on the fauna and taxonomy of *Corynoneura* Winnertz (Diptera, Chironomidae, Orthocladiinae) for the Russian Far East and bordering territories. *Euroasian Entomological Journal, 9*(3), 353−370 + II.

Some additional characters should be added: Tentorium 125 μm long, 23 μm wide. Lengths and proportions of legs as in Tables 3.40 and 3.41. Posterior margin of tergite IX almost straight. Sternapodeme inverted U-shaped, coxapodeme 38 μm long, small attachment point with phallapodeme placed caudal and caudally directed; phallapodeme short, weakly curved, not extending beyond margin of tergite IX, 35 μm long. Gonocoxite 68 μm long, with 2 long setae apically. Gonostylus 35 μm long; megaseta 5 μm long. HR 1.9, HV 3.5. The wing and the hypopygium of the holotype are redrawn (Fig. 3.74A−C). Male and pupa imago base on species from Russian Far East (Fig. 3.75A−I).

Distribution. PA: Japan, Russia (Far East).

3.53 *CORYNONEURA KIBUNESPINOSA* SASA, 1989 (FIG. 3.76A−C)

Corynoneura kibunespinosa Sasa, 1989: 60; Sasa & Suzuki, 2000b: 98; Sasa et al., 1998: 121; Fu et al., 2009: 16.

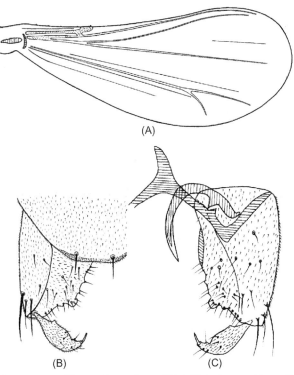

(A)

(B)   (C)

Figure 3.76 *Corynoneura kibunespinosa* Sasa, 1989; Male imago (From Fu, Y., Sæther, O. A., & Wang, X. (2009). *Corynoneura* Winnertz from East Asia, with a systematic review of the genus (Diptera: Chironomidae: Orthocladiinae). *Zootaxa, 2287*, 1—44). (A) Wing. (B) Hypopygium, dorsal view. (C) Hypopygium, ventral view.

Material examined. Japan: Honshu, Kyoto Prefecture, Kibune River, sweep net, male holotype (No. 166: 50B), 11.x.1988, M. Sasa.

Diagnostic characters. The species can be separated from other East Asian *Corynoneura* by having an antenna with 10 flagellomeres, AR 0.27; posterior margin of tergite IX slightly bilobed, sternapodeme inverted V-shaped, and inferior volsella digitiform.

Additional description and corrections. AR 0.27 while the original description by Sasa (1989) mentions AR 0.41. Some additional characters should be added: tentorium 113 μm long, 15 μm wide. Lengths and proportions of legs as in Table 3.42. Posterior margin of tergite IX slightly bilobed. Sternapodeme inverted V-shaped; coxapodeme 38 μm long; attachment point with phallapodeme placed in caudal third of lateral sternapodeme and directed caudally; phallapodeme strongly curved with projection for joint with sternapodeme placed prelateral, extending

Table 3.42 Lengths (in μm) and proportions of legs segments of male *Corynoneura kibunespinosa* Sasa, 1989

	fe	ti	ta$_1$	ta$_2$	ta$_3$	ta$_4$	ta$_5$	LR	BV	SV	BR
p$_1$	210	250	140	70	45	23	30	0.56	3.6	3.3	2.2
p$_2$	300	265	155	65	35	20	28	0.58	4.9	3.6	2.1
p$_3$	220	235	130	73	30	18	33	0.55	3.8	3.5	2.0

beyond margin of tergite IX, 88 μm long. Gonocoxite 65 μm long with 3 long setae apically. Gonostylus bent, 30 μm long; megaseta 8 μm long. HR 2.2, HV 3.4. The wing and the hypopygium of the holotype have been redrawn (Fig. 3.76A−C).

Distribution. PA: Japan.

3.54 *CORYNONEURA KISOGAWA* SASA & KONDO, 1993 (FIG. 3.77A−C)

Corynoneura kisogawa Sasa & Kondo, 1993: 102; Fu et al., 2009: 19.

Material examined. Japan: Honshu, Aichi, Bisai, sweep net, male holotype (No. 222: 100), 15.xi.1988, M. Sasa.

Diagnostic characters. The species can be separated from other East Asian *Corynoneura* by having antenna with 12 flagellomeres, AR 0.33; phallapodeme straight and erect; sternapodeme inverted U-shaped, and no inferior volsella.

Additional description and corrections. The holotype had 12 flagellomeres with AR 0.33, while the original description by Sasa and Kondo (1993) mentions antenna with 10 flagellomeres and AR 0.46. The total length is 1.12 mm and the wing length 0.83 mm, while original description mentions a total length of 1.20 mm and a wing length of 0.74 mm. The leg ratio of the mid leg is 0.52, of the hind leg 0.54, while the original description mentions 0.64 and 0.46 respectively. Lengths and proportions of legs as in Table 3.43. Some additional characters should be added: tentorium 125 μm long, 15 μm wide. Sternapodeme inverted U-shaped, coxapodeme 20 μm long, attachment point with phallapodeme placed caudal and caudally directed; phallapodeme short and straight, 33 μm long. Gonocoxite 73 μm long. Gonostylus 38 μm long. HR 2.1, HV 3.0. The wing and hypopygium of the holotype are redrawn (Fig. 3.77A−C).

Distribution. PA: Japan.

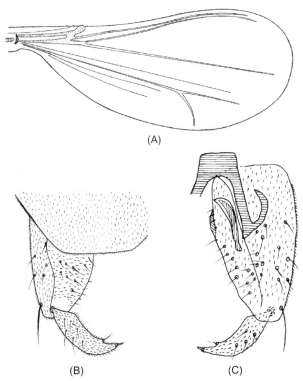

(A)

(B)   (C)

Figure 3.77 *Corynoneura kisogawa* Sasa & Kondo, 1993, Male imago (From Fu, Y., Sæther, O. A., & Wang, X. (2009). *Corynoneura* Winnertz from East Asia, with a systematic review of the genus (Diptera: Chironomidae: Orthocladiinae). *Zootaxa, 2287*, 1–44). (A) Wing. (B) Hypopygium, dorsal view. (C) Hypopygium, ventral view.

Table 3.43 Lengths (in μm) and proportion of leg segments of male *Corynoneura kisogawa* Sasa & Kondo, 1993

	fe	ti	ta₁	ta₂	ta₃	ta₄	ta₅	LR	BV	SV	BR
p_1	240	300	155	88	53	25	35	0.52	3.5	3.5	2.2
p_2	345	300	193	88	45	23	35	0.64	4.4	3.4	2.8
p_3	275	290	158	88	35	20	35	0.54	4.1	3.6	1.9

3.55 *CORYNONEURA KOREMA* FU ET AL., 2009 (FIG. 3.78A−G)

Corynoneura korema Fu et al., 2009: 19.

Material examined. Holotype male (BDN No. 1180), P.R. China: Hainan Province, Bawangling County, 19°5′N, 109°8′E, alt. 350−1438 m, sweep net, 10.v.1988, X. Wang.

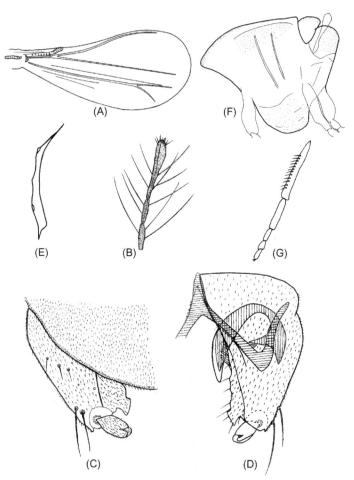

Figure 3.78 *Corynoneura korema* Fu et al., 2009, Male imago (From Fu, Y., Sæther, O. A., & Wang, X. (2009). *Corynoneura* Winnertz from East Asia, with a systematic review of the genus (Diptera: Chironomidae: Orthocladiinae). *Zootaxa*, *2287*, 1–44). (A) Wing. (B) Antenna. (C) Hypopygium, dorsal view. (D) Hypopygium, ventral view. (E) Tentorium. (F) Thorax. (G) Hind tarsomeres 1–5.

Diagnostic characters. Similar to *Corynoneura prima* in having antenna with 7 flagellomeres and AR about 0.5; but can be separated by the thorax with obvious longitudinal vittae, VR 5.0, hind ta_1 with row of stout setae, and sternapodeme inverted V-shaped.

Male (*n* = 1)

Total length 0.74 mm. Wing length 0.46 mm. Total length/wing length 1.6. Wing length/length of profemur 3.0.

Table 3.44 Lengths (in µm) and proportion of leg segments of male *Corynoneura korema* Fu, Sæther *et* Wang, 2009

	fe	ti	ta$_1$	ta$_2$	ta$_3$	ta$_4$	ta$_5$	LR	BV	SV	BR
p$_1$	153	170	88	53	28	13	20	0.52	3.6	3.7	—
p$_2$	205	185	95	45	25	15	23	0.51	3.6	4.1	1.3
p$_3$	163	180	88	40	15	13	20	0.49	4.9	3.9	2.0

Coloration. Head dark brown. Antenna and palpomere yellow. Thorax and legs yellowish brown. Tergites yellow to brown. Wing hyaline, with pale yellow clava.

Head. Antenna (Fig. 3.78B) with 7 flagellomeres; ultimate flagellomere 83 µm long, apex with short sensilla chaetica subapically, apically slightly expanded, thickening 43 µm long, AR 0.5. Tentorium (Fig. 3.78E) 98 µm long, 8 µm wide. Stipes 35 µm long. Palpomeres length (in µm): 8, 10, 13, 15, 35. Palpomere 2, 3 and 4 ellipsoid, 5 long and slender. Palpomere 5/3 ratio 2.8.

Thorax (Fig. 3.78F). With two obvious longitudinal vittae.

Wing (Fig. 3.78A). VR 5.0. C length 100 µm, C/wing length 0.22. Cu length 325 µm. Cu/wing length 0.7. Wing width/wing length 0.46. C with 5 setae.

Legs. Spur of fore tibia 20 µm long, spurs of mid tibia 13 µm long, of hind tibia 23 µm long. Width at apex of fore tibia 13 µm, of mid tibia 13 µm, of hind tibia 23 µm. Apex of hind tibia expanded, with comb of 15 setae and 1 seta near spur strongly S-shaped, hind ta$_1$ (Fig. 3.78G) with row of stout setae. Lengths and proportions of legs as in Table 3.44.

Hypopygium (Fig. 3.78C and D). Posterior margin straight, and with many short setae, laterosternite IX with 1 long seta. Anal point absent. Inferior volsella developed. Sternapodeme inverted V-shaped, coxapodeme 25 µm long, small attachment point with phallapodeme placed in caudal third of lateral sternapodeme and directed caudally; phallapodeme strongly curved and broad with lateral apex bifid and embracing knob on lateral sternapodeme, 50 µm long, base bifurcate. Gonocoxite 63 µm long with 2 setae apically. Gonostylus small and strongly curved, 13 µm long; megaseta 3 µm long. HR 5, HV 5.9.

Distribution. OR: China (Hainan).

3.56 *CORYNONEURA LACUSTRIS* EDWARDS, 1924 (FIG. 3.79A–C)

Corynoneura lacustris Edwards, 1924: 187; Schlee, 1968b: 21.

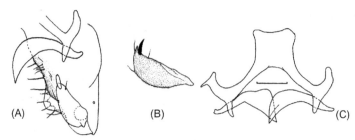

Figure 3.79 *Corynoneura lacustris* Edwards, 1924; Male imago (From Makarchenko, E. A., & Makarchenko, M. A. (2006b). Chironomids of the genera *Corynoneura* Winnertz, 1846 and *Thienemanniella* Kieffer, 1919 (Diptera, Chironomidae, Orthocladiinae) of the Russian Far East. *Euroasian Entomological Journal*, *5*(2), 151–162). (A) Part of gonocoxite and gonostylus. (B) Gonostylus. (C) Part of hypopygium (tergite IX removed).

Diagnostic characters. Male antenna with 12–13 flagellomeres; AR for 13 flagellomeres 0.32–0.42, for 12 flagellomeres 0.44–0.55. Inferior volsella rectangular.

Male

Head. Antenna with 13 flagellomeres, of which show the last two flagellomeres significant fused.

Wing. Wing length 0.9–1.2 mm. Cu/wing length 0.60–0.67. Costa length/wing length 0.26–0.30, wing width/wing length 0.37–0.41.

Hypopygium (Fig. 3.79A–C). Basal joint. The posterior margin of tergite IX medially concave, each carrying 2–3 setae. Inferior volsella rectangular, in distal half of gonocoxite. Gonostylus slightly curved.

Distribution. NE: Canada (Manitoba, Northwest Territories, Ontario), USA (South Carolina). **PA:** Austria, Balearic Islands, Corsica, Croatia, Cyprus, Czech Republic, Denmark, Estonia, Faroe Islands, Finland, France, Germany, Great Britain, Greece, Ireland, Italy, Morocco, Netherlands, Norway, Portugal, Romania, Russia (CET, NET, Far East), Spain, Sweden, Switzerland, Turkey.

3.57 *CORYNONEURA LAHULI* SINGH & MAHESHWARI, 1987 (FIG. 3.80A–G)

Corynoneura lahuli Singh & Maheshwari, 1987: 14.

Diagnostic characters. Male antenna with 7 flagellomeres. AR 1.32. This species comes near to *Corynoneura chandertali*, but differs in having

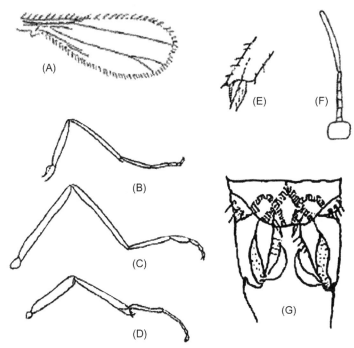

Figure 3.80 *Corynoneura lahuli* Singh & Maheshwari, 1987, Male imago (From Singh, S. & Maheshwari, G. (1987). Chironomidae (Diptera) of Chandertal Lake, Lahul Valley (Northwest Himalaya). *Annals of Entomology*, 5(2), 11–20). (A) Wing. (B–D) Fore, mid, and hind leg. (E) Hind tibial apex. (F) Antenna. (G) Hypopygium.

gonapophysis, and in differences in antennal ratio, leg ratio and in palpal proportion; ratio of gonocoxa and gonostylus being 1.5.

Male (Description from Singh & Maheshwari, 1987)

Total length 1.43–1.51, 1.47 mm. Wing length 1.32 mm.

Coloration. Antenna brown, palpi light brown. Antepronotal lobe light brown; scutum, scutellum and postnotum dark brown. Abdomen brown.

Head. Ratio of maximum head width and width between eyes 1.53. Antenna (Fig. 3.80F) with 7 flagellomeres, AR 1.32. Lengths of palpomeres (in μm): 27, 67, 100, 108.

Wing (Fig. 3.80A). Costa with 9 setae, R_{4+5} fused with costa and forms clavus. Costa ends about one-third of the length of wing.

Legs (Fig. 3.80B–D). Fore trochanter with dorsal keel, apex of fore tibia with single spur, fourth tarsomere cordiform and smaller than fifth. Middle tibial apex without spur. Apex of hind tibia enlarged and with a

Table 3.45 Lengths (in μm) and proportions of legs segments of male *Corynoneura lahuli* Singh & Maheshwari, 1987

	fe	ti	ta$_1$	ta$_2$	ta$_3$	ta$_4$	ta$_5$	LR
p$_1$	280	330	170	?	60	20	40	0.51
p$_2$	370	340	180	?	30	20	40	0.50
p$_3$	370	390	170	100	30	20	41	0.40

Note: ? = unknown.

comb, size of tibial enlargement 0.29 mm. Lengths and proportions of legs as in Table 3.45.

Hypopygium (Fig. 3.80G). Gonostylus with apical tooth. Ratio of gonocoxa and gonostylus 1.50.

Distribution. OR: India (Himachal Pradesh).

3.58 *CORYNONEURA LATUSATRA* FU ET AL., 2009 (FIG. 3.81A−E)

Corynoneura latusatra Fu et al., 2009: 21.

Material examined. Holotype male (BDN No. 22248), P.R. China: Guizhou Province, Daozhen County, Dashahe, 29°00′N−29°13′N, 107° 22′E−107°48′E, alt. 1300−1900 m, sweep net, 25.v.2004, H. Tang. Paratypes: 8 males, Guizhou Province, Daozhen County, Dashahe, 29° 00′N−29°13′N, 107°21′E−107°47′E, alt. 1300−1900 m, sweep net, 25. v.2004, H. Tang.

Diagnostic characters. Similar to *Corynoneura secunda* in having sternapodeme inverted strongly V-shaped, and phallapodeme well developed and strongly curved, but the two species can easily be separated as *Corynoneura latusatra* has 7 flagellomeres and an AR of 0.31.

Male (*n* = 9)

Total length 0.78−0.91, 0.85 mm. Wing length 0.55−0.60, 0.51 mm. Total length/wing length 1.2−1.59, 1.42. Wing length/length of profemur 3.14−3.43, 3.30.

Coloration. Head yellowish brown with brown frontal vertex. Antenna pale yellow with black apex, palpomere pale yellow. Thorax pale brown to hyaline with vittae, scutum, scutellum and postnotum brown, anterior anepisternum II, preepisternum, epimeron II light brown. Tergites I−IV yellow, tergites V−IX brown. Legs yellow. Wing yellowish to hyaline.

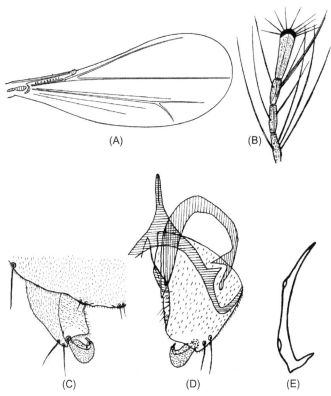

Figure 3.81 *Corynoneura latusatra* Fu et al., 2009, Male imago (From Fu, Y., Sæther, O. A., & Wang, X. (2009). *Corynoneura* Winnertz from East Asia, with a systematic review of the genus (Diptera: Chironomidae: Orthocladiinae). *Zootaxa*, *2287*, 1−44). (A) Wing. (B) Antenna. (C) Hypopygium, dorsal view. (D) Hypopygium, ventral view. (E) Tentorium.

Head. Antenna (Fig. 3.81B) with 7 flagellomeres; ultimate flagellomere 53−60, 57 μm long, antenna with rosette of apical sensilla chaetica, apically expanded, thickening 50−56, 53 μm long, with maximum width 18−25, 20 μm; AR 0.28−0.33, 0.31 (7). Clypeus with 8−12, 9 (6) setae. Tentorium (Fig. 3.81E) 93−105, 97 μm long. Stipes 40−50, 45 μm long. Palpomeres length (in μm): 10−15, 12; 10−15, 13; 13−18, 14; 13−20, 15; 28−38, 35. Palpomere 2, 3 and 4 ellipsoid, palpomere 5 rectangular. Palpomere 5/3 ratio 1.7−3, 2.5.

Thorax. Dorsocentrals 1−5, 3 (5), prealars 2−5, 3. Scutellum with 1−2 setae.

Wing (Fig. 3.81A). VR 2.9−3.6, 3.2. C length 120−150, 140 μm, C/wing length 0.22−0.25, 0.24. Cu length 295−330, 321 μm. Cu/wing

Table 3.46 Lengths (in μm) and proportion of leg segments of male *Corynoneura latusatra* Fu et al., 2009

	fe	ti	ta_1	ta_2	ta_3	ta_4
p_1	165–185, 176	213–235, 226	95–105, 101	45–58, 51	20–35, 22	18–23, 20
p_2	223–245, 234	203–233, 218	108–120, 115	50–58, 54	25–30, 28	15–20, 18
p_3	175–200, 192	155–225, 208	88–103, 98	48–58, 54	20–25, 23	13–18, 14

	ta_5	LR	BV	SV	BR
p_1	25–30, 28	0.43–0.47, 0.45	3.6–4.5, 3.9	3.8–4.2, 3.9	1.3–1.8, 1.5
p_2	25–28, 26	0.52–0.55, 0.53	4.4–4.7, 4.5	3.8–4.1, 4.0	1.2–1.5, 1.3
p_3	23–30, 26	0.43–0.58, 0.48	3.7–4.6, 4.1	3.7–4.3, 4.0	0.9–1.2, 1.0

length 0.54−0.58, 0.56. Wing width/wing length 0.4−0.43, 0.42. C with 3 setae.

Legs. Spur of fore tibia 13−20, 16 μm and 9−15, 11 (5) μm long, spurs of mid tibia 6−10, 8 (7) μm and 5−8, 6 (3) μm long, of hind tibia 25−30, 27 μm and 23 (2) μm long. Width at apex of fore tibia 13−20, 15 μm, of mid tibia 10−18, 13 μm, of hind tibia 28−33, 30 μm. Apex of hind tibia expanded, with comb of 14−17, 15 setae and one hook-like seta near spur. Lengths and proportions of legs as in Table 3.46.

Hypopygium (Fig. 3.81C and D). Tergite IX large, covering almost the whole gonocoxite, with weakly bilobed posterior margin, and many short setae. Laterosternite IX with 1 long seta. Superior volsella fused anteriomedially. Inferior volsella absent. Sternapodeme inverted strongly V-shaped, apically acute, 63−68, 65 (6) μm long; coxapodeme 13−20, 15 (6) μm long; attachment point with phallapodeme placed in caudal third of lateral sternapodeme and directed caudally; phallapodeme well developed, strongly curved with projection for joint with sternapodeme placed prelateral, not extending beyond tergite IX, 85−90, 86 μm long, 8−13, 10 (6) μm wide. Gonocoxite 45−55, 50 μm long, with 2 setae apically. Gonostylus strongly curved, 18−25, 20 μm long; megaseta 5−8, 6 μm long. HR 2−2.9, 2.3, HV 3.4−4.4, 3.9 (6).

Distribution. OR: China (Guizhou).

3.59 *CORYNONEURA LOBATA* EDWARDS, 1924 (FIGS. 3.82A−C AND 3.83A−E)

Corynoneura lobata Edwards, 1924: 186; Cranston & Oliver, 1988: 430; Sasa, 1988: 56; Sasa & Suzuki, 1998: 17; Sasa & Suzuki, 1999: 106; Schlee, 1968b: 43; Tokunaga, 1936: 42; Fu et al., 2009: 23; Fu & Sæther, 2012: 25.

Corynoneura oxfordana Boesel & Winner, 1980: 505.

Corynoneura sp. C Epler, 2001: 7.46.

Corynoneura sp. G Epler, 2001: 7.47.

Material examined. China: Liaoning Province, Kuandian County, 40°45′N, 124°46′E, alt. 400−1336 m, light trap, 1 male (BDN No. 07197), 22.iv.1992, X. Wang; Ningxia Hui Autonomous Region, Liupanshan County, 35°14′N−39°14′N, 104°17′E−109°39′E, alt. 1100−1200 m, light trap, 1 male (BDN No. 05345), 7.viii.1987, X. Wang; Sichuan Province, Daocheng County, Sangdui, 29°11′N, 100° 06′E, alt. 3943 m, light trap, 1 male (BDN No. 13410), 11.vi.1996, X.

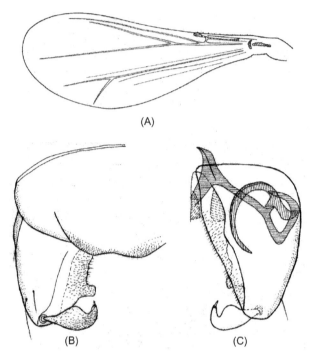

(A)

(B)   (C)

Figure 3.82 *Corynoneura lobata* Edwards, 1924; Male imago. (A) Wing. (B) Hypopygium, dorsal view. (C) Hypopygium, ventral view.

Wang. USA: Minnesota, Miller Creek, 1 mile west of Lake City, 3 males, 1.x.1972 and 25.iii.1973, P. L. Hudson; Ohio, Oxford, 1 male, holotype of *Corynoneura oxfordana* (No. YPM ENTO 500000), 8.xi.1978, R. W. Winner; 1 female, allotype (No. YPM ENTO 500001) and paratypes (8 males and 7 females) of *C. oxfordana*, as holotype except for 7.xi.1978 and 3.xi.1978; Ohio, Highland County, Rocky Fork, 1 male, 1 pupal and 1 larval exuviae, 30.ix.1986, M. J. Bolton; Ohio, Delaware County, Olentangy River upstream of Hyatts Rd., 2 females, 2 pupal and 2 larval exuviae, 31.vii.1986, M. J. Bolton; Ohio, Summit County, Sand Run Park, spring–stream, 2 males, 1 female, 3 pupal and 3 larval exuviae, 12. iv.1987, M. J. Bolton. CANADA: Alberta, Bigoray River, 2 males, 21. v.1973, H. Boerger; Manitoba, Lake Winnipeg, Victoria Beach, light trap, 1 male, 9.vii.1969, P. S. S. Chang; Grand Rapids Government Wharf, light trap, 1 male, 28.vii.1969, P. S. S. Chang; McBeth Harbour, light trap, 1 male, 7.ix.1969, P. S. S. Chang; Beaver Point, light trap, 1 male, 25.vi.1971, E. Johnson & S. Flam; Calder's Dock, light trap, 2 males, 28. vii. & 31.viii.1971, Freshwater Institute L. Winnipeg project; Calder's

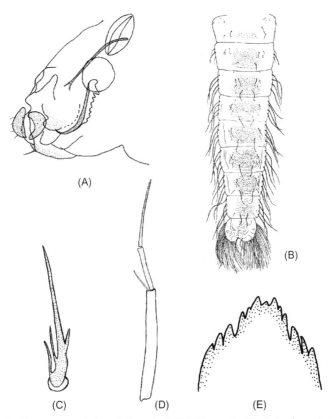

Figure 3.83 *Corynoneura lobata* Edwards, 1924 (From Fu, Y., & Sæther, O. A. (2012). *Corynoneura* Winnertz and *Thienemanniella* Kieffer from the Nearctic region (Diptera: Chironomidae: Orthocladiinae). *Zootaxa, 3536,* 1−61. www.mapress.com/j/zt). (A) Female imago: genitalia. (B) Pupa: tergites I − IX. (C−E) Larva imago. (C) Subbasal seta of posterior parapod. (D) Antenna. (E) Mentum.

Dock, rearing samples, 12 males, 4 females, 30.vi.-17.viii.1971, Freshwater Institute L. Winnipeg project (ZMBN, MJB, JHE).

Diagnostic characters. Total length 1.34−1.65, 1.50 (2) mm. Wing length 0.85−1.05, 0.95 mm. Antenna with 10 flagellomeres, apex with short sensilla chaetica subapically, apically acute. AR 0.35−0.43, 0.39. C/wing length 0.25−0.29. Posterior margin of tergite IX weakly bilobed, inferior volsella digitiform, sternapodeme inverted V-shaped, attachment point with phallapodeme placed in caudal third of lateral sternapodeme and directed caudally; phallapodeme strongly curved with projection for joint with sternapodeme placed prelateral. Gonostylus curved apically. The female has WL of 0.64−0.72 mm, only 4−6 dorsocentrals,

48−55 μm long anapleural suture, 1 seta on gonocoxite IX, 4 on tergite IX and coxosternapodeme with 7−8 lateral lamellae. The pupa is 1.6−1.8 mm long, has no taeniate L-setae on tergite I and only one on II, and 31−36 taeniae in fringe of anal lobe. The larvae are separable by AR 1.0−1.1, mentum with three median tooth, with the central tooth shorter than the lateral median one; the first lateral teeth smaller than the adjacent teeth; head capsule sculptured by reticulate pattern.

Male (*n* = 3)

Total length 1.34−1.65, 1.5(2) mm. Wing length 0.85−1.05, 0.95 mm. Total length/wing length 1.5−1.57, 1.53(1). Wing length/length of profemur 3.1−3.4, 3.2.

Coloration. Head dark brown. Antenna and palpomere yellow. Thorax brown with yellow anterior anepisternum II, median anepisternum II, posterior anepisternum II. Abdominal segments brown. Legs yellow-brown. Wings light yellow to hyaline, with pale yellow clava.

Head. Eyes bare, reniform. Antenna with 10 flagellomeres; ultimate flagellomere 98−148, 128 μm long, apex with short sensilla chaetica which extends a little way back from tip, apically acuate. AR 0.35−0.43, 0.39. Clypeus with 8 setae. Temporal setae lacking. Tentorium 125−150, 135 μm long, 20−23, 21 μm wide. Stipes: 60−65, 63 μm long. Palpomeres length (in μm): 15−25, 19; 12.5−20, 16; 23−28, 25; 28−38, 32; 55−65, 60(2). Palpomere 2 and 3 ellipsoid, 4 rectangular, 5 longer. Palpomere 5/3 ratio 2.1.

Thorax. Dorsocentrals 3−6, 5 setae. Scutellum with 1−2 seta.

Wing (Fig. 3.82A). VR 3.25(1). C length 230−300, 255 μm, C/wing length 0.25−0.29. Cu length 650 μm. Cu/wing length 0.62(1). Wing width/wing length 0.39−0.43, 0.41. C with 8−10, 9 setae.

Legs. Fore trochanter with keel. Spur of front tibia 18−23, 21 μm and 10−13, 11 μm (2) long, spurs of middle tibia 13−15, 14 μm and 10−13, 11 μm long, of hind tibia 23−33, 28 μm long. Width at apex of fore tibia 23−30, 25 μm, of mid tibia 23−25, 24 μm, of hind tibia 35−50, 42 μm. Tip of hind tibia expanded, with comb of 15−17, 16 setae and 1 seta near spur developed as hook. Lengths and proportions of legs as in Table 3.47.

Hypopygium (Figs. 3.82B). Posterior margin weakly bilobed, and with many short setae, laterosternite IX with 1 long seta. Anal point absent. Inferior volsellae digitiform. Sternapodeme inverted V-shaped, coxapodeme 43−48, 46 μm long. Phallapodeme strongly curved not extending beyond tergite IX, 30−40, 33 μm long, with basal projection.

Gonocoxite 60−83, 72 μm long with 2 setae apically. Gonostylus curved apically, 28−40, 32 μm long, median broaden. Megaseta 5−6, 5.5 μm long. HR 2.5−3.6, 2.6, HV 2.9−5.1, 3.9.

Female ($n = 2-3$)

Total length 0.95−1.17 mm. Wing length 0.64−0.72 mm. Total length/wing length 1.47−1.63. Wing length/profemur length 3.62.

Head. AR 0.32−0.36. Flagellomeres length (in μm): 33−36, 28−32, 26−32, 24−36, 37−45. Temporals absent. Clypeus with 8 setae. Tentorium 56−74 μm long. Stipes 40−64 μm long. Palp lengths (in μm): 10, 13−14, 18−22, 24−28, 44−51.

Thorax. Antepronotum with 1 seta. Dorsocentrals 4−6, prealars 2. Scutellum with 2 setae. Anapleural suture 48−55 μm long.

Wing. VR 2.50−2.83, C 282−337 μm long, C/wing length 0.44−0.47, Cu/wing length 0.57−0.59. Brachiolum with 1 seta.

Legs. Fore trochanter with distinct dorsal keel. Spur of fore tibia 12−14 μm, spurs of mid tibia 14 μm and 6−9 μm, of hind tibia 28−35 μm and 7−10 μm. Width at apex of fore tibia 18 μm, of mid tibia 16−20 μm, of hind tibia (*a*) 29−38 μm. Width of hind tibia 1/3 from apex (*d*) 19−20 μm, elongation length (*b*) 33−40 μm, length of maximum thickening (c_1) 52−60 μm, total length of thickening (c_2) 73−86 μm, a/d 1.53−1.90, b/d 1.65−2.11, c_1/d 2.70−3.16, c_2/d 3.80−4.53. Hind tibia expanded, with comb of 14−15 setae, shortest seta 14−18 μm, longest seta 20−24 μm. Sensilla chaetica absent. Lengths of leg segments and their proportions as in Table 3.48.

Genitalia (Fig. 3.83A). Cercus 22−36 μm long. Gonocoxite IX with 1 seta. Tergite IX with 4 setae. Notum 44−60 μm long. Coxosternapodeme with 7−8 lateral lamellae. Seminal capsule 42−50 μm long, 33−40 μm wide.

Pupa ($n = 3-4$)

Total length 1.64−1.79, 1.72 mm (4). Exuviae pale, transparent with pale grayish yellow thorax.

Cephalothorax. Third precorneal seta (PcS_3) 28−34 μm (3) long. PcS_{1-3} in a line; PcS_1 4−6, 5 μm (4) from PcS_2; PcS_2 3−6, 4 μm (4) from PcS_3. Frontal setae taeniate, 36−48 μm (3) long. Postorbitals 2, taeniate, 52−80, 66 μm (4) long. Second dorsocentral (Dc_2) 10−24 μm (2) long, Dc_3 50−52 μm (2) long, Dc_4 26−30 μm (3) long. Dc_1 located 8−10 μm (3) from Dc_2, Dc_2 located 56−98 μm (3) from Dc_3, Dc_3 located 14 (1) μm from Dc_4.

Table 3.47 Lengths (in μm) and proportions of legs segments of male *Corynoneura lobata* Edwards, 1924 (based on Chinese specimens)

	fe	ti	ta$_1$	ta$_2$	ta$_3$	ta$_4$
p$_1$	275–310, 300	335–375, 355	173–195, 185	100–113, 108	63–70, 65	25–35, 30
p$_2$	400–435, 413	360–395, 373	225–233, 229	95–113, 102	53–63, 56	25–30, 27
p$_3$	320–335, 326	345–380, 360	205–238, 216	108–125, 112	43–55, 47	25–33, 28

	ta$_5$	LR	BV	SV	BR
p$_1$	38–43, 40	0.50–0.53, 0.52	3.4–3.6, 3.5	3.4–3.6, 3.5	1.8–2.2, 2.0
p$_2$	35–40, 38	0.59–0.63, 0.62	4.3–4.6, 4.4	3.3–3.6, 3.4	1.6–2.5, 2.1
p$_3$	33–40, 38	0.57–0.63, 0.59	3.7–4.2, 4.0	3.0–3.3, 3.2	1.7–2.6, 2.2

Table 3.48 Lengths (in μm) and proportions of legs segments of female *Corynoneura lobata* Edwards, 1924 ($n = 2$–3)

	fe	ti	ta$_1$	ta$_2$	ta$_3$	ta$_4$
p$_1$	178–198	200–246	117–134	55–62	26–33	11–12
p$_2$	238–300	232–275	138–166	59–68	32	12–14
p$_3$	220–248	212–266	124–148	68–81	23–30	10–13

	ta$_5$	LR	BV	SV	BR
p$_1$	18–20	0.54–0.59	4.46–4.55	3.13–3.31	2.6–3.2
p$_2$	18–20	0.59–0.60	5.02–5.53	3.41–3.46	2.7–3.4
p$_3$	18–25	0.51–0.58	4.50–4.67	3.47–3.79	1.8–2.1

Abdomen (Fig. 3.83B). Shagreen and chaetotaxy as illustrated. No taeniate L-setae on Ⅰ, only 1 on Ⅱ. PSA and PSB absent. Anal lobe with 31—36, 33 (4) setae in fringe; all taeniate, not hooked apically, about 200 μm long. Median setae taeniate 80—140, 97 μm (4) long.

Larva (*n* = 3—6)

Coloration. Head yellowish. Antenna with basal segment almost transparent, other segments yellowish. Abdomen yellowish.

Head. Capsule length 256—260, 259 μm, width 172—184, 179 μm; sculptured by reticulate pattern. Postmentum 220—228, 223 μm long. Sternite Ⅱ large, Ⅰ and Ⅲ not visible. Premandible 28—32, 30 μm long, brush of premandible not observed. Mentum as in Fig. 3.83E. Mandibles 48—55, 52 μm long. Antenna as in Fig. 3.83D, AR 1.0—1.1. Lengths of flagellomeres Ⅰ—Ⅳ (in μm): 156—178, 165; 50—55, 52; 104—110, 106; 3—6, 4. Basal segment width 14—17 μm; length of blade at apex of basal segment 35—38, 36 μm; ring organ at 76—90, 82 μm from the base of antenna.

Abdomen. Length of anal setae 131—140, 135 μm. Procercus 10—13, 12 μm long, 7—8 μm wide. Subbasal seta of posterior parapods split as in Fig. 3.83C.

Remarks. Edwards (1924) established this species based on specimens from England. Sasa and Suzuki (1998, 1999) recorded the species from Japan. Schlee (1968b) redescribed the male in detail based on material from Germany. Tokunaga (1936) redescribed the male and female based on Japanese specimen.

Distribution. NE: Canada (Alberta, New Brunswick, Northwest Territories, Ontario), USA(Florida, Georgia, Michigan, Mississippi, New Jersey, New York, North Carolina, Ohio, South Carolina, Virginia). **PA:** Algeria, Austria, Balearic Islands, Belgium, China (Liaoning, Ningxia), Corsica, Denmark, Estonia, Faroe Islands, Finland, France, Germany, Great Britain, Greece, Ireland, Italy, Japan, Lebanon, Luxembourg, Macedonia, Morocco, Netherlands, Norway, Poland, Portugal, Romania, Russia (CET, NET, Far East), Slovakia, Spain, Sweden, Switzerland, Turkey. **OR:** China (Sichuan).

3.60 *CORYNONEURA LONGIPENNIS* TOKUNAGA, 1936

Corynoneura longipennis Tokunaga, 1936: 50.

Diagnostic characters. Antenna with 11 flagellomeres, AR 0.89, apex of antenna not swollen apically but not sharply pointed. Inferior volsella small and digitiform. Gonostylus large and distinctly curved, with a basal triangular projecting.

Male (Description according to Tokunaga, 1936)

Total length 1.1 mm.

Coloration. Head dark brown, thorax dark brown; legs pale brown in ground color, with knee–joints black, distal ends of tibiae black, three distal segments of tarsi brown; abdomen dark brown.

Head. Eyes bare. Antenna with 11 flagellomeres; AR 0.89, ultimate flagellomere as long as preceding eight flagellomeres together, apex of antenna not swollen apically but not sharply pointed. Clypeus with about 12 setae.

Wing. Wing length 1.0 mm, very narrow and long, with clavus pale brown.

Hypopygium. Inferior volsella small and digitiform, bare. Gonostylus large and distinctly curved, with a basal triangular projecting.

Distribution. PA: Finland, Japan.

3.61 *CORYNONEURA LONGIANTENNA* WIEDENBRUG ET AL., 2012 (FIGS. 3.84A–E AND 3.85A–G)

Corynoneura longiantenna Wiedenbrug et al., 2012: 37.

Diagnostic characters. The male of *C. longiantenna* can be differentiated from other species with phallapodeme attachment placed in the caudal apex of sternapodeme and apex of hind tibia with a S-shaped seta by the antenna plumose with 10 flagellomeres, superior volsella low only with one long seta, aedeagal lobe triangular basally narrow and long. The pupa cannot be differentiated from *C. hermanni* and *C. humbertoi*, it has rounded anal lobe separated through a straight perpendicular margin wider than half of the lobe width, shagreen on tergites fine and homogeneous. The larva is recognized by the mentum with two median teeth, first lateral teeth adpressed to median teeth. Antenna longer than 720 μm. Second antennal segment longer than the third and about half as long as the first, or shorter.

Male (Description from Wiedenbrug et al., 2012)

Total length 0.83 mm. Wing length 0.54–0.69 mm.

Coloration. Thorax brownish. Abdominal tergites I–IV whitish, V–IX brownish. Legs whitish.

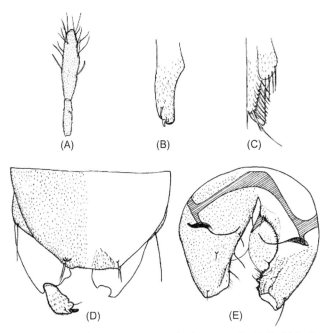

(A)   (B)   (C)

(D)   (E)

Figure 3.84 *Corynoneura longiantenna* Wiedenbrug et al., 2012; Male imago (From Wiedenbrug, S., Lamas, C. J. E., & Trivinho-Strixino, S. (2012). A review of the genus *Corynoneura* Winnertz (Diptera: Chironomidae) from the neotropical region. *Zootaxa, 3574*, 1−61). (A) Terminal flagellomeres. (B and C) Apex of hind tibia. (D and E) Hypopygium. (D) Tergite IX; left is dorsal view and gonostylus, right is ventral view of tergite IX. (E) Tergite IX and gonostylus removed, sclerites hatched; right is dorsal view, left is ventral view.

Head. AR = 0.41−0.47. Antenna with 10 flagellomeres, apical flagellomere 97−130 μm (Fig. 3.84A). Flagellomeres with more than one row of setae each. Eyes pubescent.

Thorax. Antepronotal lobes dorsally tapering.

Wing. Clavus/wing length 0.21−0.22. Anal lobe absent.

Legs. Hind tibial scale 35−45 μm long, with 1 s-seta (Fig. 3.84B and C).

Hypopygium (Fig. 3.84D and E). Tergite IX with 4 setae. Laterosternite IX with 1 seta. Superior volsella small rounded with one long seta. Inferior volsella low and apical. Aedeagal lobe long with narrow base. Transverse sternapodeme rounded, orally straight, phallapodeme caudal attached. Phallapodeme with posterior margin sclerotized, slightly curved.

Pupa (Description from Wiedenbrug et al., 2012)

Total length 1.23−1.66 mm.

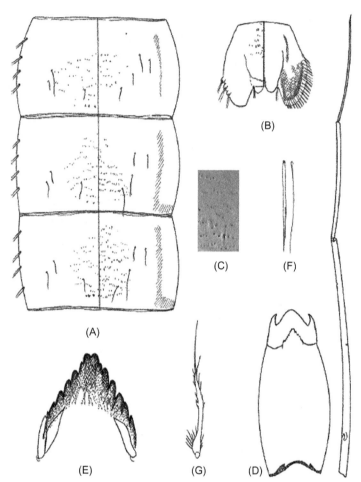

Figure 3.85 *Corynoneura longiantenna* Wiedenbrug et al., 2012; Immatures imago (From Wiedenbrug, S., Lamas, C. J. E., & Trivinho-Strixino, S. (2012). A review of the genus *Corynoneura* Winnertz (Diptera: Chironomidae) from the neotropical region. *Zootaxa, 3574*, 1–61). (A–C) Pupa imago. (A) Segments $III-V$; left is ventral view, right is dorsal view. (B) Tergite IX and anal lobe; left is dorsal view, right is ventral view. (C) Detail of posterior shagreen of tergite IV. (D–G) Larva imago. (D) Head, ventral view and separated antenna. (E) Mentum. (F) Abdominal setae. (G) Subbasal seta of posterior parapod.

Coloration (exuviae). Cephalothorax light gray, abdomen transparent except grayish muscle markings, lateral margin and anal lobe.

Cephalothorax. Frontal apotome slightly rugose. Thorax suture smooth except scutal tubercle region rugose. All Dc-setae thin taeniate. Dc_1 displaced ventrally. Wing sheaths with up to four rows of pearls or apparently absent.

Abdomen (Fig. 3.85A—C). Sternites I , II and tergite I bare, tergite II with few small shagreen points, shagreen on tergites III—IX fine, quite homogeneous. Sternites III—VIII with fine shagreen. Conjunctives sternites III/IV—VII/VIII with small spinules. Segment I with 1 seta, II with 3 L-setae and III—VIII with 4 long taeniate L-setae. Anal lobe rounded (Fig. 3.85B). Anal lobe with fringe not complete, 3 taeniate macrosetae and inner setae taeniate.

Larva (Description from Wiedenbrug et al., 2012)

Head. Postmentum 215—245 μm long. Head capsule integument smooth. Mentum with 2 median teeth, first lateral teeth small and adpressed to median, 5 additional lateral teeth (Fig. 3.85E). Antenna 722—745 μm long, segments two and three slightly darker. First segment longer than postmentum length (Fig. 3.85D).

Abdomen. Ventral setae modified, wider apically split (Fig. 3.85F). Subbasal seta on posterior parapod serrated at the base and both margins (Fig. 3.85G).

Remarks. The larva of *C. longiantenna* was collected with artificial substrate (leaves) at a slow flowing lowland stream and also in a reservoir in São Paulo State, Brazil.

Distribution. NT: Brazil.

3.62 *CORYNONEURA MACDONALDI* FU ET AL., 2009 (FIG. 3.86A—E)

Corynoneura macdonaldi Fu et al., 2009: 23.

Material examined. Holotype male (BDN No. 12635) P.R. China: Sichuan Province, Ganzi State, Yajiang County, 31°38′N, 99°58′E, alt. 2000 m, light trap, 14.vi.1996, X. Wang.

Diagnostic characters. The species is similar to *Corynoneura lobata* in having antenna with 10 flagellomeres and AR about 0.35, but can easily be separated by the subrectangular inferior volsella with round corner; and the inverted M-shaped sternapodeme.

Male (n = 1)

Total length 1.05 mm. Wing length 0.80 mm. Total length/wing length 1.30. Wing length/length of profemur 3.27.

Coloration. Head dark brown. Antenna and palpomere yellow. Thorax brown with yellow anterior anepisternum II . Legs yellowish brown. Tergites I—V yellow, tergites VI—IX brown. Wings yellowish to hyaline, with pale yellow clava.

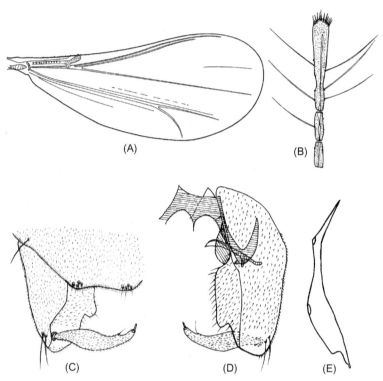

Figure 3.86 *Corynoneura macdonaldi* Fu et al., 2009, Male imago (From Fu, Y., Sæther, O. A., & Wang, X. (2009). *Corynoneura* Winnertz from East Asia, with a systematic review of the genus (Diptera: Chironomidae: Orthocladiinae). *Zootaxa, 2287*, 1—44). (A) Wing. (B) Antenna. (C) Hypopygium, dorsal view. (D) Hypopygium, ventral view. (E) Tentorium.

Head. Antenna (Fig. 3.86B) with 10 flagellomeres; ultimate flagellomere 113 µm long, apex with rosette of apical sensilla chaetica, slightly expanded apically. AR 0.36. Clypeus with 9 setae. Tentorium (Fig. 3.86E) 118 µm long, 18 µm wide. Stipes 45 µm long. Palpomeres length (in µm): 13, 13, 18, 20, 53. Palpomere 2, 3 and 4 ellipsoid, 5 long. Palpomere 5/3 ratio 3.0.

Thorax. Dorsocentrals 5.

Wing (Fig. 3.86A). VR 3.2. C length 190 µm, C/wing length 0.24. Cu length 450 µm. Cu/wing length 0.56. Wing width/wing length 0.43. C with 5 setae.

Legs. Spur of fore tibia 20 µm long, spur of mid tibia 8 µm long, of hind tibia 15 µm long. Width at apex of fore tibia 20 µm, of mid tibia 15 µm, of hind tibia 38 µm. Apex of hind tibia expanded, with comb of

Table 3.49 Lengths (in μm) and proportion of leg segments of male *Corynoneura macdonaldi* Fu et al., 2009

	fe	ti	ta₁	ta₂	ta₃	ta₄	ta₅	LR	BV	SV	BR
p_1	233	290	155	80	50	25	28	0.53	3.7	3.4	1.7
p_2	330	280	193	83	40	20	28	0.69	4.7	3.2	2.2
p_3	265	280	155	80	38	23	30	0.55	4.1	3.5	2.0

16 setae, 1 seta near spur strongly S-shaped. Lengths and proportions of legs as in Table 3.49.

Hypopygium (Fig. 3.86C and D). Tergite IX large, almost covering gonocoxite. Posterior margin of tergite IX slightly bilobed, with 4 setae on each side; laterosternite IX with 1 long seta. Superior volsella broad, anteriomedially fused, partly united with inferior volsella. Inferior volsella subrectangular, with round corners. Sternapodeme inverted M-shaped; coxapodeme 25 μm long, small attachment point with phallapodeme placed caudal and ventrally directed; phallapodeme short, weakly curved, 24 μm long, with medial projection. Gonocoxite 70 μm long with 2 setae apically. Gonostylus slender, 28 μm long, with single setae near apex; megaseta 3 μm long. HR 2.5, HV 3.8.

Distribution. OR: China (Sichuan).

3.63 *CORYNONEURA MACULA* FU & SÆTHER, 2012 (FIGS. 3.87A–G AND 3.88A–E)

Corynoneura sp. 8 Bolton, 2007: 26.

  Corynoneura sp. H Epler, 2001: 7.48.

  Corynoneura macula Fu & Sæther, 2012: 38.

Material examined. Holotype male, with associated larval and pupal exuviae, USA, Ohio, Geauga County, Fern Lake Bog, 2.viii.1992, M. J. Bolton (ZMBN Type No. 458). Allotype: female with associated pupal and larval exuviae, as holotype except for 12.vii.1992 (ZMBN). Paratypes: 1 female, 1 larval and 1 pupal exuviae as holotype; 1 female, 1 larval and 1 pupal exuviae as holotype except for 12.vii.1992. USA, Ohio, Hocking County, 3 larvae, 29.viii.2001, M. J. Bolton (MJB).

Diagnostic characters. The adult male is characterized by the yellowish antenna with dark brown apical spot; anterior margin of cibarial pump strongly concave; superior volsella low and conspicuously projecting; inferior volsella large, digitiform, placed caudally of gonocoxite; and

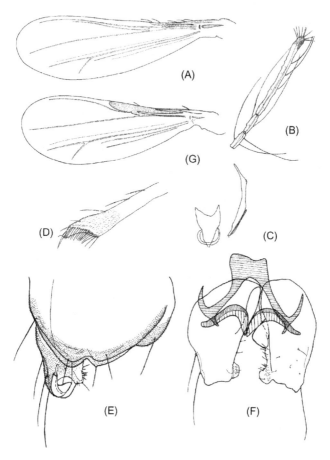

Figure 3.87 *Corynoneura macula* Fu & Sæther, 2012 (From Fu, Y., & Sæther, O. A. (2012). *Corynoneura* Winnertz and *Thienemanniella* Kieffer from the Nearctic region (Diptera: Chironomidae: Orthocladiinae). *Zootaxa, 3536*, 1−61. www.mapress.com/j/ zt). (A−F) Male imago. (A) Wing. (B) Antenna. (C) Tentorium, stipes and cibarial pump. (D) Hind tibial apex. (E) Hypopygium, dorsal view. (F) Hypopygium, ventral view. (G) Female imago: wing.

sternapodeme U-shaped. The adult female has AR 0.27−0.37; 10 transparent fused bead-like lateral lamellae on coxosternapodeme, and very short notum. The pupa has no taeniate L-setae and shagreen on tergites I−II, only tergites IV−VI with 4−6 thick hooklets, and 26−29 taeniae in fringe of anal lobe. The larvae are separable by mentum with three median teeth, the lateral tooth adjacent to median tooth not small, all antennal segments of the same color, and head capsule weakly sculptured.

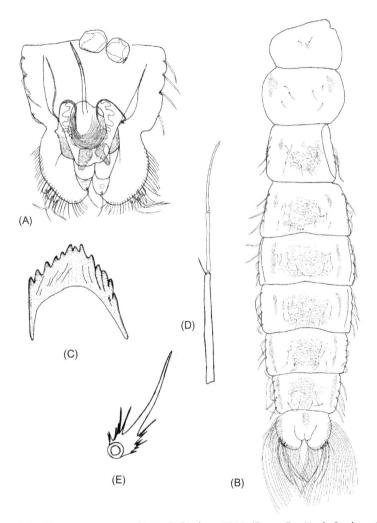

Figure 3.88 *Corynoneura macula* Fu & Sæther, 2012 (From Fu, Y., & Sæther, O. A. (2012). *Corynoneura* Winnertz and *Thienemanniella* Kieffer from the Nearctic region (Diptera: Chironomidae: Orthocladiinae). *Zootaxa, 3536*, 1–61. www.mapress.com/j/zt). (A) Pharate female imago: genitalia, ventral view. (B) Pupa: tergites I–IX. (C–E) Larva. (C) Mentum. (D) Antenna. (E) Subbasal seta of posterior parapod.

Male (*n* = 1)

Total length 1.02 mm. Wing length 0.65 mm. Total length/wing length 1.56. Wing length/profemur length 3.0.

Coloration. Head and thorax brown, antenna yellowish with dark brown apical spot, legs yellowish, abdomen light brown.

Head. Antenna (Fig. 3.87B) with 10 flagellomeres, ultimate flagellomere 132 μm, AR 0.55; antenna slightly expanded apically, with about 17 apical sensilla chaetica, longest antennal seta 202 μm. Temporals absent. Tentorium, stipes and cibarial pump as in Fig. 3.87C; tentorium 112 μm long, 12 μm wide; stipes 58 μm long, 3 μm wide. Anterior margin of cibarial pump strongly concave. Clypeus with 6 setae. Lengths of palpomeres (in μm): 10, 12, 17, 22, 46. Palpomere 5/3 ratio 2.7.

Thorax. Dorsocentrals 5, prealars 2. Scutellum with 2 setae. Anapleural suture 91 μm long.

Wing (Fig. 3.87A). VR 3.3. Cu/wing length 0.60, C length 160 μm long, C length/wing length 0.25, Cu 392 μm long, wing width/wing length 0.38. Costa with 5 setae.

Legs. Fore trochanter with dorsal keel. Spurs of fore tibia 18 μm and 8 μm long, of mid tibia 12 μm and 10 μm long, of hind tibia 33 μm and 11 μm long. Width at apex of fore tibia 17 μm, of mid tibia 14 μm, of hind tibia (a) 30 μm. Width of hind tibia 1/3 from apex (d) 14 μm, elongation length (b) 30 μm, length of maximum thickening (c_1) 62 μm, total length of thickening (c_2) 110 μm; a/d 2.1; b/d 2.1; c_1/d 4.4; c_2/d 7.9. Hind tibia (Fig. 3.87D) expanded, with comb of 16 setae, 1 seta near spur S-shaped. Lengths and proportions of legs as in Table 3.50.

Hypopygium (Fig. 3.87E and F). The posterior margin of tergite IX medially concave, with 4 long setae. Laterosternite IX with 1 long seta. Superior volsella low and conspicuously projecting. Inferior volsella small, placed caudally. Phallapodeme curved, 41 μm long, placed in caudal position of sternapodeme. Transverse sternapodeme very thick longitudinally, 21 μm wide. Large attachment point of lateral sternapodeme with phallapodeme placed and directed caudally. Gonostylus strongly curved, 17 μm long, megaseta 6 μm long. HR 3.6, HV 6.0.

Pharate female ($n = 1-2$)

Total length 0.93 mm. Wing length 0.58−0.65 mm. Total length/wing length 1.60. Wing length/profemur length 3.4.

Table 3.50 Lengths (in μm) and proportions of legs segments of male *Corynoneura macula* Fu & Sæther, 2012 ($n = 1$)

	fe	ti	ta₁	ta₂	ta₃	ta₄	ta₅	LR	BV	SV	BR
p₁	220	248	139	71	38	18	30	0.56	3.9	3.4	2.5
p₂	288	264	162	65	36	16	28	0.60	4.9	3.4	3.2
p₃	280	260	145	81	30	14	24	0.56	4.6	3.7	2.6

Source: From Fu, Y., & Sæther, O. A. (2012). *Corynoneura* Winnertz and *Thienemanniella* Kieffer from the Nearctic region (Diptera: Chironomidae: Orthocladiinae). *Zootaxa*, 3536, 1−61. www.mapress.com/j/zt.

Coloration. Head brown. Thorax yellowish brown, medially yellowish. Antenna and legs yellowish, abdomen yellowish brown.

Head. AR 0.27−0.37. Lengths of flagellomeres (in μm): 25, 26, 28, 32, 39. Ultimate flagellomere with 8 apical sensilla chaetica. Tentorium 83 μm long, 7 μm wide. Clypeus with 8−10 setae. Palpomere lengths (in μm): 19, 19, 21, 22, 50. Palpomere 5/3 ratio 2.4.

Thorax. Dorsocentrals 8−9; prealars 2−3. Scutellum with 3−4 setae.

Wing (Fig. 3.87G). Wing broader than in male. VR 2.2−2.6, two anal veins present, Cu 340−380 μm long, Cu/wing length 0.58−0.59, C 272−280 μm long; C/wing length 0.43−0.47, wing width/ wing length 0.42−0.47. Costa with 11−14 setae.

Legs. Fore trochanter with keel. Spurs of fore tibia 14 μm and 10 μm long, of mid tibia 11 μm and 7 μm long, of hind tibia 32 μm and 12 μm long. Width at apex of fore tibia 24 μm wide, of mid tibia 18 μm, of hind tibia (a) 33−35 μm. Width of hind tibia 1/3 from apex (d) 12−19 μm, elongation length (b) 25−35 μm, length of maximum thickening (c_1) 76−80 μm, total length of thickening (c_2) 116−117 μm, a/d 1.7−2.1, b/d 1.3−2.1, c_1/d 4.0−4.7, c_2/d 6.2−6.8. Apex of hind tibia expanded, with comb of 16 setae, with one S-shaped spur. Lengths and proportions of legs as in Table 3.51.

Abdomen. Number of setae on tergites II −VIII as: 1, 1, 1, 1, 1, 3, 3.

Genitalia (Fig. 3.88A). Tergite IX with 2 long caudal setae. Cercus 32−34 μm long, 18−24 μm wide. Notum length 29−30 μm. Gonocoxite with 1 long seta. Coxosternapodeme with 10 transparent fused bead-like lateral lamellae on inner side of coxosternapodeme. Seminal capsule 40−43 μm long; neck 8 μm long, 6 μm wide.

Pupa ($n = 2-3$)

Table 3.51 Lengths (in μm) and proportions of legs segments of female *Corynoneura macula* Fu & Sæther, 2012 ($n = 1-2$)

	fe	ti	ta_1	ta_2	ta_3	ta_4
p_1	172−200	208−236	111−129	57−67	32−40	16−20
p_2	280	248	152	73	38	16
p_3	204−244	224−252	123−133	61−73	24−30	16−18

	ta_5	LR	BV	SV	BR	
p_1	22−28	0.53−0.55	3.7−3.8	3.4	1.5−1.8	
p_2	28	0.62	4.4	3.5	2.1	
p_3	26−30	0.53−0.55	4.2−4.3	3.5−3.7	1.0−1.5	

Source: From Fu, Y., & Sæther, O. A. (2012). *Corynoneura* Winnertz and *Thienemanniella* Kieffer from the Nearctic region (Diptera: Chironomidae: Orthocladiinae). *Zootaxa*, 3536, 1−61. www.mapress.com/j/zt.

Total length 1.80 mm. Exuviae yellowish brown, abdomen yellowish with brown belt on each side.

Cephalothorax. Frontal setae $28-38\,\mu m$ long. Median antepronotals $24\,\mu m$ and $28\,\mu m$ long. Lateral antepronotals $13\,\mu m$ and $15\,\mu m$ long. Anterior precorneal seta (PcS$_1$) $7\,\mu m$ long, PcS$_2$ $8-10\,\mu m$ long, PcS$_3$ $23\,\mu m$ long. PcS$_{1-3}$ in a line; PcS$_1$ $6-7\,\mu m$ from PcS$_2$; PcS$_2$ $4-7$, $6\,\mu m$ from PcS$_3$; PcS$_1$ $10-14$, $12\,\mu m$ from PcS$_3$; PcS$_3$ $20-28$, $25\,\mu m$ from thoracic horn. Anterior dorsocentral (Dc$_1$) $15\,\mu m$ long; Dc$_2$ $10\,\mu m$ long; Dc$_3$ $14\,\mu m$ long; Dc$_4$ $11\,\mu m$ long. Dc$_1$ located $5-6\,\mu m$ from Dc$_2$, Dc$_2$ located $7-11\,\mu m$ from Dc$_3$. Wing sheath with 4 rows of pearls.

Abdomen (Fig. 3.88B). Shagreen and chaetotaxy as illustrated. No taeniate L-seta on tergites I−II. Anal lobe $111\,\mu m$ long. Anal lobe fringe with $26-29$ setae, $280\,\mu m$ long. Three taeniate anal macrosetae $200-260\,\mu m$ long; median setae shorter than anal macrosetae, about $100-120\,\mu m$ long.

Larva ($n = 1-3$)

Coloration. Head brown. Antenna with all segments yellowish. Abdomen yellowish brown.

Head. Capsule length $232-264$, $249\,\mu m$; width $112-120\,\mu m$; weakly sculptured. Postmentum $196-216$, $204\,\mu m$ long. Sternite II obvious, rising from small tubercle, I simple, III not observed. Premandible $28-35$, $34\,\mu m$ long; brush of premandible not observed. Mentum as in Fig. 3.88C. Mandibles $41-51$, $48\,\mu m$ long. Antenna as in Fig. 3.88D, AR 0.88. Lengths of flagellomeres I−IV (in μm): 232, 121, 137, 6. Basal segment width $15-18$, $16\,\mu m$; length of blade at apex of basal segment $28-32$, $30\,\mu m$.

Abdomen. Length of anal setae $90-131\,\mu m$. Procercus $44\,\mu m$ long, $18\,\mu m$ wide. Subbasal seta of posterior parapods split (as in Fig. 3.88E), $43-47\,\mu m$ long.

Distribution. NE: USA (Ohio).

3.64 *CORYNONEURA MAGNA* BRUNDIN, 1949

Corynoneura magna Brundin, 1949: 833.

Diagnostic characters. Total length 2.0 mm. Wing length 1.8 mm. Male antenna with 12 flagellomeres. The posterior margin of tergite IX medially concave, with 2 short setae. Inferior volsella small rectangular, placed caudally of gonocoxite.

Distribution. PA: Finland, Sweden.

3.65 *CORYNONEURA MAKARCHENKORUM* KRASHENINNIKOV, 2012 (FIG. 3.89A–E)

Corynoneura makarchenkorum Krasheninnikov, 2012: 83.

Diagnostic characters. Total length 2.4–2.7 mm. Wing length 1.4 mm. Antenna with 12 flagellomeres. Apex of terminal flagellomere with two groups of short setae. t_3 with straight seta in apex; $a/d = 1.2–1.6$; $b/d = 0.6–1.0$. Superior volsella roundish-rectangular, connected anteromedially. Inferior volsellae located in second part of gonocoxite, with roundish or angular edge, and bare apex. Sternapodeme inverted V-shaped. Phallapodeme crescent-shaped.

Male (Description from Krasheninnikov, 2012)

Total length 2.6–2.7 mm. Total length/wing length 1.7–1.9.

Coloration. Thorax and abdomen brown.

Head. Eyes bare. Antenna with 12 flagellomeres, AR 0.92–1.0; ultimate flagellomere with apical two groups of light hairs (Fig. 3.89A). Lengths of palpomeres (in μm): 20, 31–35, 47–55, 55–59, 71.

Thorax. Dorsocentrals 7–9.

Wing. Wing length 1.4 mm.

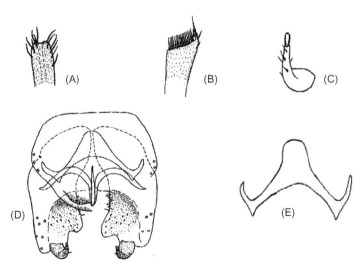

Figure 3.89 *Corynoneura makarchenkorum* Krasheninnikov, 2012; Male imago (From Krasheninnikov, A. B. (2012). A new and little known Chironomidae of subfamily Orthocladiinae (Diptera, Chironomidae) from the Urals. *Eurasian Entomological Journal, 11*(1), 83–86). (A) Distal part of terminal flagellomere. (B) Apex of hind tibia. (C) Gonostylus. (D) Total view of hypopygium, from above. (E) Sternapodema.

Table 3.52 Lengths (in μm) and proportions of legs segments of male *Corynoneura makarchenkorum* Krasheninnikov, 2012

	fe	ti	ta$_1$	ta$_2$	ta$_3$	ta$_4$
p$_1$	390	452−468	250−265	140	78	31
p$_2$	530	515	296	140−156	62−78	31
p$_3$	452	484−499	328	172	62−78	31

	ta$_5$	LR	BV	SV	BR	
p$_1$	47	0.55−0.57	3.68−3.79	3.24−3.38	2.0−2.2	
p$_2$	47	0.58	4.53	3.53	1.9−2.8	
p$_3$	47	0.66−0.68	3.90−4.05	2.86−2.90	2.1−3.3	

Legs. Fore trochanter with dorsal keel. Spurs of fore tibia 31−39 μm long, of mid tibia 12 μm and 16 μm long, of hind tibia 43 μm and 12−16 μm long. Hind tibia (Fig. 3.89B) expanded, with comb of 16 setae. Lengths and proportions of legs as in Table 3.52.

Hypopygium (Fig. 3.89C−E). The posterior margin of tergite IX medially concave, with 4−5 long setae. Laterosternite IX with 1−4 long setae. Superior volsella rounded-rectangular. Inferior volsella located in the second half gonocoxite, apical naked small. Gonocoxite 161−169 μm long. Gonostylus apically curved, narrowed and elongated, 50 μm long, with a 8 mm long apical spur, transverse sternapodeme curved into V-shaped.

Distribution. PA: Russia (CET).

3.66 *CORYNONEURA MARINA* KIEFFER, 1924

Corynoneura magna Kieffer, 1924: 43.

Diagnostic character. Male antenna with 11 flagellomeres.

Distribution. PA: Germany.

3.67 *CORYNONEURA MEDIASPICULA* WIEDENBRUG & TRIVINHO-STRIXINO, 2011 (FIGS. 3.90A−H, 3.91A−D, AND 3.92A−M)

Corynoneura mediaspicula Wiedenbrug & Trivinho-Strixino, 2011: 12.

Diagnostic characters. The adult males are separable from other species by the antenna with 7 flagellomeres, 6 distinct and the last composed of 5 fused flagellomeres, not very plumose with one row of seta each flagellomere, AR about 1.2, eyes pubescent, abdominal segments

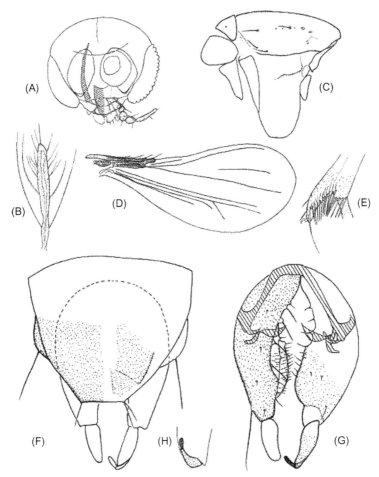

Figure 3.90 *Corynoneura mediaspicula* Wiedenbrug & Trivinho-Strixino, 2011, Male imago (From Wiedenbrug, S., & Trivinho-Strixino, S. (2011). New species of the genus *Corynoneura* Winnertz (Diptera, Chironomidae) from Brazil. *Zootaxa, 2822,* 1−40). (A) Head, dorsal view, left is sclerites hatched. (B) Terminal flagellomeres. (C) Thorax. (D) Wing. (E) Apex of hind tibia. (F−H) Hypopygium. (F) Tergite IX; left is dorsal view, right is ventral view. (G) Tergite IX removed, sclerites hatched; right is dorsal view, left is ventral view. (H) Gonostylus.

I−IV whitish, V usually whitish, not as dark as following segments, VI−IX brownish, attachment of the phallapodeme caudal on the lateral sternapodeme, phallapodeme posterior margin sclerotized, rounded and curved to anterior, gonostylus slender tapering apically. Adult females are separable from other species except from *C. hermanni* by the shape of the labia funnel, seminal capsules size subequal, abdominal segments III−IX

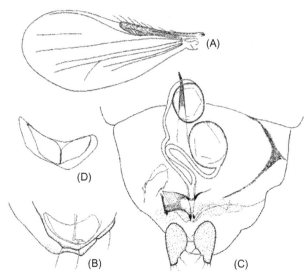

Figure 3.91 *Corynoneura mediaspicula* Wiedenbrug & Trivinho-Strixino, 2011, Female imago (From Wiedenbrug, S., & Trivinho-Strixino, S. (2011). New species of the genus *Corynoneura* Winnertz (Diptera, Chironomidae) from Brazil. *Zootaxa, 2822,* 1−40). (A) Wing. (B−D) Genitalia. (B) Dorsal with view of labia. (C) Ventral view of genitalia; left is gonapophysis removed. (D) Ventral view of labia.

brownish. The pupae are distinguished from other species by the shagreen of relatively strong spines even distributed, on tergite IV median spines slightly longer than on posterior field, 4 small taeniate lateral setae on segments III−VIII, anal lobe rectangular separated from each other through a perpendicular margin, anal lobe fringe restricted to posterior margin. The larvae can be recognized by the head integument without sculptures, the mentum with 3 median and 6 lateral teeth, antenna longer than postmentum, AR 0.81−0.97, second antennal segment longer than the third, basal seta on posterior parapods split from the base.

Male (Description from Wiedenbrug & Trivinho-Strixino, 2011)

Total length 1.00−1.27 mm. Wing length 0.52−0.58 mm. Total length/wing length 1.73−2.29. Wing length/length of profemur 2.65−3.11.

Coloration. Thorax brownish; legs whitish; tergites I−IV whitish, V usually whitish, if brown not as dark as following segments, VI−IX brownish.

Head (Fig. 3.90A). AR = 1.16−1.23. Antenna with 7 flagellomeres, flagellomeres 1−6 distinctly separated from each other, ultimate

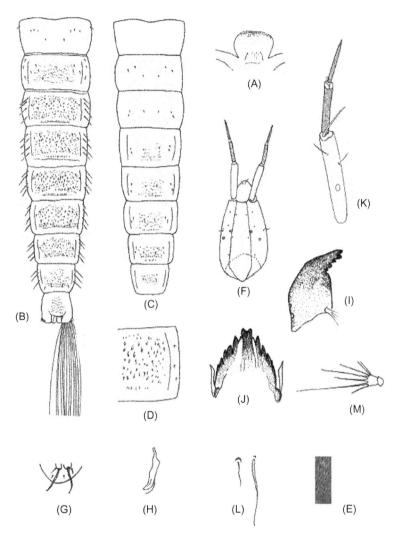

Figure 3.92 *Corynoneura mediaspicula* Wiedenbrug & Trivinho-Strixino, 2011, Immatures imago (From Wiedenbrug, S., & Trivinho-Strixino, S. (2011). New species of the genus *Corynoneura* Winnertz (Diptera, Chironomidae) from Brazil. *Zootaxa, 2822*, 1–40). (A–E) Pupa imago. (A) Frontal apotome. (B) Tergites $I-IX$ and anal lobe, left is without fringe. (C) Sternites $I-VIII$. (D) Detail of tergite IV. (E) Detail of posterior shagreen of tergite IV. (F–M) Larva imago. (F) Dorsal view of the head and antenna. (G) Dorsal view of labrum. (H) Premandible. (I) Mandible. (J) Mentum. (K) Antenna. (L) Abdominal seta. (M) Subbasal seta of posterior parapod.

Table 3.53 Lengths (in μm) and proportions of legs segments of male *Corynoneura mediaspicula* Wiedenbrug & Trivinho-Strixino, 2011

	fe	ti	ta_1	ta_2	ta_3	ta_4
p_1	185−187	235−240	90−93	62−65	32−35	17
p_2	255−280	237−245	130	65−68	33	17
p_3	197−200	210−225	125	67−−70	27	15

	ta_5	LR	BV	SV	BR
p_1	22	0.37−0.43	3.57	4.40−4.72	2.25−2.50
p_2	20−25	0.53−0.55	4.45−4.76	3.78−4.04	2.25−2.50
p_3	25	0.55−0.59	3.87−4.07	3.26−3.40	4.25−4.75

flagellomere 105−160 μm long, composed of 5 distinct but fused flagellomeres, two sensilla chaetica at the last flagellomere (Fig. 3.90B). All flagellomeres, except the first and last, with one row of seta each. Temporal setae absent. Clypeus 32 μm long, with 6−7 setae. Tentorium 102−107 μm long, 7−10 μm wide. Palpomere lengths (in μm): 7−10, 10, 15, 17, 35−45. Third palpomere with one sensilla clavata. Eyes pubescent.

Thorax (Fig. 3.90C). Antepronotum with 1 lateral seta. Dorsocentrals 4, prealars 2. Scutellars 2. Antepronotal lobes dorsally reduced, tapering.

Wing (Fig. 3.90D). VR 3.17−3.36. Clavus 25−27 μm wide, ending 130−132 μm from arculus. Clavus/wing length 0.23−0.25. Anal lobe absent.

Legs. Spur of front tibia 20−22 μm long; spur of middle tibia 10 μm and 7−10 μm long; spurs of hind tibia 37 μm long, second spur S-shaped. Width of apex of front tibia 15−17 μm, of middle tibia 12−15 μm, of hind tibia 27−35 μm. Front tibial scale 7 μm long; hind tibial scale 32−37 μm long (Fig. 3.90E). Leg measurements (in μm) and ratios as in Table 3.53.

Abdomen setation. Tergite VI 0−1, VII 3−4, VIII 1.

Hypopygium (Fig. 3.90F−H). Tergite IX with 2−4 setae. Laterosternite IX with 2 setae. Superior volsella absent. Inferior volsella smooth with longer microtrichia directed to median. Phallapodeme 20 μm long, rounded and sclerotized on posterior margin; transverse sternapodeme 15 μm long. Gonocoxite 55−60 μm long. Gonostylus tapering apically 22 μm long; megaseta 6 μm long. HR 2.44−2.67; HV 4.43−4.76.

Female (*n* = 1) (Description from Wiedenbrug & Trivinho-Strixino, 2011)

Table 3.54 Lengths (in µm) and proportions of legs segments of female *Corynoneura mediaspicula* Wiedenbrug & Trivinho-Strixino, 2011

	fe	ti	ta$_1$	ta$_2$	ta$_3$	ta$_4$	ta$_5$	LR	BV	SV	BR
p$_1$	165	217	87	55	27	15	25	0.40	3.84	4.37	2.40
p$_2$	250	235	120	62	27	12	20	0.51	4.94	4.04	2.50
p$_3$	195	205	112	65	25	15	22	0.55	4.02	3.56	2.20

Total length 1.01 mm. Wing length 0.56 mm. Total length/wing length 1.82. Wing length/length of profemur 3.37.

Coloration. Thorax light brown, brownish vittae, postnotum and anepisternum and preepisternum; legs whitish except for brownish ring anterior on tibia; tergites I−II whitish, III−IX brownish.

Head. AR 0.48. Antenna with 5 flagellomeres, apical flagellomere 50 µm long; two sensilla chaetica on flagellomeres 3−5. Temporal setae absent. Clypeus 37 µm long, with 7 setae. Tentorium 70 µm long, 5 µm wide. Palpomere lengths (in µm): 12, 12, 17, —, 35. Sensilla clavata not possible to see. Eyes pubescent.

Thorax. Antepronotal 1. Dorsocentrals 5, prealars 2. Scutellars 2. Antepronotal lobes dorsally tapering.

Wing (Fig. 3.91A). VR 2.45. Clavus 30 µm wide, ending 225 µm from arculus. Clavus/wing length 0.40.

Legs. Spur of front tibia 10 µm long; spur of middle tibia 7 µm and 7 µm long; spurs of hind tibia 35 µm long, second spur small S-shaped. Front tibial scale 10 µm long. Hind tibial scale 40 µm long. Width of apex of front tibia 15 µm, of middle tibia 15 µm, of hind tibia 30 µm. Leg measurements (in µm) and ratios as in Table 3.54.

Abdomen. One seta on tergite VI, 3 setae on tergite VII, 1 seta on tergite VIII.

Genitalia (Fig. 3.91B−D). Gonocoxite IX with 1 seta. Tergite IX with 2 setae. Two seminal capsules respectively 32 µm and 37 µm long, one spermathecal duct with a loop, second straighter, both ducts join together short distance before seminal eminence, which has sclerotized outer borders. Notum 70 µm long. Membrane well sclerotized. Apodeme lobe well sclerotized, apically pointed. Coxosternapodeme strongly curved, with one tiny lateral lamellae, coxosternapodeme with one end at the roof of copulatory bursa, the latter semicircle-shaped. Labia membranous, bare, separated, dorsal funnel shaped, apically building the accessory gonopore, ventral divided in two lobes (Fig. 3.91D). Gonocoxapodeme straight, gonapophyses medially slightly lobed. Postgenital plate inconspicuous. Cercus 37 μm long.

Pupa (Description from Wiedenbrug & Trivinho-Strixino, 2011)
Total length 1.32−1.42 mm.

Cephalothorax. Frontal apotome with granulation (Fig. 3.92A). Frontal setae 55−60 μm long. Median antepronotals 40 μm. Dorsocentrals Dc_1 27 μm long; Dc_2 25 μm long; Dc_3 30 μm long; Dc_4 27 μm long. Distance between Dc_1 and Dc_2 67−77 μm; Dc_2 and Dc_3 32−37 μm; Dc_3 and Dc_4 12−17 μm. Dc_1 displaced ventrally. Wing sheath with about 4 rows of pearls.

Abdomen (Fig. 3.92B−E). Tergite I and sternites I−II without shagreen. Tergites II, VIII−IX with fine shagreen. Tergites III−VII with relative strong points evenly distributed, median field of shagreen with slightly stronger shagreen points as posterior (Fig. 3.92D and E). Sternite III with very fine shagreen (not drawn); sternites IV−VIII with fine shagreen. Conjunctive tergite II/III with 0−9 spinules; tergite III/IV with 18−20 hooklets; tergite IV/V with 15−19 hooklets; tergite V/VI with 10−13 hooklets; tergite VI/VII without hooklets; Sternite IV/V with 10−11 spinules; sternite V/VI with 8−12 hooklets; sternite VI/VII with 8−12 hooklets, sternites VII/VIII with 7−8 hooklets. segment I with 3 D-setae, 1 L-setae, without V-setae; segment II 4 D-setae, 3 L-setae and 3V-setae; segments III−VII with 4 D-setae, 3V-setae and 4 small taeniate L-setae; segment VIII with 2 D-setae, 1 V-setae and 3 or 4 taeniate L-setae. Anal lobe rectangular 90−105 μm long. Anal lobe fringe without lateral setae, posterior with 11−13 taeniate setae 375−400 μm long and median 1−3 shorter and thinner fringe setae; 3 macrosetae taeniate; inner setae taeniate.

Larva (Description from Wiedenbrug & Trivinho-Strixino, 2011)
Head (Fig. 3.92F). Head capsule integument smooth. Frontal apotome 160−185 μm long; head width 110−137 μm; postmentum 130−135 μm; postmentum/head width 0.98−1.18. Sternite I simple (Fig. 3.92G). Premandible with an outer lateral lamella with a brush (Fig. 3.92H). Mentum with three median teeth, intermediate teeth minute, and six adjacent teeth (Fig. 3.92J). Distance between setae submenti 37−42 μm. Mandible length 40−42 μm with outer sclerotized protuberance (Fig. 3.92I). Antennae (Fig. 3.92K): AR 0.81−0.97. Length of segment I 75−82 μm, II 42−50 μm, III 37−40 μm, IV 2−5 μm; basal segment width 12 μm; antennal blade 30−42 μm long; ring organ at 17−25 μm from the base of first antennal segment. Antennal segments 2 and 3 brown.

Abdomen. Ventral setae modified, wide and longer (Fig. 3.92L). Subbasal seta on posterior parapod split from the base (Fig. 3.92M).

Remarks. The larvae of *C. mediaspicula* were collected from litter laying near the water surface from small mountain streams. This species was found in Minas Gerais and São Paulo states.

Distribution. NT: Brazil.

3.68 *CORYNONEURA MEDICINA* FU ET AL., 2009 (FIG. 3.93A–C)

Corynoneura medicina Fu et al., 2009: 25.

Material examined. Holotype male (BDN No. 05368), P.R. China: Sichuan Province, Emeishan County, Medicine School, 29°21′N, 103° 17′E, alt. 1500 m, sweep net, 13.v.1986, X. Wang.

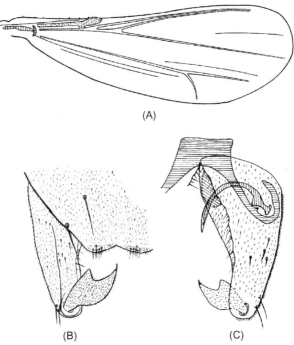

(A)

(B)   (C)

Figure 3.93 *Corynoneura medicina* Fu et al., 2009, Male imago (From Fu, Y., Sæther, O. A., & Wang, X. (2009). *Corynoneura* Winnertz from East Asia, with a systematic review of the genus (Diptera: Chironomidae: Orthocladiinae). *Zootaxa, 2287,* 1–44). (A) Wing. (B) Hypopygium, dorsal view. (C) Hypopygium, ventral view.

Diagnostic characters. Similar to *C. scutellata* in having antenna with 10 flagellomeres and AR 0.74, but can be separated by the inverted U-shaped sternapodeme and lack of inferior volsella.

Male (*n* = 1)

Total length 1.08 mm. Wing length 0.87 mm. Total length/wing length 1.24. Wing length/length of profemur 3.35.

Coloration. Head dark brown. Antenna and palpomere yellow. Thorax yellowish brown with brown vittae, scutum, scutellum, preepisternum, postnotum and yellow anterior anepisternum II, median anepisternum II, posterior anepisternum II. Abdominal segments brown. Legs yellowish brown. Wing light yellow to hyaline, with pale yellow clava.

Head. Antenna with 10 flagellomeres; ultimate flagellomere 215 μm long, apex with short sensilla chaetica subapically, apically acute. AR 0.74. Clypeus with 4 setae. Tentorium 120 μm long, 13 μm wide. Stipes 55 μm long. Palpomeres length (in μm): 13, 13, 23, 33, 48. Palpomere 2 and 3 ellipsoid, 4 rectangular, 5 curved. Palpomere 5/3 ratio 2.1.

Wing (Fig. 3.93A). VR 2.6. C length 240 μm, C/wing length 0.3. Cu length 520 μm. Cu/wing length 0.6. Wing width/wing length 0.37. C with 8 setae.

Legs. Spurs of fore tibia 20 μm and 13 μm long, spur of mid tibia 10 μm long, of hind tibia 33 μm long. Width at apex of fore tibia 18 μm, of mid tibia 18 μm, of hind tibia 30 μm. Apex of hind tibia expanded, with comb of 12 setae and 1 seta near spur near straight. Lengths and proportions of legs as in Table 3.55.

Hypopygium (Fig. 3.93B and C). Posterior margin of tergite IX bilobed, 5—6 setae on each side; laterosternite IX with 2 setae. Anal point absent. Superior volsella fused anteriomedially. Inferior volsella absent. Sternapodeme inverted U-shaped, 33 μm wide; coxapodeme 25 μm long, attachment point with phallapodeme placed in caudal third of lateral sternapodeme and directed caudally; phallapodeme well developed, strongly curved with projection for joint with sternapodeme placed prelateral, not

Table 3.55 Lengths (in μm) and proportion of leg segments of male *Corynoneura medicinea* Fu et al., 2009

	fe	ti	ta_1	ta_2	ta_3	ta_4	ta_5	LR	BV	SV	BR
p_1	260	325	185	90	48	20	32	0.57	4.1	3.0	2.0
p_2	365	335	195	88	44	22	34	0.58	4.8	3.8	2.0
p_3	285	340	175	85	30	15	35	0.51	4.8	3.6	1.7

extending beyond margin of tergite IX, 55 μm long, 3 μm wide. Gonocoxite 63 μm long with 2 setae apically. Gonostylus hooked, 30 μm long; megaseta 5 μm long. HR 2.1, HV 3.6.

Distribution. OR: China (Sichuan).

3.69 *CORYNONEURA MINEIRA* WIEDENBRUG & TRIVINHO-STRIXINO, 2011 (FIGS. 3.94A−F, 3.95A−D, AND 3.96A−L)

Corynoneura-group spec. 9: Wiedenbrug, 2000: 107.

Corynoneura mineira Wiedenbrug & Trivinho-Strixino, 2011: 17.

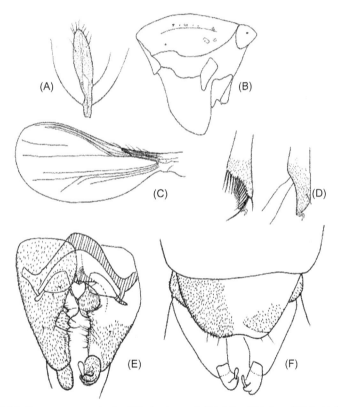

Figure 3.94 *Corynoneura mineira* Wiedenbrug & Trivinho-Strixino, 2011, Male imago (From Wiedenbrug, S., & Trivinho-Strixino, S. (2011). New species of the genus *Corynoneura* Winnertz (Diptera, Chironomidae) from Brazil. *Zootaxa, 2822*, 1−40). (A) Terminal flagellomeres. (B) Thorax. (C) Wing. (D) Apex of hind tibia. (E and F) Hypopygium. (E) Tergite IX removed, sclerites hatched; left is ventral view, right is dorsal view. (F) Tergite IX; left is dorsal view, right is ventral view.

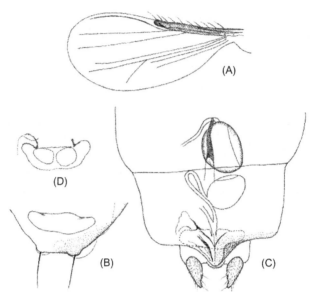

Figure 3.95 *Corynoneura mineira* Wiedenbrug & Trivinho-Strixino, 2011, Female imago (From Wiedenbrug, S., & Trivinho-Strixino, S. (2011). New species of the genus *Corynoneura* Winnertz (Diptera, Chironomidae) from Brazil. *Zootaxa, 2822*, 1—40). (A) Wing. (B—D) Genitalia. (B) Dorsal view of copulatory bursa. (C) Ventral view of genitalia; left is gonapophysis removed. (D) Detail of labia.

Diagnostic characters. The adult males are separable from other species by the antenna with 10 flagellomeres, AR about 0.3, eyes pubescent, aedeagal lobe triangular, attachment of the phallapodeme caudal on the lateral sternapodeme, phallapodeme posterior margin sclerotized, curved to anterior and rounded, gonocoxite with a median lobe with longer microtrichia. Adult females are separable from other species by the labia not strongly sclerotized, apparently separated not building together a funnel, copulatory burse with posterior margin not strongly invaginated, two seminal capsules subequal in size, one more sclerotized than the other. The pupae are distinguished from other species by tergites III and IV usually with median shagreen field separated from posterior shagreen, spines on posterior row relatively wide and with rounded lateral margin, anal lobe not joined medially as an inverted "V" but separated by a straight margin, posterior margin rounded with fringe. The larvae can be recognized by the head integument without sculptures, the mentum with 3 median and 6 lateral teeth, head elongate (postmentum length/head width 1.5), AR 0.88, antenna longer than frontoclypeal apotome, basal seta on posterior parapods split from the base.

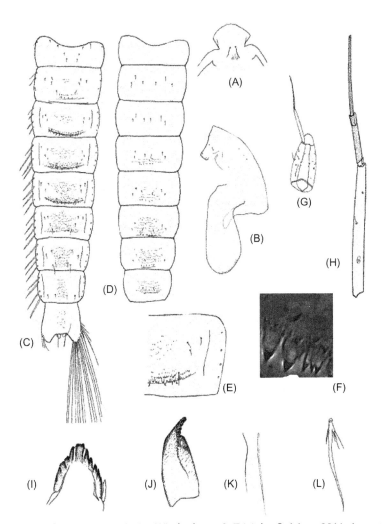

Figure 3.96 *Corynoneura mineira* Wiedenbrug & Trivinho-Strixino, 2011, Immatures imago (From Wiedenbrug, S., & Trivinho-Strixino, S. (2011). New species of the genus *Corynoneura* Winnertz (Diptera, Chironomidae) from Brazil. *Zootaxa, 2822*, 1−40). (A−F) Pupa imago. (A) Frontal apotome. (B) Thorax, lateral view. (C) Tergites I−IX and anal lobe, left is without fringe. (D) Sternites I−VIII. (E) Detail of tergite IV. (F) Detail of posterior shagreen of tergite IV. (G−L) Larva imago. (G) Dorsal view of the head and antenna. (H) Antenna. (I) Mentum. (J) Mandible. (K) Abdominal setae. (L) Subbasal seta of posterior parapod.

Male (Description from Wiedenbrug & Trivinho-Strixino, 2011)
Total length 0.83 mm.
Coloration. Thorax brownish; legs whitish; tergites I−IV whitish, V−IX brownish; gonocoxite and gonostylus brownish.

Head. AR = 0.28. Antenna with 10 flagellomeres, sensilla chaetica not possible to see. Flagellomeres 2−9, with more than one row of setae. Apical flagellomere composed of two fused flagellomeres, with one row of setae (Fig. 3.94A). Temporal setae absent. Clypeus 30 µm long, with 8 setae. Tentorium 112 µm long, 10 µm wide. Palpomere lengths (in µm): 10, 12, 17, 20, 35. One sensilla clavata on the third palpomere. Eyes pubescent.

Thorax (Fig. 3.94B). Antepronotum apparently without seta. Dorsocentrals 3, prealars 2. Scutellars 2. Antepronotal lobes dorsally reduced.

Wing (Fig. 3.94C). VR 2.94. Clavus 25 µm wide, ending 150 µm from arculus. Clavus/wing length 0.24.

Legs. Spur of front tibia 17 µm long; spur of middle tibia 10 µm long; spurs of hind tibia 37 µm long, second spur S-shaped. Width of apex of front tibia 12 µm, of middle tibia 15 µm, of hind tibia 32 µm. Front tibial scale 17 µm long; mid tibial scale 7 µm, hind tibial scale 35 µm long (Fig. 3.94D). Leg measurements (in µm) and ratios as in Table 3.56.

Abdomen setation. Three setae on tergite VII, 3 Setae on tergite VIII.

Hypopygium (Fig. 3.94E and F). Tergite IX with 4 small setae. Laterosternite IX with 1 seta. Superior volsella small with small seta directed to median. Aedeagal lobe triangular. Gonocoxite with a median lobe with longer microtrichia. Inferior volsella smooth anteriorly with longer setae medially directed. Phallapodeme 30 µm long; transverse sternapodeme rounded. Gonocoxite 55 µm long. Gonostylus 15 µm long, tapering apically; megaseta 5 µm long. HR 3.67. HV 5.53.

Female (Description from Wiedenbrug & Trivinho-Strixino, 2011)

Thorax. length 0.42 mm. Wing length 0.66 mm. Wing length/length of profemur 3.61.

Coloration. Thorax brownish; legs whitish; tergites I−II whitish, III−IX brownish.

Head. AR 0.42. Antenna with 5 flagellomeres, apical flagellomere 45 µm long; one sensilla chaetica on flagellomere 1, two sensilla chaetica on flagellomeres 2−5. Temporal setae absent. Clypeus 40 µm long, with 8

Table 3.56 Lengths (in µm) and proportions of legs segments of male *Corynoneura mineira* Wiedenbrug & Trivinho-Strixino, 2011

	fe	ti	ta_1	ta_2	ta_3	ta_4	ta_5	LR	BV	SV	BR
p_1	202	240	82	62	35	15	25	0.34	3.82	5.36	1.60
p_2	280	255	140	70	25	15	17	0.55	5.29	3.82	2.00
p_3	200	215	120	67	25	17	25	0.56	3.96	3.46	2.75

Table 3.57 Lengths (in μm) and proportions of legs segments of female *Corynoneura mineira* Wiedenbrug & Trivinho-Strixino, 2011

	fe	ti	ta_1	ta_2	ta_3	ta_4	ta_5	LR	BV	SV	BR
p_1	182	212	92	57	30	27	25	0.44	3.48	4.27	1.80
p_2	257	242	122	52	25	12	22	0.48	5.53	4.08	2.25
p_3	205	222	110	65	25	12	27	0.49	4.13	3.89	2.20

setae. Tentorium 70 μm long, 5 μm wide. Palpomere lengths (in μm): 10, 10, 17, 20, 37. One sensilla clavata on third palpomere. Eyes pubescent.

Thorax. Antepronotal not possible to observe. Dorsocentrals 3, prealars 2. Scutellars 2. Antepronotal lobes dorsally tapering.

Wing (Fig. 3.95A). VR 2.4. Clavus 32 μm wide, ending 290 μm from arculus. Clavus/wing length 0.44.

Legs. Spur of front tibia 17 μm long; spur of mid tibia 10 μm; spurs of hind tibia 32 μm long, second spur small S-shaped. Front tibial scale 10 μm long, mid tibial scale 7 μm long, hind tibial scale 37 μm long. Width of apex of front tibia 17 μm, of mid tibia 17 μm, of hind tibia 35 μm. Leg measurements (in μm) and ratios as in Table 3.57.

Abdomen setation. Tergite VII with 3 setae, tergite VIII with 1 seta.

Genitalia (Fig. 3.95B−D). Gonocoxite IX with 1 seta. Tergite IX with 2 setae. Two seminal capsules respectively 32 μm and 40 μm long, one more sclerotized than the other, one spermathecal duct with a loop, second not discernable, both ducts join together short distance before seminal eminence, which has sclerotized outer borders. Notum 30 μm long. Membrane well sclerotized V-shaped. Apodeme lobe anterior margin curved to posterior, apically pointed. Coxosternapodeme strongly curved, with a tiny lateral lamella, copulatory bursa semicircle-shaped, with rounded lateral corners. Two labia membranous, bare, separated and rounded. Gonocoxapodeme straight, gonapophyses triangular medially. Postgenital plate with microtrichia. Cercus 25 μm long.

Pupa (Description from Wiedenbrug & Trivinho-Strixino, 2011) Total length 1.22−1.32 mm.

Cephalothorax. Frontal apotome smooth (Fig. 3.96A). Frontal setae 62−67 μm long. Median antepronotals 30 μm and 80 μm. Pc_3 35 μm. Dorsocentrals Dc_3 32 μm long; Dc_4 42 μm long. Distance between Dc_1 and Dc_2 87−90 μm; Dc_2 and Dc_3 30−35 μm; Dc_3 and Dc_4 22 μm. Dc_1 displaced ventrally. Wing sheath with 1−4 row of pearls (Fig. 3.96B).

Abdomen (Fig. 3.96C−F). Tergite I and sternite I without shagreen. Tergite II with few median and posterior shagreen points.

Tergites III–IV with median shagreen field separated from posterior shagreen field, last with spines increasing in size, posterior spines wider than longer (Fig. 3.96E and F). Tergite V similar to IV but median shagreen field not separated from posterior shagreen. Tergites VI–IX with shagreen. Sternites II–III with very fine shagreen (not drawn), IV–VIII with fine shagreen. Conjunctive tergite II/III with 8–9 spinules; tergite III/IV with 11–12 hooklets; tergite IV/V with 7–10 hooklets; tergite V/VI with 0–2 spinules; sternite IV/V with 8–9 spinules; sternite V/VI with 8–9 spinules; sternite VI/VII with 8–9 spinules, sternite VII/VIII with 7–8 spinules. Segment I with 4 D-setae, 1 L-seta, without V-seta; segment II with 3 D-setae, 3 L-setae and 3 V-setae; segments III–VII with 4 D-setae, 3V-seta and 4 taeniate L-setae; segment VIII with 2 D-setae, 1 V-seta and 4 taeniate L-setae. Anal lobe distally rounded 82–105 μm long. Anal lobe fringe without lateral setae, posterior with 24 taeniate setae 300 μm long; 3 taeniate macrosetae; inner setae taeniate.

Larva (Description from Wiedenbrug & Trivinho-Strixino, 2011)

Head (Fig. 3.96G). Head capsule integument smooth. Frontal apotome 187 μm long; head width 107 μm; postmentum 162–167 μm; postmentum/head width 1.56. Sternite I simple. Premandible with outer lateral lamella with a brush. Mentum with three median teeth and six adjacent teeth (Fig. 3.96I). Distance between setae submenti 35 μm. Mandible (Fig. 3.96J) length 30–37 μm. Antennae (Fig. 3.96H): AR 0.88. Length of segment I 140 μm, II 72 μm, III 80 μm, IV 5 μm; basal segment width 10 μm; antennal blade not visible; ring organ at 30 μm from the base of first antennal segment. Lauterborn organs inserted on second antennal segment not overreaching the distal border of this segment. Antennal segments two and three brown.

Abdomen. With ventral setae modified, wider and longer (Fig. 3.96K). Subbasal seta on posterior parapod as Fig. 3.96L.

Remarks. The larvae from *Corynoneura mineira* were collected from litter laying near the water surface from small mountain streams. This species was found in Minas Gerais, São Paulo and Rio Grande do Sul states.

Distribution. NT: Brazil.

3.70 *CORYNONEURA NANKAIENSIS* FU ET AL., 2009 (FIG. 3.97A–E)

Corynoneura nankaiensis Fu et al., 2009: 25.

Material examined. Holotype male (BDN No. 05361), China: Tianjin City, Nankai University, 39°8′N, 117°1′E, alt. 2–5 m, sweep net,

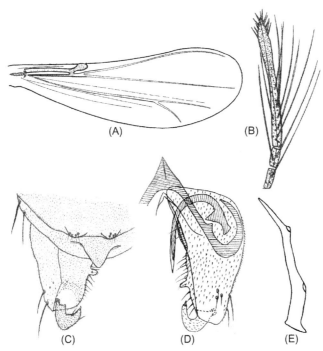

Figure 3.97 *Corynoneura nankaiensis* Fu et al., 2009, Male imago (From Fu, Y., Sæther, O. A., & Wang, X. (2009). *Corynoneura* Winnertz from East Asia, with a systematic review of the genus (Diptera: Chironomidae: Orthocladiinae). *Zootaxa, 2287*, 1–44). (A) Wing. (B) Antenna. (C) Hypopygium, dorsal view. (D) Hypopygium, ventral view. (E) Tentorium.

20.iv.1986, X. Wang. Paratypes: 3 males as holotype (BDN slides No. 05356, 05351, 05357).

Diagnostic characters. Similar to *C. scutellata* in having antenna with 10 flagellomeres, inferior volsella digitiform, and sternapodeme inverted V-shaped, but can be separated by tergite IX with a posterior hyaline structure and anal point.

Male (*n* = 4)

Total length 1.32–1.82, 1.63 mm. Wing length 0.95–1.05, 1.02 mm. Total length/wing length 1.3–1.9, 1.6. Wing length/length of profemur 2.6–3.3, 3.0.

Coloration. Head blackish brown. Antenna and palpomere pale yellowish brown. Thorax and abdomen brown. Legs yellowish brown. Wings yellowish to hyaline.

Head. Antenna with 10 flagellomeres; ultimate flagellomere 235–310, 268 μm long, apex with short sensilla chaetica subapically, apically acute (Fig. 3.97B). AR 0.72–0.94, 0.84. Clypeus with 6–10, 8 (3) setae.

Table 3.58 Lengths (in μm) and proportion of leg segments of male *Corynoneura nankaiensis* Fu et al., 2009

	fe	ti	ta_1	ta_2	ta_3	ta_4
p_1	315–360, 339	360–430, 385	215–260, 239	155–133, 123	65–80, 70	28–30, 28
p_2	440–510, 461	410–450, 420	240–255, 245	100–165, 123	58–63, 61	23–25, 24
p_3	365–425, 384	390–450, 409	223–260, 235	113–138, 123	48–58, 51	23–27, 26

	ta_5	LR	BV	SV	BR
p_1	38–48, 42	0.6–0.7, 0.64	3.4–3.9, 3.7	2.9–3.2, 3.0	2.0–2.3, 2.2
p_2	40–48, 43	0.57–0.6, 058	3.7–5.0, 4.5	3.5–3.8, 3.6	2.4–3.0, 2.6
p_3	40–50, 44	0.56–0.6, 0.58	4.0–4.6, 4.3	3.3–3.4, 3.4	2.0–2.3, 2.2

Tentorium (Fig. 3.97E) 103−148, 129 (3) μm long, 13−20, 17 (3) μm wide. Stipes 65−75, 70 (3) μm long. Palpomeres length (in μm): 13−18, 15; 20−23, 21; 23−28, 26; 18−35, 30; 68 (2). Palpomere 2 and 3 ellipsoid, 4 rectangular, 5 long and slender. Palpomere 5/3 ratio 2.45−3.0, 2.7.

Thorax. Dorsocentrals 6−7.

Wing (Fig. 3.97A). VR 2.6−3.0, 2.8. C length 300−400 μm, C/wing length 0.29−0.4, 0.33(3). Cu length 600−710, 697 μm. Cu/wing length 0.57−0.71, 0.64. Wing width/wing length 0.32−0.34, 0.33 (2). C with 7 setae.

Legs. Spurs of fore tibia 25−35, 31 μm and 15−18, 16 μm long, spurs of mid tibia 13−20, 15 (3) μm long, of hind tibia 28−35, 31 μm long. Width at apex of fore tibia 25−30, 27 μm, of mid tibia 23−28, 25 μm, of hind tibia 45−50, 48 μm. Apex of hind tibia expanded, with comb of 12−18, 14 setae and 1 seta near spur strongly S-shaped. Lengths and proportions of legs as in Table 3.58.

Hypopygium (Fig. 3.97C and D). Tergite IX with posterior margin slightly bilobed, with a hyaline structure, each lobe with about 4 setae, anal point present. Laterosternite IX with 2 setae. Inferior volsella digitiform, with glandular setae; sternapodeme inverted V-shaped; coxapodeme 45−55, 50 μm long, attachment point with phallapodeme placed in caudal third of lateral sternapodeme and directed caudally; phallapodeme strongly curved with projection for joint with sternapodeme placed prelateral, almost extending beyond tergite IX, 80−85, 83 μm long, with projection basally. Gonocoxite 88−100, 91 μm long with 2 setae apically. Gonostylus strongly curved, 18−25, 22 (4) μm long, with protuberance basally; megaseta 8−11 μm long. HR 2.9−4.0, 3.3, HV 4.4−4.6, 4.5 (2).

Distribution. PA: China (Tianjin).

3.71 *CORYNONEURA NASUTICEPS* HAZRA ET AL., 2003 (FIGS. 3.98A−E, 3.99A−D, 3.100A−D, AND 3.101A−F)

Corynoneura nasuticeps Hazra et al., 2003: 76.

Diagnostic characters. The adult male is characterized by AR 0.49, hind tibia with comb of 14−15 setae and 1 seta near spur developed as hook; inferior volsella nose-like, with several setae located near the apex of gonocoxite; gonostylus short, simple, little tapered with a pointed; sternapodeme inverted V-shaped. The adult female is characterized by AR 0.34; gonocoxite IX with single seta; tergite IX undivided with 4 setae. The pupa has shagreen of small points only on tergites II, VIII and IX,

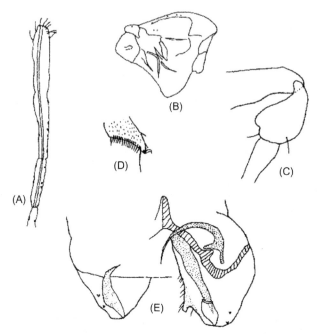

Figure 3.98 *Corynoneura nasuticeps* Hazra et al., 2003, Male imago (From Hazra, N., Nath, S., & Chaudhuri, P. K. (2003). The genus *Corynoneura* Winnertz (Diptera. Chironomidae) from the Darjeeling-Sikkim Himalayas of India, with description of three new species. *Entomologist's Monthly Magazine, 139,* 69–82). (A) The last two flagellomeres of antenna. (B) Thorax. (C) Fore trochanter. (D) Hind tibial comb. (E) Hypopygium.

shagreen of numerous small points and posteriorly located spinules of about similar size on tergites III–VII, tergites III–VII with 4–5 spinules arranged posteriorly in a transverse row; anal lobe with 24–25 filaments in fringe covering the lobes completely. The larva has AR 0.92–1.01; mandible with 1 apical and 4 inner teeth, apical tooth shorter than first inner one.

Male (Description from Hazra et al., 2003)

Total length 1.22 mm. Wing length 0.58 mm. Total length/wing length 2.10. Wing length/profemur length 3.13.

Head. AR 0.49, length of flagellomeres (I–VIII): 26, 22, 30, 37, 41, 44, 41, 118; apex with rosette of short hairs (Fig. 3.98A), ultimate flagellomere slightly clubbed and about equal to preceding three together with few plume hairs basally for nearly half of its length. Tentorium 103 μm long. Clypeus with seta. Length of palpomeres (I–V): 7, 13, 18, 22, 52. Cibarial pump 74 μm long. CA 0.67; CP 1.96.

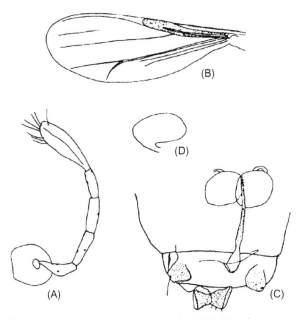

Figure 3.99 *Corynoneura nasuticeps* Hazra et al., 2003, Female imago (From Hazra, N., Nath, S., & Chaudhuri, P. K. (2003). The genus *Corynoneura* Winnertz (Diptera. Chironomidae) from the Darjeeling-Sikkim Himalayas of India, with description of three new species. *Entomologist's Monthly Magazine*, *139*, 69−82). (A) Antenna. (B) Wing. (C) Genitalia. (D) Seminal capsule.

Thorax (Fig. 3.98B). Dorsocentrals 5, prealars 2. Scutellum with 2 setae.

Wing. Membrane with fine punctuation; clavus terminates about 0.26 wing length; Cu_1 weakly sinuous; haltere light brown; VR 3.66; CR 0.26.

Legs. Fore trochanter as in Fig. 3.98C. Spur of fore tibia 18 μm long, of mid tibia 11 μm long, of hind tibia 25 μm long. Width of the apex of fore tibia 15 μm, mid tibia 18 μm and of hind tibia 26 μm. Hind tibia with comb of 14−15 setae and 1 seta near spur developed as hook; hind apex as in Fig. 3.98D, lengths and proportions of legs as in Table 3.59.

Hypopygium (Fig. 3.98E). Tergite IX with straight margin occupying much of the gonocoxites. Gonocoxite 51 μm long; short, stout; nose-like inferior volsella with several setae located near the apex of gonocoxite; gonostylus 37 μm long, short, simple, little tapered with a pointed; mega-seta 4 μm long; sternapodeme inverted V-shaped, phallapodeme strongly curved not extending beyond tergite IX; HR 1.4; HV 2.83.

Female (Description from Hazra et al., 2003)

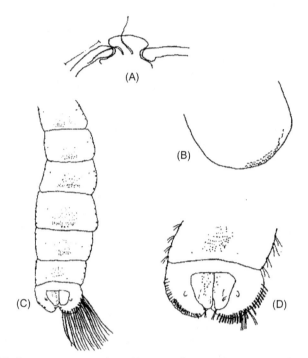

Figure 3.100 *Corynoneura nasuticeps* Hazra et al., 2003, Pupa imago (From Hazra, N., Nath, S., & Chaudhuri, P. K. (2003). The genus *Corynoneura* Winnertz (Diptera. Chironomidae) from the Darjeeling-Sikkim Himalayas of India, with description of three new species. *Entomologist's Monthly Magazine, 139,* 69–82). (A) Frontal apotome. (B) Tip of wing sheath. (C) Tergites. (D) Tergite VIII, anal lobe and male genital sac.

Similar to male with usual sex difference. Antenna (Fig. 3.99A): AR 0.34, length of flagellomeres $(I-V)$: 33, 30, 33, 33, 44.

Wing (Fig. 3.99B). Wing length 0.73 mm, clavus terminates 0.46 wing length.

Legs. LR_1 0.51, LR_2 0.57 and LR_3 0.56.

Genitalia (Fig. 3.99C and D). Gonocoxite IX with single seta; tergite IX undivided with 4 setae; seminal capsule (Fig. 3.99D) measuring 46×37; notum 99 µm long; cercus 33 µm long.

Pupa (Description from Hazra et al., 2003)

Total length of exuviae 1.53 mm.

Cephalothorax. Frontal setae 137 µm long, frontal apotome lamelliform (Fig. 3.100A). Postorbital setae 16 µm long, lamelliform; median antepronotals 22 µm and 18 µm long each; distance between anterior and median precorneals 5, between median and posterior precorneals 5; distance

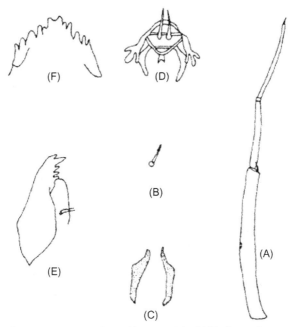

Figure 3.101 *Corynoneura nasuticeps* Hazra et al., 2003, Larva imago (From Hazra, N., Nath, S., & Chaudhuri, P. K. (2003). The genus *Corynoneura* Winnertz (Diptera. Chironomidae) from the Darjeeling-Sikkim Himalayas of India, with description of three new species. *Entomologist's Monthly Magazine, 139,* 69–82). (A) Antenna. (B) S I of labrum. (C) Premandible. (D) Palatum. (E) Mandible. (F) Mentum.

Table 3.59 Lengths (in μm) and proportions of legs segments of male *Corynoneura nasuticeps* Hazra et al., 2003

	fe	ti	ta$_1$	ta$_2$	ta$_3$	ta$_4$	ta$_5$	LR	BV	SV	BR
p$_1$	185	248	122	74	44	18	26	0.49	3.42	4.92	1.42
p$_2$	296	248	148	66	37	16	29	0.59	4.67	6.63	2.33
p$_3$	203	222	122	70	30	18	29	0.55	5.52	5.52	2.00

between Dc$_2$ and Dc$_3$ 7 μm long, between Dc$_3$ and Dc$_4$ 11 μm long. Wing sheath (Fig. 3.100B) apically with double rows of 30 pearls.

Abdomen (Fig. 3.100C). Shagreen of small points only on tergites II, VIII and IX, shagreen of numerous small points and posteriorly located spinules of about similar size on tergites III–VII, tergites III–VII with 4–5 spinules arranged posteriorly in a transverse row. Segment I with 1 L-seta, segment II 4 lateral setae of which 1 and 2 L-setae, 3 and 4 L-setae; segments III–VIII with 4 setae each. Anal lobe (Fig. 3.100D)

111 μm long, 148 μm wide with 24—25 filaments in fringe covering the lobes completely, and 3 long lamelliform lateral anal macrosetae; genital sac 74 μm long; anterior macroseta located at lateral filamentous setae 9—10, G/F 0.66; ALR 1.5.

Larva (Description from Hazra et al., 2003)

Total length of larva 1.48—2.22 mm long.

Antenna (Fig. 3.101A). AR 0.92—1.01, length of antennal segments (I—IV): 118—177, 65—81, 59—89, 4—5; distance of ring organ from the base 67—92 μm long; blade 14—16 μm long.

Labrum. All S-setae simple; S I (Fig. 3.101B) arising from socket.

Premandible (Fig. 3.101C). 26—28 μm long, simple. Palatum as in Fig. 3.101D. Mandible (Fig. 3.101E) 44—48 μm long with 1 apical and 4 inner teeth, apical tooth shorter than first inner one.

Mentum (Fig. 3.101F). 29—33 μm wide, triangular with 2 median teeth.

Body. Each segment bearing 2 setae ventrally. Procercus 11—15 μm long with 4 apical setae. Basiventral spine of posterior parapod dark brown, plumose 44—57 μm long; anterior parapods 111—120 μm long and posterior parapods 121—132 μm long.

Distribution. OR: India (West Bengal).

3.72 *CORYNONEURA PORRECTA* FU & SÆTHER, 2012 (FIG. 3.102A—G)

Corynoneura sp. A: Sæther, 1977; Fig. 61D, E.

Corynoneura porrecta Fu & Sæther, 2012: 41.

Material examined. Holotype male, CANADA: Lake Winnipeg, Victoria Beach, light trap, 9.vii.1969, P. S. S. Chang (CNC). Paratypes: 2 females as holotype; 1 male as holotype except for 25.vii.1969; Lake Winnipeg, 5 km off Selkirk (Horse) Island, 1 male, 25.vii.1969, P. S. S. Chang (ZMBN).

Diagnostic characters. The male imago is characterized by the wing length of 0.710.87 mm, AR of 0.5—0.6, 10 flagellomeres with "normal" whorl of setae, shortened palp, phallapodeme attached laterally on sterna-podeme, and both superior and inferior volsella strongly protruding. The female apparently lacks sensilla chaetica, lacks setae on gonocoxite IX, has 6—8 setae on tergite IX, and 8—9 well-developed lateral lamellae on the coxosternapodeme. The species is very closely related to *C. edwardsi* Brundin (Schlee, 1968b), but differs in the smaller size, lower AR, more

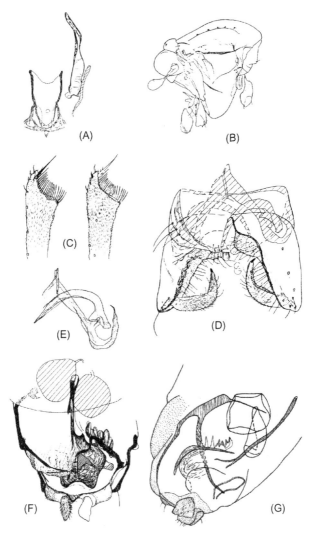

Figure 3.102 *Corynoneura porrecta* Fu & Sæther, 2012 (From Fu, Y., & Sæther, O. A. (2012). *Corynoneura* Winnertz and *Thienemanniella* Kieffer from the Nearctic region (Diptera: Chironomidae: Orthocladiinae). *Zootaxa*, *3536*, 1−61. www.mapress.com/j/zt). (A−E) Male imago. (A) Tentorium, stipes and cibarial pump. (B) Thorax. (C) Apex of hind tibia. (D) Hypopygium. (E) Sternapodeme and phallapodeme. (F and G) Female imago: genitalia.

numerous setae on tergite IX, more projecting inferior volsella, and differently shaped inferior and superior volsella. The short palpal segments of the male together with 10 flagellomeres with "normal" whorl of setae will probably also separate this species.

Male ($n = 2-3$)

Total length 1.19—1.45 mm. Wing length 0.71—0.87 mm. Total length/wing length 1.64—1.70. Wing length/profemur length 2.94—3.16.

Coloration. Brownish black, with brown legs and basal part of tarsi slightly lighter.

Head. AR 0.52—0.58. Ten distinct flagellomeres, longest antennal seta 220—240 μm. Sensilla chaetica and shape of ultimate flagellomere as in *C. edwardsi* Brundin (Schlee, 1968b). Temporals absent. Clypeus with 9—10 setae. Cibarial pump, tentorium and stipes as in Fig. 3.102A. Palp reduced, palpomere lengths (in μm): 10—13, 18—22, 27, 30—33, 50—51.

Thorax (Fig. 3.102B). Antepronotum with 2 setae. Dorsocentrals 5—6, prealars 2. Scutellum with 2 setae. Anapleural suture 97—104 μm long.

Wing. VR 2.10—2.16. C 216—272 μm long, C/wing length 0.30—0.32, Cu/wing length 0.62—0.63. Brachiolum with 1 seta.

Legs. Spur of fore tibia 22—25 μm, spurs of mid tibia 10—12 μm and 10 μm, and of hind tibia 35—40 μm and 8—10 μm. Width at apex of fore tibia 23—25 μm, of mid tibia 20—25 μm, and of hind tibia (a) 40—42 μm. Width of hind tibia 1/3 from apex (d) 25 μm, elongation length (b) 33—37 μm, length of maximum thickening (c_1) 55—65 μm, total length of thickening (c_2) 110—116 μm; a/d 1.60—1.68, b/d 1.32—1.48, c_1/d 2.20—2.60, c_2/d 4.40—4.64. Hind tibia (Fig. 3.102C) expanded, with a slightly S-shaped and relatively short apical seta, and comb with 17—19 setae, shortest seta 13—18 μm, longest seta 22 μm. Sensilla chaetica absent. Lengths of leg segments and their proportions as in Table 3.60.

Hypopygium (Fig. 3.102D and E). Tergite IX with 9—12 setae; laterosternite IX with 2—3 setae. Phallapodeme 70—80 μm long, attached laterally to sternapodeme. Superior volsella tongue-shaped, strongly

Table 3.60 Lengths (in μm) and proportions of legs segments of male *Corynoneura porrecta* Fu & Sæther, 2012 ($n = 2-3$)

	fe	ti	ta_1	ta_2	ta_3	ta_4
p_1	252—282	299—354	163—185	83—87	38—43	15
p_2	333—394	323—381	173—200	78—83	37—43	17—18
p_3	279—337	306—354	167—190	90—105	35—38	15

	ta_5	LR	BV	SV	BR
p_1	25—32	0.52—0.55	4.22—4.83	3.31—3.44	2.0—2.2
p_2	25—29	0.52—0.54	5.25—5.67	3.79—3.93	2.0—2.1
p_3	30—32	0.53—0.55	4.42—4.64	3.50—3.64	2.1—2.5

Source: From Fu, Y., & Sæther, O. A. (2012). *Corynoneura* Winnertz and *Thienemanniella* Kieffer from the Nearctic region (Diptera: Chironomidae: Orthocladiinae). *Zootaxa*, 3536, 1—61. www.mapress.com/j/zt.

Table 3.61 Lengths (in μm) and proportions of legs segments of female *Corynoneura porrecta* Fu & Sæther, 2012 (*n* = 1−2)

	fe	ti	ta_1	ta_2	ta_3	ta_4
p_1	198−230	254−284	140−150	62−80	31−38	13−18
p_2	300−342	288−314	160−180	60−86	30−38	14−16
p_3	260−300	268−247	150−165	64−98	36	11−13

	ta_5	LR	BV	SV	BR	
p_1	22−26	0.53−0.55	4.10−4.63	3.23−3.43	2.6−3.0	
p_2	22−25	0.56−0.57	5.07−5.94	3.64−3.68	2.4−2.5	
p_3	24−25	0.56−0.57	4.43−5.02	3.52−3.62	2.3−2.4	

Source: From Fu, Y., & Sæther, O. A. (2012). *Corynoneura* Winnertz and *Thienemanniella* Kieffer from the Nearctic region (Diptera: Chironomidae: Orthocladiinae). *Zootaxa*, 3536, 1−61. www.mapress.com/j/zt.

projecting together with basal lobe. Gonocoxite 100−120 μm long, gonostylus 50−57 μm long. HR 1.74−2.00, HV 2.08−2.38.

Female (*n* = 1−2)

Total length 1.23−1.24 mm. Wing length 0.69−0.79 mm. Total length/wing length 1.55−1.79. Wing length/profemur length 3.44−3.49.

Coloration. Yellowish brown with blackish brown vittae, scutellum, postnotum, preepisternum and abdomen. Femora, tibiae, apices of ta_1−ta_4 and ta_5 dark brown; remaining part of tarsi pale.

Head. AR 0.38−0.45. Flagellomeres lengths (in μm): 38−42, 30−33, 29−33, 28, 48−56. Temporals absent. Clypeus with 8−9 setae. Tentorium 70−100 μm long. Stipes 50 μm long. Palpomere lengths (in μm): 11−12, 16−22, 14−18, 31−36, 45−55.

Thorax. Antepronotum with 1 seta. Dorsocentrals 7−8, prealars 2. Scutellum with 2 setae. Anapleural suture 95−96 μm long.

Wing. VR 3.00−3.25. C 362−380 μm long, C/wing length 0.48−0.52, Cu/wing length 0.60−0.64. Brachiolum with 1 seta.

Legs. Spur of fore tibia 14 μm, spurs of mid tibia 12−20 μm and 10−14 μm, and of hind tibia 33−39 μm and 7−10 μm. Width at apex of fore tibia 22 μm, of mid tibia 20−24 μm, and of hind tibia (*a*) 34−35 μm. Width of hind tibia 1/3 from apex (*d*) 18−21 μm, elongation length (*b*) 27−32 μm, length of maximum thickening (c_1) 49−54 μm, total length of thickening (c_2) 66−80 μm; *a/d* 1.62−1.92, *b/d* 1.50−1.52, c_1/*d* 2.57−2.72, c_2/*d* 3.67−3.81. Hind tibia expanded, with a comb of 14 setae, shortest seta 14−22 μm, longest seta 23−30 μm. Sensilla chaetica absent. Lengths of leg segments and their proportions as in Table 3.61.

Genitalia (Fig. 3.102F and G). Cercus 31−36 μm long. Gonocoxite IX with 1−2 setae. Tergite IX with 6−8 setae. Notum 70−80 μm long.

Coxosternapodeme with 8—9 lateral lamellae. Seminal capsule 60—70 μm long, 48—50 μm wide.

Remarks. The female illustrated by Sæther (1977) as *Corynoneura* sp. A probably belongs to *Corynoneura porrecta*, but it is not included in the type material.

Distribution. NE: Canada.

3.73 *CORYNONEURA PRIMA* MAKARCHENKO & MAKARCHENKO, 2006B (FIGS. 3.103A—C AND 3.104A—C)

Corynoneura prima Makarchenko & Makarchenko, 2006b: 154.

Diagnostic characters. Male: antenna with 7—8 flagellomeres, apical flagellomere with rosette of 20—24 long and pale setae. AR 0.45—0.48. Inferior volsella reduced and looks like low roundish protrusion. Transverse sternapodeme disposed low, with concave apex. Phallapodemes short and situated low. Pupa (Fig. 3.104A—C): wing case with 2—3 rows of pearls. Tergite II with 26 medial and small spinules which not form the row. Segment II with 4 pairs of lateral setae, 2 anterior setae are simple.

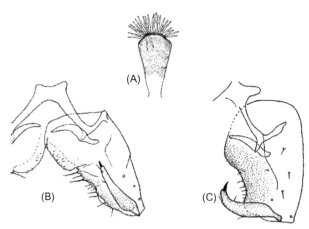

(A)

(B)   (C)

Figure 3.103 *Corynoneura prima* Makarchenko & Makarchenko, 2006b, Male imago (From Makarchenko, E. A., & Makarchenko, M. A. (2006b). Chironomids of the genera *Corynoneura* Winnertz, 1846 and *Thienemanniella* Kieffer, 1919 (Diptera, Chironomidae, Orthocladiinae) of the Russian Far East. *Euroasian Entomological Journal*, 5(2), 151—162). (A) Apex of apical flagellomere with rosette of setae. (B and C) Part of hypopygium (tergite IX removed).

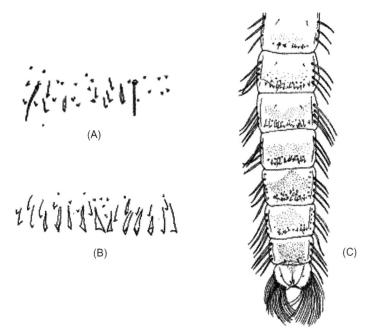

Figure 3.104 *Corynoneura prima* Makarchenko & Makarchenko, 2006b, Pupa imago (From Makarchenko, E. A., & Makarchenko, M. A. (2006b). Chironomids of the genera Corynoneura Winnertz, 1846 and Thienemanniella Kieffer, 1919 (Diptera, Chironomidae, Orthocladiinae) of the Russian Far East. *Euroasian Entomological Journal, 5*(2), 151−162). (A) Spinules and spines in posterior side of tergite II. (B) Spinules and spines in posterior side of tergite III. (C) Anal segments.

Male (Description from Makarchenko & Makarchenko, 2006b**)**
Total length 1.0−1.2 mm. Total length/wing length 1.56−1.74.

Head. Eyes bare. Antenna (Fig. 3.103A) with 7−8 flagellomeres, AR 0.45−0.48. Lengths of palpomeres (in μm): —, 10, 14, 16, 52.

Thorax. Dorsocentrals 3−4, prealars 2.

Wing. Wing length 0.62−0.69 mm.

Legs. Fore trochanter with dorsal keel. Spurs of fore tibia 20 μm long, of mid tibia 8 μm long, of hind tibia 28 μm and 15 μm long. Hind tibia with comb of 15 setae. Lengths and proportions of legs as in Table 3.62.

Hypopygium (Fig. 3.103B and C). Tergite IX with 6 long setae. Inferior volsella with small rounded corner, located in caudal of gonocoxite. Gonostylus apically curved, narrowed and elongated, transverse sternapodeme curved into U-shaped.

Distribution. PA: Russia (Far East).

Table 3.62 Lengths (in μm) and proportions of legs segments of male *Corynoneura prima* Makarchenko & Makarchenko, 2006b

	fe	ti	ta$_1$	ta$_2$	ta$_3$	ta$_4$
p$_1$	208−228	252−272	108−132	64−80	40−44	20
p$_2$	284−308	252−280	144−160	68−76	36−40	20
p$_3$	232−252	232−260	120−128	72−84	28−36	16−20

	ta$_5$	LR	BV	SV	BR
p$_1$	24−28	0.43−0.48	3.67−3.84	3.79−4.26	2.4
p$_2$	24−30	0.57	4.51−4.59	3.67−3.72	2.8
p$_3$	28	0.49−0.52	3.81−4.05	3.87−4.0	2.1

3.74 *CORYNONEURA PROMINENS* FU ET AL., 2009 (FIG. 3.105A−E)

Corynoneura prominens Fu et al., 2009: 28.

Material examined. Holotype male (BDN No. 24734), P.R. China: Jiangxi Province, Wuyi Mountain, 27°28′N, 118°1′E, alt. 400 m, sweep net, 15.vi.2004, C. Yan. Paratypes: 2 males, as holotype (BDN No. 24583, 24735); 1 male (BDN No. 05363), Chongqi City, Nanchuan County, peak of Jinfo Mountain, 29°9′N, 107°8′E, alt. 1200 m, sweep net, 9.v.1986, X. Wang; 1 male (BDN No. 25152), Hunan Province, Hengyang County, Heng Mountain, 112°52′E, 27°15′N, alt. 1290 m, sweep net, 20.vii.2004, C. Yan.

Diagnostic characters. Similar to *C. prima* in having antenna with 8 flagellomeres, but can be separated by having an anal point.

Male (*n* = 5)

Total length 0.82−0.95, 0.91 (4) mm. Wing length 0.50−0.64, 0.57 (4) mm. Total length/wing length 1.5−1.55, 1.53 (3). Wing length/length of profemur 2.7−3.0, 3.0 (3).

Coloration. Head yellow-brown with brown frontal vertex. Antenna and palpomere pale yellowish brown. Thorax yellowish brown, abdomen brown. Legs yellow. Wings light yellow to hyaline.

Head. Antenna with 8 flagellomeres; ultimate flagellomere 83−133, 108 (4) μm long, antenna apically expanded, with rosette of apical sensilla chaetica(Fig. 3.105B), thickening 55−60, 57 μm long, with maximum width 15−20, 18 μm; AR 0.42−0.60, 0.52 (4). Clypeus with 4−6, 5 (3) setae. Tentorium (Fig. 3.105E) 92−115, 105 μm long, 10−15, 13 μm wide. Stipes 35−40, 38 (3) μm long. Palpomeres length (in μm): 10; 10−13, 11; 13−18, 15; 18−20, 19; 33−43, 39 (4); Palpomere 2,

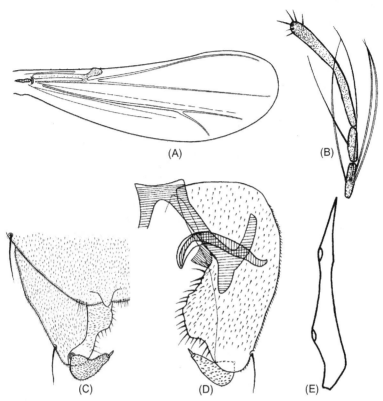

Figure 3.105 *Corynoneura prominens* Fu et al., 2009, Male imago (From Fu, Y., Sæther, O. A., & Wang, X. (2009). *Corynoneura* Winnertz from East Asia, with a systematic review of the genus (Diptera: Chironomidae: Orthocladiinae). *Zootaxa, 2287,* 1–44). (A) Wing. (B) Antenna. (C) Hypopygium, dorsal view. (D) Hypopygium, ventral view. (E) Tentorium.

3 and 4 ellipsoid, palpomere 5 long and slender. Palpomere 5/3 ratio 2.1−2.8, 2.6 (4).

Thorax. Dorsocentrals 5.

Wing (Fig. 3.105A). VR 3.2−3.4, 3.3 (3). C length 100−140 μm, C/wing length 0.20−0.23, 0.22 (4). Cu length 310−375, 337 (3) μm. Cu/wing length 0.53−0.59, 0.57. Wing width/wing length 0.44−0.48, 0.46. C with 5−7, 6 setae.

Legs. Spur of fore tibia 10−18, 13 (4) μm and 8−10, 8 (2) μm long; spurs of mid tibia 8−10, 8 (2) μm long; of hind tibia 25−35, 29 μm long. Width at apex of fore tibia 13−18, 15 μm; of mid tibia 13−20, 15 μm; of hind tibia 23−33, 28 μm. Apex of hind tibia expanded, with comb of 14−16, 15 setae and 1 seta near spur strongly S-shaped. Lengths and proportions of legs as in Table 3.63.

Table 3.63 Lengths (in μm) and proportion of leg segments of male *Corynoneura prominens* Fu et al., 2009

	fe	ti	ta_1	ta_2	ta_3	ta_4
p_1	175−218, 197	210−240, 227	105−125, 114	55−75, 68	28−43, 35	13−25, 19
p_2	235−290, 260	210−260, 229	110−165, 135	58−73, 64	30−38, 33	15−23, 24
p_3	198−245, 215	200−275, 230	100−130, 110	60−70, 64	25−33, 28	15−20, 18

	ta_5	LR	BV	SV	BR
p_1	20−30, 26	0.48−0.52, 0.5	3.3−4.6, 3.8	3.6−3.9, 3.7	1.5−2.3, 1.8
p_2	25−28, 26	0.51−0.63, 069	4.2−4.9, 4.5	3.3−3.7, 3.5	1.7−2.0, 1.85
p_3	20−30, 25	0.48−0.51, 0.49	3.8−4.2, 4.0	3.8−4.0, 3.9	1.6−1.8, 1.7

Hypopygium (Fig. 3.105C and D). Tergite IX posterior margin slightly bilobed, laterosternite IX with 1 long seta. Anal point present. Superior volsella anteriomedially fused; inferior volsella broad and protruding, with glandular setae. Sternapodeme inverted U-shaped; coxapodeme 18 μm long, attachment point with phallapodeme placed caudal and caudally directed; phallapodeme weakly curved, not extending beyond tergite IX, 30–35, 33 μm long, with projection basally. Gonocoxite 50–65, 58 μm long with 1 seta apically. Gonostylus strongly curved, 18–25, 22 (4) μm long; megaseta 3–5, 4 μm long. HR 2–3.4, 2.7 (4), HV 3.8–4.2, 4.0.

Distribution. OR: China (Jiangxi and Hunan).

3.75 *CORYNONEURA RENATA* WIEDENBRUG ET AL., 2012 (FIGS. 3.106A AND B AND 3.107A–H)

Corynoneura renata Wiedenbrug et al., 2012: 40.

Diagnostic characters. Male of *Corynoneura renata* differs from other species with phallapodeme attached in the caudal apex of the sternapodeme and S-seta on apex of hind tibia, by the antenna with 7 flagellomeres, aedeagal lobe with a large base and inferior volsella apical and large. The pupa of this species can be recognized by the conjunctive of tergite and sternite IV/V armed with long and thin spines and the anal

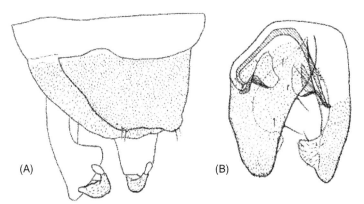

(A)   (B)

Figure 3.106 *Corynoneura renata* Wiedenbrug et al., 2012; Male imago (From Wiedenbrug, S., Lamas, C. J. E., & Trivinho-Strixino, S. (2012). A review of the genus *Corynoneura* Winnertz (Diptera: Chironomidae) from the neotropical region. *Zootaxa*, *3574*, 1–61). (A) Tergite IX and gonostylus. (B) Tergite IX and gonostylus removed, sclerites hatched; right is dorsal view, left is ventral view.

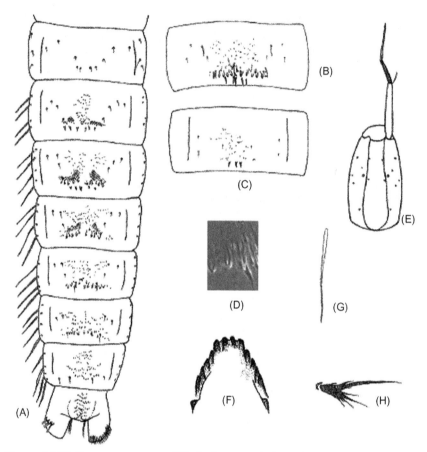

Figure 3.107 *Corynoneura renata* Wiedenbrug et al., 2012; Immatures imago (From Wiedenbrug, S., Lamas, C. J. E., & Trivinho-Strixino, S. (2012). A review of the genus *Corynoneura* Winnertz (Diptera: Chironomidae) from the neotropical region. *Zootaxa, 3574*, 1–61). (A–D) Pupa imago. (A) Tergites II–IX. (B) Tergite IV. (C) Sternite IV. (D) Detail of posterior shagreen of tergite IV. (E–H) Larva imago. (E) Head, dorsal view and separated antenna. (F) Mentum. (G) Abdominal seta. (H) Subbasal seta of posterior parapod.

lobes laterally straight, posteriorly rounded joined medially through a straight perpendicular margin. The larva has postmentum length/head width 1.16, first antennal segment shorter than postmentum length, second antennal segment shorter than the third, basal seta on posterior parapods split from the base and dark brown.

Male (Description from Wiedenbrug et al., 2012)

Thorax length 0.37–0.39 mm.

Coloration. Thorax brownish. Legs whitish.

Head. AR = 0.4−0.5. Antenna with 7 flagellomeres, apical flagellomere 130−145 µm. Flagellomeres, except first and last, with more than one row of setae each. Eyes pubescent.

Thorax. Antepronotal lobes dorsally tapering.

Wing. Not possible to measure.

Legs. Hind tibial scale 30−37 µm long, with one subapical S-seta.

Hypopygium (Fig. 3.106A and B). Tergite IX with 4 setae. Laterosternite IX with 1 seta (not drawn). Superior volsella absent. Inferior volsella apical and large. Aedeagal lobe with a broad base and distally rounded. Transverse sternapodeme straight. Phallapodeme with posterior margin sclerotized, attached in the caudal apex of the sternapodeme. Gonostylus curved.

Pupa (Description from Wiedenbrug et al., 2012)

Total length 1.18−1.29 mm.

Color (exuviae). Cephalothorax light brown, abdomen transparent except brownish muscle markings, lateral margin and anal lobe.

Cephalothorax. Frontal apotome slightly granulated. Thorax suture smooth except scutal tubercle region rugose. All Dc-setae thin taeniate. Dc_1 displaced ventrally. Wing sheaths with one to two rows of pearls.

Abdomen (Fig. 3.107A−D). Sternite and tergite I bare, tergite II with few thin shagreen points, tergites III−VII with very fine shagreen increasing in size posteriorly. Sternites II and III with scarse shagreen of elongate spinules, also on parasternites. Sternites IV−VII with fine shagreen with stronger posterior spines. Conjunctives tergites II/III−VI/VII and Sternites IV/V−VI/VII with spinules, longer on tergite IV/V and Sternite IV/V. Segment I with 1, II with 3 L-setae and III−VIII with 4 long taeniate L-setae. Anal lobe laterally straight, posteriorly rounded (Fig. 3.107A). Anal lobe with fringe not complete, 3 taeniate macrosetae and inner setae taeniate.

Larva (Description from Wiedenbrug et al., 2012)

Head. Postmentum 125 µm long. Head capsule integument smooth. Mentum of observed specimen frayed. Possibly with two median teeth, first lateral teeth almost as large as median teeth (Fig. 3.107F). Antenna 233 µm long, segments two and three darker (Fig. 3.107E). First segment shorter than postmentum length, second antennal segment shorter than the third.

Abdomen. Ventral setae modified, longer and wider (Fig. 3.107G). Subbasal seta on posterior parapod dark, split from the base (Fig. 3.107H).

Remarks. Larva of *C. renata* was collected from the surface of stones, of a fast flowing small river in the Atlantic Forest in São Paulo State. Pupal exuviae of this species were also collected in Brazil in small rivers in the same biome at Rio Grande do Sul State (Wiedenbrug, 2000).

Distribution. NT: Brazil.

3.76 *CORYNONEURA SALVINIATILIS* WIEDENBRUG ET AL., 2012 (FIGS. 3.108A−E, 3.109A−C, AND 3.110A−G)

Corynoneura salviniatilis Wiedenbrug et al., 2012: 42.

Diagnostic characters. Male of *C. salviniatilis* is very similar to *C. diogo*, but the apex of the antenna bears about 10 sensilla chaetica, the gonocoxite is laterally rounded and the aedeagal lobe is small and rectangular. The female is characterized by the antennal apex rounded, genitalia with two Sca about 30 µm long, copulatory bursa semi-circled, dorsal with membranous oral extension, not median invaginated and laterosternite IX without seta. Pupae with abdominal shagreen composed of very fine shagreen, points of the sternites are slightly elongated; anal lobe rounded, both together wider than longer. The larval head integument smooth, first antennal segment shorter than postmentum length, second antennal segment subequal or larger than the third, mentum with 2 median teeth, first pair of lateral teeth larger than the second.

Male (Description from Wiedenbrug et al., 2012)

Total length 0.83 mm. Wing length 0.58 mm.

Coloration. Thorax brownish. Abdominal tergites I−IV whitish, other tergites and genitalia brownish. Legs whitish.

Head. AR = 0.32. Antenna with 9 flagellomeres, apical flagellomere 140 µm. Flagellomeres, except first and last, with more than one row of setae each. Antennal apex with 10 sensilla (Fig. 3.108A). Eyes pubescent.

Thorax. Antepronotal lobes dorsally tapering.

Wing. Clavus/wing length 0.22. Anal lobe absent (Fig. 3.108C).

Legs. Hind tibial scale 27 µm long, with one S-seta (Fig. 3.108B).

Hypopygium (Fig. 3.108D and E). Tergite IX with 2 setae. Laterosternite IX without seta. Superior and inferior volsella absent. Gonocoxite laterally rounded, posterior region between inner and outer margin with sclerotized longitudinal ridge. Aedeagal lobe small and rectangular. Lateral and transverse sternapodeme straight. Phallapodeme elongated, with posterior margin sclerotized, attached in the caudal apex of the sternapodeme.

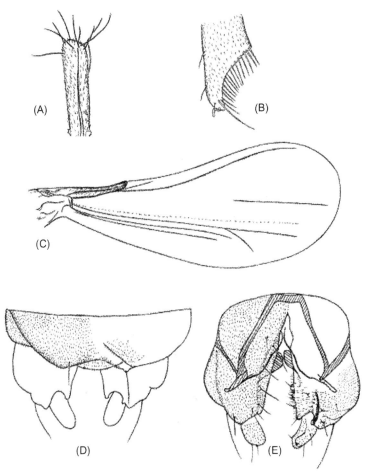

Figure 3.108 *Corynoneura salviniatilis* Wiedenbrug et al., 2012; Male imago (From Wiedenbrug, S., Lamas, C. J. E., & Trivinho-Strixino, S. (2012). A review of the genus *Corynoneura* Winnertz (Diptera: Chironomidae) from the neotropical region. *Zootaxa*, *3574*, 1−61). (A) Terminal flagellomere. (B) Apex of hind tibia. (C) Wing. (D and E) Hypopygium. (D) Tergite IX; left is dorsal view, right is ventral view. (E) Tergite IX removed, sclerites hatched; right is dorsal view, left is ventral view.

Female (Description from Wiedenbrug et al., 2012)

Wing length 0.56 mm.

Coloration. Thorax brownish. Abdominal tergites II−IX brownish. Legs whitish.

Head. AR = 0.37. Antenna with 5 flagellomeres, apical flagellomere 40 μm long. Flagellomeres with more than one row of setae each. Antennal apex with about 15 sensilla (Fig. 3.109A). Eyes pubescent.

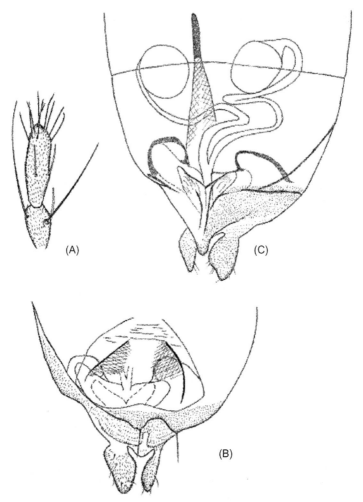

Figure 3.109 *Corynoneura salviniatilis* Wiedenbrug et al., 2012; Female imago (From Wiedenbrug, S., Lamas, C. J. E., & Trivinho-Strixino, S. (2012). A review of the genus *Corynoneura* Winnertz (Diptera: Chironomidae) from the neotropical region. *Zootaxa*, *3574*, 1–61). (A) Terminal flagellomeres. (B and C) Genitalia. (B) Dorsal view. (C) Ventral view.

Thorax. Antepronotal lobes dorsally tapering.

Wing. Clavus/wing length 0.40. Anal lobe absent.

Legs. Hind tibial scale 30 μm long, with one small S-seta.

Genitalia (Fig. 3.109B and C). Tergite IX with 2 setae. Laterosternite IX without seta. Two seminal capsules 30 μm long, both spermathecal ducts with a loop, ducts join together shortly before seminal eminence,

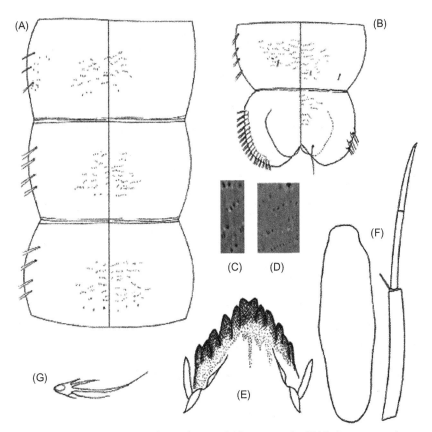

Figure 3.110 *Corynoneura salviniatilis* Wiedenbrug et al., 2012; Immatures imago (From Wiedenbrug, S., Lamas, C. J. E., & Trivinho-Strixino, S. (2012). A review of the genus *Corynoneura* Winnertz (Diptera: Chironomidae) from the neotropical region. *Zootaxa, 3574*, 1−61). (A−D) Pupa. (A) Segments II−IV; left is ventral view, right is dorsal view. (B) Tergites VIII, IX and anal lobe; left is ventral, right is dorsal view. (C) Detail of posterior shagreen of tergite IV. (D) Detail of posterior shagreen of sternite IV. (E−G) Larva. (E) Mentum. (F) Frontal apotome and antenna. (G) Subbasal seta of posterior parapod.

which has sclerotized outer borders. Membrane well sclerotized. Apodeme lobe with median border sclerotized. Coxosternapodeme curved, with one end at roof of copulatory bursa, last with sclerotized dorsal-oral extension, oral median invaginated. Labia difficult to define. Gonocoxapodeme straight, gonapophyses median pointed. Cercus 20 μm long.

Pupa (Description from Wiedenbrug et al., 2012)

Total length 1.35 mm.

Coloration (exuviae). Cephalothorax and abdomen almost transparent.

Cephalothorax. Frontal apotome almost smooth. Thorax suture smooth except scutal tubercle region rugose. Dc₃ taeniate much longer than the others, longer than 100 µm. Dc₁ displaced ventrally. Wing sheaths with two rows of pearls.

Abdomen (Fig. 3.110A–D). Sternite I and tergites I and II bare, tergites III–IX with very fine shagreen of short points. Sternites II–VIII with very fine shagreen of slightly elongated points. Conjunctives tergites III/IV–V/VI and Sternites III/IV–VII/VIII with spinules. Segment I with 1, II with 3 L-setae and IV–VIII with 4 long taeniate L-setae. Anal lobe rounded (Fig. 3.110B). Anal lobe with fringe almost complete, 3 taeniate macrosetae and inner setae taeniate.

Larva (Description from Wiedenbrug et al., 2012)

Head. Postmentum 185–188 µm long. Head capsule integument of the exuviae ventral slightly rugose. Mentum with two median teeth, first lateral teeth larger than the second (Fig. 3.110E). Antenna 240–248 µm long, segments two and three darker. First segment shorter than postmentum length (Fig. 3.110F).

Abdomen. Ventral setae modified, wider apically split. Subbasal seta on posterior parapod split from the base with one spine longer (Fig. 3.110G).

Remarks. This species was collected from *Salvinia* sp. (Salviniaceae), as well as with artificial substrate, left between the aquatic macrophytes in marginal lakes of the Mogi-Guaçú River in São Paulo State, Brazil. See *C. diogo* Wiedenbrug et al., 2012.

Distribution. NT: Brazil.

3.77 *CORYNONEURA SCHLEEI* MAKARCHENKO & MAKARCHENKO, 2010 (FIG. 3.111A–F)

Corynoneura schleei Makarchenko & Makarchenko, 2010: 363.

Diagnostic characters. Antenna with 12 flagellomeres; terminal flagellomere with long setae and with light sensitive hairs in 1/6 of apical part. t₃ with straight apical seta; $a/d = 1.5$; $b/d = 0.75-1.0$. Inferior volsella long and wide, with apex not far from distal edge of gonocoxite; superior volsella is triangular and bare; under superior volsella in basal part of gonocoxite fan-shaped long setae are situated. Sternapodeme inverted U-shaped. Phallapodemes low and large and situated near of connecting of trunnion with lateral part of sternapodeme.

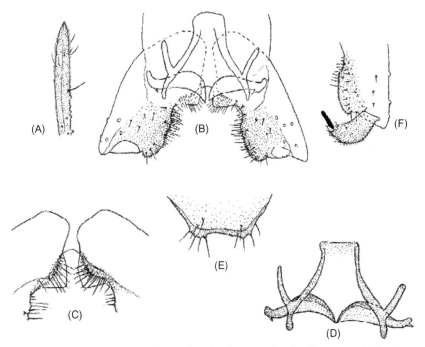

Figure 3.111 *Corynoneura schleei* Makarchenko & Makarchenko, 2010, Male imago (From Makarchenko, E. A., & Makarchenko, M. A. (2010). New data on the fauna and taxonomy of *Corynoneura* Winnertz (Diptera, Chironomidae, Orthocladiinae) for the Russian Far East and bordering territories. *Euroasian Entomological Journal*, *9*(3), 353–370 + II). (A) Distal part of terminal antennal flagellomere. (B) Total view of hypopygium, from above (tergite IX deleted); (C) Basal part of gonocoxite, from below. (D) Sternapodema and phallapodemes. (E) Posterior margin of tergite IX. (F) Distal part of gonocoxite and gonostylus.

Male (Description from Makarchenko & Makarchenko, 2010)

Coloration. Head brown, thorax dark brown, abdomen brown, legs brownish.

Total length 1.6–1.8 mm. Total length/wing length 1.31–1.50.

Head. Eyes bare. Antenna with 12 flagellomeres, ultimate flagellomere covered with long setae, its 1/6 apical portion with light sensitive hairs (Fig. 3.111A); AR 0.98–1.05. Lengths of palpomeres (in µm): 16, 32, 30, 40–44, 56–64.

Thorax. Dorsocentrals 3–6, prealars 1–3. Scutellum with 2 setae.

Wing. Wing length 1.17–1.28 mm.

Legs. Fore trochanter with dorsal keel. Spurs of fore tibia 28 µm long, of mid tibia 8–16 µm and 16 µm long, of hind tibia 8–32 µm and

Table 3.64 Lengths (in μm) and proportions of legs segments of male *Corynoneura schleei* Makarchenko & Makarchenko, 2010

	fe	ti	ta₁	ta₂	ta₃	ta₄
p₁	336−352	416−448	224−248	138−144	72−80	32−36
p₂	496−512	504−528	240−256	148−152	68−76	28−32
p₃	416−432	464−496	268−292	152−172	68−72	28−36

	ta₅	LR	BV	SV	BR	
p₁	48−52	0.52−0.57	3.25−3.46	3.16−3.48	1.5−1.8	
p₂	52	0.48	4.15−4.19	4.06−4.17	3.0	
p₃	52	0.58−0.59	3.67−3.90	3.18−3.28	2.2	

12−36 μm long. Hind tibia expanded, with comb of 10−12 setae, lengths and proportions of legs as in Table 3.64.

Hypopygium (Fig. 3.111B−F). The posterior margin of tergite IX medially concave, with 6−12 short setae. Laterosternite IX with 3−4 short setae. Gonocoxite 120−132 μm, superior volsella triangular, bare. Inferior volsella long and wide. Sternapodeme curved into U-shaped, transverse sternapodeme 28 μm. Gonostylus 36−44 μm long.

Distribution. PA: Russia (Far East).

3.78 *CORYNONEURA SCUTELLATA* WINNERTZ, 1846 (FIGS. 3.112A−C AND 3.113A−C)

Corynoneura scutellata Winnertz, 1846: 13; Boesel & Winner, 1980: 502; Hirvenoja & Hirvenoja, 1988: 217; Schlee, 1968b: 40; Fu et al., 2009: 30; Fu & Sæther, 2012: 44.

Material examined. P.R. China: Fujian Province, Liancheng County, Guanzhi Mountain, 25°43′N, 116°45′E, alt. 660 m, sweep net, 1 male, 4.v.1993, X. Wang; Yunnan Province, Lijiang County, Heilongtan, 26°52′N, 100°15′E, alt. 2400 m, sweep net, 1 male, 28.v.1996, X. Wang; Tianjin City, Ji County, 40°3′N, 117°24′E, alt. 857 m, sweep net, 1 male, 15.ix.1987, X. Wang; Tianjin City, Nankai University, 39°8′N, 117°12′E, alt. 2−5 m, sweep net, 6 males, 20.iv.1986, X. Wang. Canada: Manitoba, Lake Winnipeg, Victoria Beach, light trap, 17 males, 2 females, 9.vii. & 25.vii.1969, P. S. S. Chang; Pine Dock, light trap, 1 male, 10. vii.1969, P. S. S. Chang; 20 Mile Creek, light trap, 4 males, 26.viii.1971, N. Hooper. USA: Ohio, Knox County, Ballfield Marsh, 1 mile south of Bladensburg, 1 larval exuviae, 18.v.2001, M. Micacchion; Summit

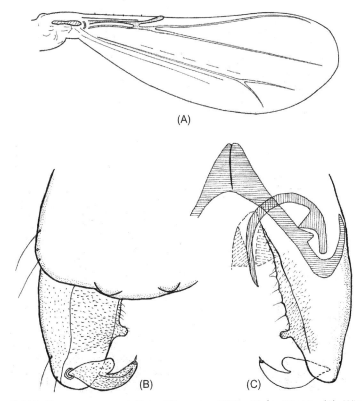

Figure 3.112 *Corynoneura scutellata* Winnertz, 1846; Male imago. (A) Wing. (B) Hypopygium, dorsal view. (C) Hypopygium, ventral view.

County, Steels Corners Wetland, 3 larval exuviae, 22.v.2001, M. Micacchion (ZMBN, MJB).

Diagnostic characters. Total length 1.07−1.82, 1.50 mm. Wing length 0.85−1.25, 1.03 mm. Antenna with 10 flagellomeres; apically acute; AR 0.74−1.1, 0.82. Posterior margin of tergite IX bilobed. Inferior volsella digitiform, with small inner lobe in middle part. Sternapodeme inverted V-shaped, attachment point with phallapodeme placed in caudal third of lateral sternapodeme and directed caudally, phallapodeme strongly curved with projection for joint with sternapodeme placed prelateral. Gonostylus curved apically.

Male (*n* = 9)

Total length 1.07−1.82, 1.50 mm. Wing length 0.85−1.25, 1.03 mm. Total length/wing length 1.32−1.67, 1.5 (8). Wing length/length of profemur 3.13−3.53, 3.28 (8).

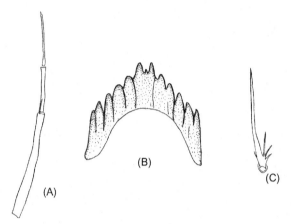

Figure 3.113 *Corynoneura scutellata* Winnertz, 1846; Larva imago (From Fu, Y., & Sæther, O. A. (2012). *Corynoneura* Winnertz and *Thienemanniella* Kieffer from the Nearctic region (Diptera: Chironomidae: Orthocladiinae). *Zootaxa, 3536,* 1−61. www. mapress.com/j/zt). (A) Antenna. (B) Mentum. (C) Subbasal seta of posterior parapod.

Coloration. Head dark brown. Antenna and palpomere pale yellow-brown. Thorax brown. Abdominal segments yellow-brown. Legs yellow-brown. Wings transparent and yellowish, with pale yellow clava.

Head. Eyes bare, reniform. Antenna with 10 flagellomeres; ultimate flagellomere 225−335, 280 µm long, apex with short sensilla chaetica which extends a little way back from tip, apically acuate. AR 0.74−1.1, 0.82. Clypeus with 5−10, 8 setae. Temporal setae lacking. Tentorium 95−170, 139 µm long, 15−25, 19 µm wide. Stipes: 62−65, 64 µm long. Palpomeres length (in µm): 12.5−20, 17; 15−25, 20; 23−28, 26; 28−40, 35; 40−65, 57. Palpomere 2 ellipsoid, 3 and 4 rectangular, 5 long and slender. Palpomere 5/3 ratio 1.5−2.4, 2.2.

Thorax. Antepronotum without lateral setae. Dorsocentrals 5. Other setae cannot be seen.

Wing (Fig. 3.112A). VR 2.1−3.2, 2.9. C length 300−360, 345 µm, C/wing length 0.26−0.33, 0.31. Cu length 600−720, 660 µm. Cu/wing length 0.22−0.35, 0.25. Wing width/wing length 0.40−0.44, 0.42. C with 6−8, 7 setae.

Legs. Fore trochanter with keel. Spur of front tibia 17.5−35, 26 µm and 15−18, 16 µm long, spurs of middle tibia 7.5−15, 13 µm long, of hind tibia 38−50, 45 µm long. Width at apex of fore tibia 23−33, 28 µm, of mid tibia 20−28, 25 µm, of hind tibia 35−50, 43 µm. Tip of hind tibia expanded, with comb of 11−17, 15 setae and 1 seta near spur strongly hooked. Lengths and proportions of legs as in Table 3.65.

Table 3.65 Lengths (in μm) and proportions of legs segments of male *Corynoneura scutellata* Winnertz, 1846 (based on Chinese specimens)

	fe	ti	ta₁	ta₂	ta₃	ta₄
p₁	250–380, 310	300–450, 376	180–280, 225	90–150, 113	50–80, 62	28–30, 28
p₂	350–530, 400	320–420, 370	175–270, 215	93–128, 115	48–78, 58	23–25, 24
p₃	285–420, 375	305–450, 385	180–270, 235	85–145, 120	38–65, 50	23–28, 26

	ta₅	LR	BV	SV	BR
p₁	38–48, 42	0.58–0.69, 0.62	3.4–3.9, 3.7	2.9–3.2, 3.0	2.0–2.3, 2.2
p₂	40–48, 43	0.57–0.59, 058	3.7–5.0, 4.5	3.5–3.8, 3.6	2.4–3.0, 2.6
p₃	40–50, 44	0.56–0.59, 0.58	4.0–4.6, 4.3	3.3–3.4, 3.37	2.0–2.3, 2.2

Hypopygium (Figs. 3.112B and C). Posterior margin bilobed, and with many short setae, laterosternite IX with 1−2 long setae. Anal point absent. Inferior volsellae digitiform. Sternapodeme inverted slightly V-shaped. Coxapodeme 40−45, 43 μm long. Phallapodeme strongly curved not extending beyond tergite IX, 80−88, 86 μm long. Gonocoxite 80−167.5, 104 μm long with 3−4 long setae apically. Gonostylus curved apically, 25−37.5, 30 μm long. Megaseta 5−7.5, 6 μm long. HR 2.2−4.4, 3.4; HV 4.1−5.8, 4.7.

Larva (*n* = 3−4)

Coloration. Head yellowish. Antenna with basal and second segments yellowish, other segments light brown. Abdomen yellowish.

Head. Capsule length 256−272, 263 μm, width 164−172, 167 μm; strongly sculptured. Postmentum 220−228, 223 μm long. Sternite II obvious, rising from small tubercle, I and III not visible. Premandible 18−28, 22 μm long. Mentum as in Fig. 3.114B. Mandible 55−69, 61 μm long. Antenna (Fig. 3.113A), AR 0.91−1.1, 1.03. Lengths of flagellomeres I−IV (in μm): 204−236, 227; 101−105, 103; 105−117, 112; 4−6, 5. Basal segment width 15−18, 17 μm; length of blade at apex of basal segment 28−35, 30 μm. Length of antenna/length of head 1.71−1.74, 1.73.

Abdomen. Length of anal setae 224−280, 245 μm. Procercus 10−12, 11 μm long, 10−12, 11 μm wide. Subbasal seta of posterior parapods split as in Fig. 3.113C, 55−63, 59 μm long.

Remarks. Winnertz (1846) established this species based on specimens from Poland. Schlee (1968b) redescribed the male based on specimens from Germany, but it referred sternapodeme inverted slightly U-shaped. Hirvenoja and Hirvenoja (1988) redescribed the male, pupa and larva in detail, it referred sternapodeme inverted V-shaped. Ouyang, Lu, and Yan (1984) recorded the larvae of this species in China. Wang (2000) listed this species from Liaoning Province, China. The Lake Winnipeg specimens are slightly smaller than mentioned by Schlee, with wing length of 0.83−1.00, 0.90 (6) mm and wing length/length of profemur of 3.32−3.58, 3.43. Two females which may belong to this species differ from *C. porrecta* by having an AR of 0.51−0.52, slightly larger size, and having about 6 lateral lamellae on the coxosternapodeme.

Distribution. NT: Argentina, Chile. **NE:** Canada (Northwest Territories, Nunavut, Ontario, Saskatchewan), Greenland, USA (Florida, Michigan, New York, Ohio). **PA:** Algeria, Austria, Belgium, Bulgaria, China (Tianjin), Czech Republic, Croatia, Denmark, Estonia, Finland,

France, Germany, Great Britain, Greece, Hungary, Ireland, Italy, Lebanon, Luxembourg, Macedonia, Moldova, Mongolia, Netherlands, Norway, Novaya Zemlya, Poland, Romania, Russia (CET, NET, SET, East Siberia, Far East, West Siberia), Slovakia, Spain, Sweden, Switzerland, Syria, Turkey. **OR:** China (Fujian, Yunnan, Zhejiang). **AU:** Australia (Victoria, Western Australia), New Zealand (North Island). The species thus is recorded from all zoogeographical regions except the Afrotropical.

3.79 *CORYNONEURA SECUNDA* MAKARCHENKO & MAKARCHENKO, 2006B (FIGS. 3.114A−C AND 3.115A−E)

Corynoneura secunda Makarchenko & Makarchenko, 2006b: 156.

Diagnostic characters. Male: antenna with 9 flagellomeres, apical flagellomere with rosette of 15−16 long and pale setae; AR 0.25. Inferior volsella absent. Transverse sternapodeme high and triangular, with

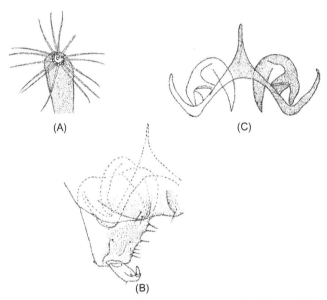

(A)　　　　　　　　　(C)

(B)

Figure 3.114 *Corynoneura secunda* Makarchenko & Makarchenko, 2006b, Male imago (From Makarchenko, E. A., & Makarchenko, M. A. (2006b). Chironomids of the genera *Corynoneura* Winnertz, 1846 and *Thienemanniella* Kieffer, 1919 (Diptera, Chironomidae, Orthocladiinae) of the Russian Far East. *Euroasian Entomological Journal, 5*(2), 151−162). (A) Apex of apical flagellomere with rosette of setae. (B) Total view of hypopygium, from above. (C) Transverse sternapodeme and phallapodemes.

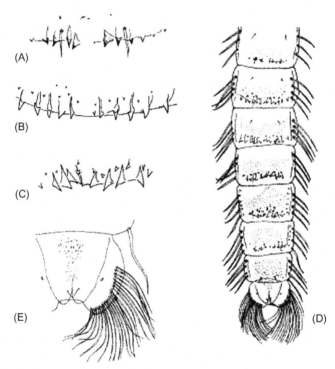

Figure 3.115 *Corynoneura secunda* Makarchenko & Makarchenko, 2006b, Pupa imago (From Makarchenko, E. A., & Makarchenko, M. A. (2006b). Chironomids of the genera *Corynoneura* Winnertz, 1846 and *Thienemanniella* Kieffer, 1919 (Diptera, Chironomidae, Orthocladiinae) of the Russian Far East. *Euroasian Entomological Journal*, 5(2), 151−162). (A) Spinules and spines in posterior side of tergite II. (B) Spinules and spines in posterior side of tergite III. (C) Spines in posterior side of sternite V. (D) Tergites II−VIII and anal segment. (E) Anal segment.

elongate apical part. Phallapodemes disposed and falciformly curved, with point apex, in basal half with short inner projection. Pupa (Fig. 3.115A−E): tergite II with 12 spinules in posterior row. Segment II with 3 pairs of taeniate lateral setae, which slightly narrower than taeniate lateral setae of 4 pairs in segments III−VIII.

Male (Description from Makarchenko & Makarchenko, 2006b)

Total length 1.1 mm. Total length/wing length 1.72.

Head. Eyes bare. Antenna with 9 flagellomeres (Fig. 3.114A), AR 0.25. Lengths of palpomeres (in μm): —, 12, 16, 20, 48.

Thorax. Dorsocentrals 3, prealars 2. Scutellum with 2 setae.

Wing. Wing length 0.64 mm.

Table 3.66 Lengths (in μm) and proportions of legs segments of male *Corynoneura secunda* Makarchenko & Makarchenko, 2006b

	fe	ti	ta$_1$	ta$_2$	ta$_3$	ta$_4$	ta$_5$	LR	BV	SV	BR
p$_1$	200	236	128	68	40	20	28	0.54	3.61	3.41	2.0
p$_2$	276	236	156	68	32	16	24	0.66	4.77	3.28	2.0
p$_3$	208	224	128	68	28	16	24	0.57	4.12	3.37	2.2

Legs. Fore trochanter with dorsal keel. Spurs of fore tibia 20 μm long, of mid tibia 12 μm long, of hind tibia 28 μm and 17 μm long. Lengths and proportions of legs as in Table 3.66.

Hypopygium (Fig. 3.114B and C). Tergite IX with 4 long setae. Inferior volsella absent. Gonostylus apically curved, narrowed and elongated, transverse sternapodeme curved into V-shaped.

Distribution. PA: Russia (Far East).

3.80 *CORYNONEURA SEIRYURESEA* SASA ET AL., 1998 (FIG. 3.116A−C)

Corynoneura seiryuresea Sasa et al., 1998: 123; Fu et al., 2009: 32.

Material examined. Japan: Shikoku, Kochi Prefecture, Ekawasaki, Nishitosa-Mura, sweep net, male holotype (No. 358: 81), 26.iv.1998, H. Suzuki.

Diagnostic characters. The species can be separated from other East Asian *Corynoneura* by having antenna 8 flagellomeres, AR 0.47; phallapodeme short, almost straight; sternapodeme inverted U-shaped; and gonostylus apically curved.

Additional description and corrections. The holotype had 8 flagellomeres, AR 0.47 and leg ratio of the midleg of 0.44; while the original description Sasa et al. (1998) mentions antenna with only 7 flagellomeres, AR 0.69 and midleg leg ratio of 0.51. The following characters should be added: Tentorium 100 μm long, 15 μm wide. Lengths and proportions of legs as in Table 3.67. Sternapodeme inverted U-shaped; coxapodeme 25 μm long; attachment point with phallapodeme placed caudal and caudally directed; phallapodeme short, straight and broad, 20 μm long. Gonocoxite 63 μm long with 2 long setae apically. Gonostylus 20 μm long; megaseta 8 μm long. HR 3.1, HV 5.75. The wing and hypopygium of the holotype are redrawn (Fig. 3.116A−C).

Distribution. PA: Japan.

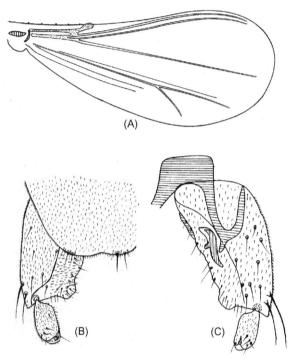

Figure 3.116 *Corynoneura seiryuresea* Sasa et al., 1998, Male imago (From Fu, Y., Sæther, O. A., & Wang, X. (2009). *Corynoneura* Winnertz from East Asia, with a systematic review of the genus (Diptera: Chironomidae: Orthocladiinae). *Zootaxa, 2287,* 1−44). (A) Wing. (B) Hypopygium, dorsal view. (C) Hypopygium, ventral view.

Table 3.67 Lengths (in μm) and proportions of legs segments of male *Corynoneura seiryuresea* Sasa et al., 1998

	fe	ti	ta$_1$	ta$_2$	ta$_3$	ta$_4$	ta$_5$	LR	BV	SV	BR
p$_1$	195	260	115	60	33	20	43	0.44	3.6	4.0	1.7
p$_2$	263	260	115	58	33	20	30	0.44	4.6	4.5	1.8
p$_3$	213	233	120	80	25	18	30	0.52	3.7	3.7	1.7

3.81 *CORYNONEURA SEPTADENTATA* WIEDENBRUG & TRIVINHO-STRIXINO, 2011 (FIGS. 3.117A−F, 3.118A−C, AND 3.119A−M)

Corynoneura-group spec. 2: Wiedenbrug, 2000: 99.

Corynoneura septadentata Wiedenbrug & Trivinho-Strixino, 2011: 21; Wiedenbrug et al., 2012: 46.

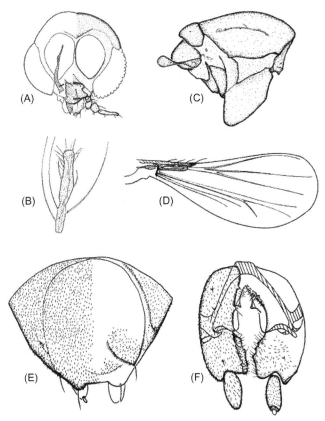

Figure 3.117 *Corynoneura septadentata* Wiedenbrug & Trivinho-Strixino, 2011, Male imago (From Wiedenbrug, S., & Trivinho-Strixino, S. (2011). New species of the genus *Corynoneura* Winnertz (Diptera, Chironomidae) from Brazil. *Zootaxa, 2822,* 1–40). (A) Head, dorsal view, left is sclerites hatched. (B) Terminal flagellomeres. (C) Thorax. (D) Wing. (E and F) Hypopygium. (E) Tergite IX; left is dorsal view, right is ventral view. (F) Tergite IX removed, sclerites hatched; right is dorsal view, left is ventral view.

Diagnostic characters. The adult males are separable from other species by the antenna with 11 flagellomeres, AR 0.2, eyes bare, attachment of the phallapodeme caudal on the sternapodeme, phallapodeme posterior margin sclerotized and strongly curved to anterior, the presence of a small aedeagal lobe and the short gonostylus. Adult females are distinguished from other species except *Corynoneura sertaodaquina* and *Corynoneura* sp. *sexadentata* by the labia funnel shaped, one large seminal capsule sclerotized, the second less sclerotized difficult to see, gonapophysis posteromedian margin straight. The pupae are separable from other

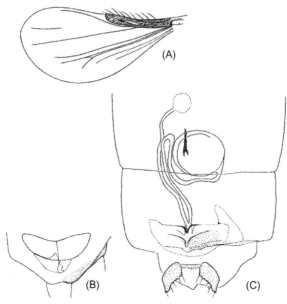

Figure 3.118 *Corynoneura septadentata* Wiedenbrug & Trivinho-Strixino, 2011, Female imago (From Wiedenbrug, S., & Trivinho-Strixino, S. (2011). New species of the genus *Corynoneura* Winnertz (Diptera, Chironomidae) from Brazil. *Zootaxa, 2822*, 1–40). (A) Wing. (B) Dorsal view of genitalia with labia, right is dorsal view of labia, left is ventral view of labia. (C) Ventral view of genitalia; left is gonapophysis removed.

species except *C.* sp. *sexadentata* by having tergites IV and V with spines of shagreen increasing in size posteriorly, sternite II without shagreen, 4 LS-setae on segments III−VIII, anal lobe rectangular, more than a half of the anal lobe lateral margin with short and thin setae and posterior margin with fringe. The larvae can be recognized by the mentum with 7 median teeth, antenna slightly smaller as postmentum, head not elongated (postmentum length/head length less than 1), frontal apotome about 170 μm long, modified abdominal seta split apically.

Male (Description from Wiedenbrug & Trivinho-Strixino, 2011)

Total length 0.73−0.80 mm. Wing length 0.53−0.57 mm. Total length/wing length 1.39−1.41. Wing length/length of profemur 3.10−3.54, 3.23.

Coloration. Thorax brownish; legs whitish; tergites I−II white, III−IV with brownish posteromedian area, V−IX brownish; sternites V−VII brownish.

Head (Fig. 3.117A). AR = 0.22−0.24. Antenna with 11 flagellomeres, ultimate flagellomere 55−65 μm long; one sensilla chaetica at first, second

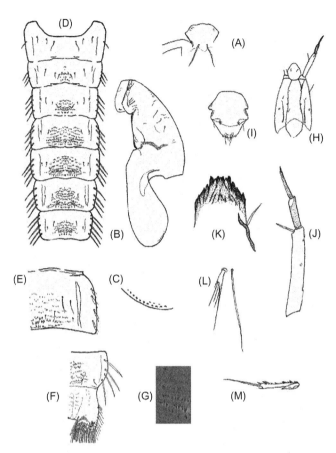

Figure 3.119 *Corynoneura septadentata* Wiedenbrug & Trivinho-Strixino, 2011, Immatures imago (From Wiedenbrug, S., & Trivinho-Strixino, S. (2011). New species of the genus *Corynoneura* Winnertz (Diptera, Chironomidae) from Brazil. *Zootaxa, 2822,* 1−40). (A−G) Pupa imago. (A) Frontal apotome. (B) Thorax, lateral view. (C) Detail of wing sheath margin. (D) Abdomen, tergites I−VII. (E) Tergite IV. (F) Tergite VIII and anal lobe. (G) Detail of posterior shagreen of tergite IV. (H−M) Larva imago. (H) Dorsal view of the head and antenna. (I) Dorsal view of labrum. (J) Antenna. (K) Mentum. (L) Abdominal setae. (M) Subbasal seta of posterior parapod.

and last flagellomeres; flagellomeres, except the first and last with more than one row of setae each (Fig. 3.117B). Temporal setae absent. Clypeus 28−38 μm long, with 6−7 setae. Tentorium 93−110 μm long; 10 μm wide. Palpomere lengths (in μm): 10, 10, 15, 15−18, 35−43. Third palpomere with one sensilla clavata. Eyes bare.

Thorax (Fig. 3.117C). Antepronotum with 1 seta. Dorsocentrals 3, prealars 2. Scutellars 2. Antepronotal lobes dorsally tapering.

Table 3.68 Lengths (in μm) and proportions of legs segments of male *Corynoneura septadentata* Wiedenbrug & Trivinho-Strixino, 2011

	fe	ti	ta$_1$	ta$_2$	ta$_3$	ta$_4$
p$_1$	165−183	195−215	92−100	57−62	27−33	12−15
p$_2$	245−275	200−225	140−148	57−70	30−35	15−18
p$_3$	190−200	175−207	110−113	57−65	25−27	12−15

	ta$_5$	LR	BV	SV	BR	
p$_1$	22	0.47−0.48	3.69−3.89	3.89−3.97	2.66−3.33	
p$_2$	25	0.64−0.70	4.37−4.58	3.17−3.44	2.50−4.33	
p$_3$	20−25	0.56−0.59	3.84−4.27	3.40−3.56	2.75−3.00	

Wing (Fig. 3.117D). VR 3.30−3.41. Clavus 25−30 μm wide, ending 125−130 μm from arculus. Clavus/wing length 0.23−0.24. Anal lobe absent.

Legs. Spur of front tibia 17−20 μm long; spur of middle tibia 13−15 μm long; spurs of hind tibia 35 μm long, 2nd spur S-shaped. Fore and mid tibial scales 5 μm long; hind tibial scale 33−43 μm long. Width of apex of front tibia 15 μm; of mid tibia 13−15 μm; of hind tibia 28−33 μm. Leg measurements (in μm) and ratios as in Table 3.68.

Abdomen. One seta on tergite VIII.

Hypopygium (Fig. 3.117E and F). Tergite IX with 4 small setae. Laterosternite IX with 1 seta. Aedeagal lobe small, triangular. Superior volsella low, with short setae directed to median. Inferior volsella prominent with longer microtrichia directed medially. Phallapodeme, posterior margin sclerotized strongly curved to anterior 17−22 μm long; transverse sternapodeme 7−12 μm long. Gonocoxite 45 μm long. Gonostylus oval, not tapering apically, 10−17 μm long; megaseta 2−5 μm long. HR 2.57−4.50; HV 4.17−5.33.

Female (Description from Wiedenbrug & Trivinho-Strixino, 2011)

Total length 0.86 mm. Wing length 0.47 mm. Total length/wing length 1.84. Wing length/length of profemur 3.45.

Coloration. Thorax and abdomen light brown, legs whitish.

Head. AR 0.49. Antenna with 5 flagellomeres, apical flagellomere 47 μm long; one sensilla chaetica on flagellomeres 2−4 and two on flagellomeres 5. Temporal setae absent. Clypeus 32 μm long, with 8 setae. Tentorium 65 μm long, 2 μm wide. Palpomere lengths (in μm): 5, 10, 12, 15, 35. Sensilla clavata not possible to see. Eyes bare.

Thorax. Dorsocentrals and prealars not possible to see. Scutellum with 2 setae. Antepronotal lobes dorsally tapering.

Table 3.69 Lengths (in μm) and proportions of legs segments of female *Corynoneura septadentata* Wiedenbrug & Trivinho-Strixino, 2011

	fe	ti	ta$_1$	ta$_2$	ta$_3$	ta$_4$	ta$_5$	LR	BV	SV	BR
p$_1$	138	172	82	47	22	12	22	0.48	3.74	3.76	2.67
p$_2$	215	195	117	50	25	12	22	0.60	4.79	3.49	2.50
p$_3$	175	162	87	50	20	12	25	0.53	3.95	3.86	2.75

Wing (Fig. 3.118A). VR 2.04. Clavus 25 μm wide, ending 207 μm from arculus. Clavus/wing length 0.44.

Legs. Spur of front tibia 7 μm long; spur of middle tibia 7 μm long; spurs of hind tibia 30–40 μm long, second spur small S-shaped. Hind tibial scale 33 μm long. Width of apex of front tibia 15 μm, of middle tibia 15 μm, of hind tibia 30 μm. Leg measurements (in μm) and ratios as in Table 3.69.

Abdomen. One seta on tergite VIII.

Genitalia (Fig. 3.118B and C). Gonocoxite IX with 1 seta. Tergite IX with 4 setae, two stronger, dorsal and two smaller, ventral. Seminal capsule 37 μm and spermathecal duct long with a loop, second seminal capsule less sclerotized apparently absent with spermathecal duct shorter and straight, both ducts join together shortly before seminal eminence, which has sclerotized outer borders. Notum 30 μm long. Membrane median curved to posterior, less sclerotized on the link of both sides. Apodeme lobe anterior margin in the same shape as membrane, apically pointed, posterior margin difficult to define. Coxosternapodeme apparently without lateral lamellae, strongly curved with one end at the roof of copulatory bursa, last semicircle-shaped. Labia separated, membranous, bare, triangle-shaped, ventrally longer than copulatory bursa posterior margin, apically funnel shaped, median apical margin with small curves building the accessory gonopore, labia dorsally with straight median margin, building the funnel posterior to the posterior margin of the copulatory burse. Gonocoxapodeme straight, gonapophyses without lobes. Post-genital plate with a straight posterior margin and few microtrichia. Cercus 30 μm long.

Pupa (Description from Wiedenbrug & Trivinho-Strixino, 2011) Total length 1.1–1.2 mm.

Cephalothorax. Frontal apotome not granulated (Fig. 3.119A). Frontal setae 30–62 μm long. Postorbitals 40–100 and 25–27 μm; Longest median antepronotals 55 and 30 μm. Lateral antepronotals 50 μm.

Precorneal setae Pc_1 20 μm long. Dorsocentrals Dc_1 32−42 μm long; Dc_2 27−30 μm long; Dc_3 20−27 μm long; Dc_4 27−35 μm long. Distance between Dc_1 and Dc_2 75 μm; Dc_2 and Dc_3 37−47 μm; Dc_3 and Dc_4 27−32 μm. Dc_1 displaced ventrally. Wing sheath with two to three rows of pearls (Fig. 3.119B and C).

Abdomen (Fig. 3.119D−G). Tergite I and sternites I−III without shagreen.

Tergite II with few shagreen points. Tergites III−VIII with fine shagreen grading to a row of slightly stronger points posteriorly (Fig. 3.127E−G). Tergite IX with shagreen. Sternites VI−VIII with very fine shagreen. Sternite VIII on female bare. Conjunctive tergite II/III with 8 hooklets; tergite III/IV with 12−14 hooklets and about 6 spinules; tergite IV/V with 12−15 hooklets; tergite V/VI with 2−4 hooklets; tergite VI/VII without hooklet, with a row of small spinules; Sternite IV/V with 8 hooklets; Sternite V/VI with 6−8 hooklets; Sternite VI/VII with 11 hooklets; Sternite VII/VIII with 7−11 hooklets.

Segment I with 3 D-setae, 1 L-seta, without V-seta; segment II 3 D-setae, 3 L-setae and 3 V-setae; segment III 3 D-setae, 3 V-setae and 4 short and taeniate L-setae; segments IV−VII with 4 D-setae, 3 V-setae and 4 taeniate L-setae; segment VIII with 2 D-setae, 1 V-seta and 4 taeniate L-setae.

Anal lobe (Fig. 3.119F) rectangular 92−100 μm long. Anal lobe fringe with 10−11 lateral fine and shorter setae and posterior 16−17 taeniate setae 250−330 μm long with about 3 thinner and shorter median setae; 3 macrosetae taeniate, 82 μm long; inner setae taeniate.

Larva (Description from Wiedenbrug & Trivinho-Strixino, 2011)

Head (Fig. 3.119H). Head capsule integument smooth. Frontal apotome 170−175 μm long, head width 125−145 μm. Postmentum 120−125 μm. Postmentum/head width 0.83−1.00. Sternite I simple (Fig. 3.119I). Premandible with an outer lateral lamellae with a brush. Mentum with seven median teeth and four adjacent teeth (Fig. 3.119K). Distance between setae submenti 40−43 μm. Mandible length 40−43 μm with outer sclerotized protuberance. Antennae (Fig. 3.119J): AR 1.2−1.3. Length of segment I 62 μm, II 22−25 μm, III 22 μm, IV 2 μm; basal segment width 10 μm; antennal blade 17 μm long; ring organ at 12 μm from the base of first antennal segment. Third antennal segment with wall from one side not sclerotized. Antennal segments two and three brown.

Abdomen. At least 3 anterior segments with a pair of ventral modified setae, pectinate (Fig. 3.119L). Anal setae length 87 μm. Subbasal seta on posterior parapod simple, 52—55 μm long, with a serrate side (Fig. 3.119M).

Remarks. Larvae from *C. septadentata* were collected from litter near the water surface of a small shallow stream. The species has been found in low order streams in the mountainous regions from Minas Gerais, Rio de Janeiro, São Paulo and Rio Grande do Sul States. See also the remarks of *C. sertaodaquina*.

Distribution. NT: Brazil.

3.82 *CORYNONEURA SERTAODAQUINA* WIEDENBRUG & TRIVINHO-STRIXINO, 2011 (FIGS. 3.120A—D, 3.121A—C; FIG. 3.122A—I)

Corynoneura sertaodaquina Wiedenbrug & Trivinho-Strixino, 2011: 26.

Diagnostic characters. The adult males are separable from other species by the antenna with 11 flagellomeres, with more than one row of seta on each flagellomere, last flagellomere subequal to preceding two, eyes bare, AR 0.2, superior volsella absent, attachment of the

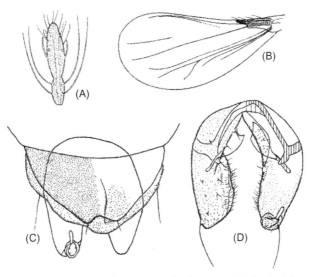

Figure 3.120 *Corynoneura sertaodaquina* Wiedenbrug & Trivinho-Strixino, 2011, Male imago (From Wiedenbrug, S., & Trivinho-Strixino, S. (2011). New species of the genus *Corynoneura* Winnertz (Diptera, Chironomidae) from Brazil. *Zootaxa, 2822*, 1—40). (A) Terminal flagellomeres. (B) Wing. (C and D) Hypopygium. (C) Tergite IX; left is dorsal view, right is ventral view. (D) Tergite IX removed, sclerites hatched; right is ventral view, left is dorsal view.

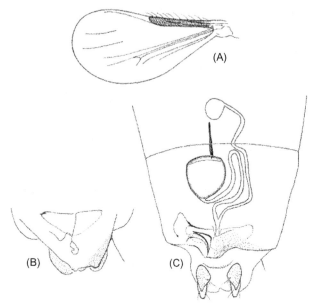

Figure 3.121 *Corynoneura sertaodaquina* Wiedenbrug & Trivinho-Strixino, 2011, Female imago (From Wiedenbrug, S., & Trivinho-Strixino, S. (2011). New species of the genus *Corynoneura* Winnertz (Diptera, Chironomidae) from Brazil. *Zootaxa, 2822*, 1−40). (A) Wing. (B) Dorsal view of genitalia with labia, right is dorsal view of labia, left is ventral view of labia. (C) Ventral view of genitalia; left is gonapophysis removed.

phallapodeme caudal on the lateral sternapodeme, phallapodeme paddle shaped, gonostylus tapering apically. Adult females are separable from other species but from *C. septadentata* and *C.* sp. *sexadentata* by the labia funnel shaped, one large seminal capsule sclerotized, the second less chitinized difficult to see, gonapophysis posterior median margin straight. The pupae are distinguished from other species except from *Corynoneura* sp. *circulimentum* by having spines of shagreen on tergites IV and V increasing in size posteriorly, sternites II and III with minute spines, 4 L-setae on segments III−VIII, anal lobe rectangular, less than one-fourth of the anal lobe lateral margin with short and thin setae, posterior margin with fringe. The larvae can be recognized by the mentum with 7 median teeth, head not elongated (postmentum length/head length < 1), antenna slightly smaller as postmentum, AR 1.3, frontal apotome about 140 μm long, modified abdominal seta apically split and basal seta on posterior parapod with one pectinate side.

Male (Description from Wiedenbrug & Trivinho-Strixino, 2011)

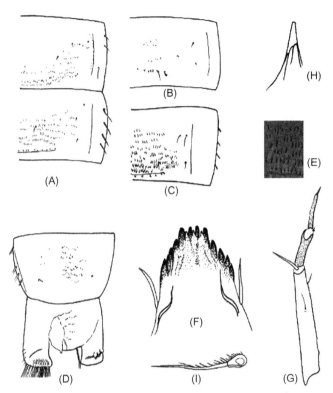

Figure 3.122 *Corynoneura sertaodaquina* Wiedenbrug & Trivinho-Strixino, 2011, Immatures imago (From Wiedenbrug, S., & Trivinho-Strixino, S. (2011). New species of the genus *Corynoneura* Winnertz (Diptera, Chironomidae) from Brazil. *Zootaxa, 2822*, 1–40). (A–E) Pupa imago. (A) Sternites II–III. (B) Tergite II. (C) Tergite IV. (D) Segment VIII and anal lobe, left is ventral view, right is dorsal view, without fringe. (E) Detail of posterior shagreen of tergite IV. (F–I) Larva imago. (F) Mentum. (G) Antenna. (H) Abdominal seta. (I) Subbasal seta of posterior parapod.

Total length 0.88 mm. Wing length 0.50 mm. Total length/wing length 1.76. Wing length/length of profemur 3.28.

Coloration. Thorax brownish; legs whitish; tergites I–III whitish, IV brownish in the median area, V–IX brownish. Head: AR = 0.2. Antenna with 11 flagellomeres, ultimate flagellomere 48 μm long; two sensilla chaetica at last flagellomere. Flagellomeres with more than one rows of setae (Fig. 3.120A). Temporal seta absent. Clypeus 27 μm long, with 8 setae. Tentorium 87 μm long. Palpomere lengths (in μm): 10, 10, 15, 17, 42. Third palpomere with one sensilla clavata. Eyes bare.

Thorax. Antepronotum apparently without seta. Dorsocentrals 3, prealars 2. Scutellars 2. Antepronotal lobes dorsally reduced.

Table 3.70 Lengths (in μm) and proportions of legs segments of male *Corynoneura sertaodaquina* Wiedenbrug & Trivinho-Strixino, 2011

	fe	ti	ta₁	ta₂	ta₃	ta₄	ta₅	LR	BV	SV	BR
p₁	153	185	88	53	25	13	22	0.47	3.77	3.86	1.50
p₂	235	200	125	53	25	13	23	0.63	4.98	3.48	3.33
p₃	175	187	93	55	23	15	23	0.49	3.96	3.92	2.50

Wing (Fig. 3.120B). Clavus 37 μm wide, ending 115 μm from arculus. Clavus/wing length 0.23. Anal lobe absent.

Legs. Spur of front tibia 17 μm long, scale 7 μm long; spur of mid tibia 10 μm long, scale 2 μm long; spurs of hind tibia 27 μm long, second spur S-shaped. Width of apex of front tibia 15 μm, of mid tibia 12 μm, of hind tibia 27 μm. Fore tibial scale 7 μm long, mid tibial scale 2 μm long, hind tibial scale 40 μm long. Leg measurements (in μm) and ratios as in Table 3.70.

Abdomen. One seta on tergite VII, 1 Seta on tergite VIII.

Hypopygium (Fig. 3.120C and D). Tergite IX with 3 small setae. Laterosternite IX with 1 seta. Superior volsella absent. Inferior volsella smooth with longer microtrichia medially directed. Phallapodeme paddle shaped, 27 μm long; transverse sternapodeme rounded. Gonocoxite 55 μm long. Gonostylus tapering apically not possible to measure; mega-seta 5 μm long. HR 5.5; HV 8.8.

Female (Description from Wiedenbrug & Trivinho-Strixino, 2011)

Total length 0.76 mm. Wing length 0.52 mm. Total length/ wing length 1.48.

Coloration. Thorax light brown; legs whitish; abdomen, light brown, tergites VI and VII apparently whitish.

Head. AR 0.42. Antenna with 5 flagellomeres; apical flagellomere 45 μm long; one sensilla chaetica on flagellomeres 2−4 and two on flagellomere 5. Temporal setae absent. Clypeus 37 μm long, with 7 setae. Tentorium 62 μm long, 2 μm wide. Palpomere lengths (in μm): 10, 10, 15, 12, 40. Sensilla clavata not possible to see. Eyes bare.

Thorax. Dorsocentrals 4, prealars 2. Scutellum with 2 setae. Antepronotal lobes dorsally tapering.

Wing (Fig. 3.121A). VR 2.00. Clavus 25 μm wide, ending 227 μm from arculus. Clavus/wing length 0.44.

Legs. Spur of front tibia 5 μm long; spur of middle tibia 5 μm long; spurs of hind tibia 27 μm long, second spur small S-shaped. Hind tibial

Table 3.71 Lengths (in μm) and proportions of legs segments of female *Corynoneura sertaodaquina* Wiedenbrug & Trivinho-Strixino, 2011

	fe	ti	ta$_1$	ta$_2$	ta$_3$	ta$_4$	ta$_5$	LR	BV	SV	BR
p$_1$	135	165	65	38	20	13	20	0.39	4.06	4.61	2.00
p$_2$	195	175	103	43	23	13	13	0.56	5.25	3.61	2.00
p$_3$	158	155	90	48	18	13	23	0.58	4.02	3.47	1.40

scale 37 μm long. Width of apex of front tibia 15 μm, of middle tibia 15 μm, of hind tibia 20 μm. Leg measurements (in μm) and ratios as in Table 3.71.

Abdomen. One seta on tergite VI, 1 seta on tergite VII.

Genitalia (Fig. 3.121B and C). Gonocoxite IX with 1 seta. Tergite IX with 4 setae. Seminal capsule 30 μm and spermathecal duct long with a loop, second seminal capsule less sclerotized apparently absent with spermathecal duct straighter, both ducts join together short before seminal eminence, which has sclerotized outer borders. Notum 25 μm long. Membrane median curved to posterior, less sclerotized on the link of both sides. Apodeme lobe anterior margin in the same shape as membrane, apically pointed. Coxosternapodeme apparently without lateral lamellae, strongly curved, one strongly sclerotized end at the roof of copulatory bursa, last semicircle-shaped. Labia, membranous, bare, separated, apparently funnel shaped as *C. septadentata*. Gonocoxapodeme straight, gonapophyses slightly medially lobed. Postgenital plate with few microtrichia. Cercus 25 μm long.

Pupa (Description from Wiedenbrug & Trivinho-Strixino, 2011)

Total length 1.06 mm.

Cephalothorax. Frontal apotome not granulated. Frontal setae 100 μm long. Postorbitals 30 and 25 μm. Median antepronotals 25 and 20 μm. Lateral antepronotals 12 μm. Dorsocentrals Dc$_1$ 25 μm long. Distance between Dc$_1$ and Dc$_2$ 70 μm; Dc$_2$ and Dc$_3$ 40 μm; Dc$_3$ and Dc$_4$ 22 μm. Dc$_1$ displaced ventrally. Wing sheath with one row of pearls.

Abdomen (Fig. 3.122A−D). Tergite I and sternite I without shagreen. Tergite II with few shagreen points. Tergites III−VIII with fine shagreen grading to a row of slightly stronger points posteriorly (Fig. 3.122C−E). Tergite IX with shagreen. Sternites II−III with shagreen of very fine spinules, IV−VIII with fine shagreen. Sternite VIII on female bare. Conjunctive tergite II/III with 11−14 hooklets; tergite III/IV with 13 hooklets; tergite IV/V with 13 hooklets; tergite V/VI with 0−8 hooklets; tergite VI/VII and tergite VII−VIII without hooklet, with a row of small spinules; Sternite IV/V with 8 hooklets; Sternite V/VI with 0−10 hooklets; Sternite VI/VII with 0−8 hooklets, Sternite VII/VIII with 0−5 hooklets.

Segment I with 3 D-setae, 1 L-seta, 1 V-seta; segment II 3 D-setae, 3 L-setae and 3 V-setae; segment III 3 D-setae, 3 V-setae and 4 short and taeniate L-setae; segments IV−VII with 4 D-setae, 3V-setae and 4 taeniate L-setae; segment VIII with 2 D-setae, 1 V-seta and 4 taeniate L-setae.

Anal lobe rectangular 75−87 μm long (Fig. 3.122D). Anal lobe fringe with 2 lateral fine and shorter setae and posterior 12−13 taeniate setae 250−275 μm long with about 1 thinner and shorter median setae; 3 macrosetae taeniate; inner setae taeniate.

Larva (Description from Wiedenbrug & Trivinho-Strixino, 2011)

Head. Head capsule integument smooth. Frontal apotome 140 μm long, head width 120 μm. Postmentum 107 μm. Postmentum/head width 0.90. Sternite I simple. Premandible with an outer lateral lamellae with a brush. Mentum with seven median teeth, intermediate teeth minute and four adjacent teeth (Fig. 3.122F). Distance between setae submenti 37 μm. Mandible length 37 μm with outer sclerotized protuberance. Antennae (Fig. 3.122G). AR 1.33. Length of segment I 60 μm, II 20 μm, III 20 μm, IV 5 μm; basal segment width 10 μm; antennal blade 20 μm long; ring organ at 12 μm from the base of first antennal segment. Third antennal segment with wall from one side not sclerotized. Antennal segments two and three brown.

Abdomen. At least 2 anterior segments with a pair of modified ventral setae, pectinate (Fig. 3.122H). Anal setae length 75 μm. Subbasal seta on posterior parapod 55 μm μm long with a serrate side (Fig. 3.122I).

Remarks. *C. septadentata* and *C. sertaodaquina* are similar to each other. More material and DNA analyses would be necessary to clear relationship between this species. Larva of *C. sertaodaquina* was collected from the surface of stones of shallow fast flowing waters and also from litter laying near the water surface of streams.

Distribution. NT: Brazil.

3.83 *CORYNONEURA SISBIOTA* WIEDENBRUG ET AL., 2012 (FIGS. 3.123A−D, 3.124A AND B, AND 3.125A−D)

Corynoneura sisbiota Wiedenbrug et al., 2012: 47.

Diagnostic characters. The male of *Corynoneura sisbiota* can be differentiated from other species with phallapodeme long and rounded,

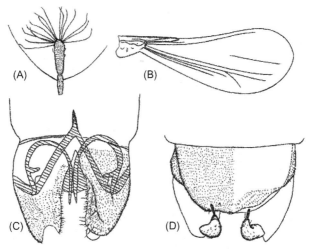

Figure 3.123 *Corynoneura sisbiota* Wiedenbrug et al., 2012; Male imago (From Wiedenbrug, S., & Trivinho-Strixino, S. (2011). New species of the genus *Corynoneura* Winnertz (Diptera, Chironomidae) from Brazil. *Zootaxa, 2822*, 1−40). (A) Terminal flagellomere. (B) Wing. (C and D) Hypopygium. (C) Tergite IX and gonostylus removed, sclerites hatched; right is dorsal view, left is ventral view. (D) Tergite IX; left is dorsal view, right is ventral view of tergite IX.

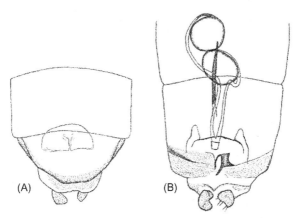

Figure 3.124 *Corynoneura sisbiota* Wiedenbrug et al., 2012; Female imago (From Wiedenbrug, S., & Trivinho-Strixino, S. (2011). New species of the genus *Corynoneura* Winnertz (Diptera, Chironomidae) from Brazil. *Zootaxa, 2822*, 1−40). (A and B) Genitalia. (A) Dorsal view. (B) Ventral view.

sternapodeme as an inverted "V" and lateral attached phallapodeme, by the gonostylus slender and apically hooked without basal hyaline lobus, antenna with sensilla at antennal apex as long as last antennal segment, 8 flagellomeres, AR 0.24−0.43 and inferior volsella absent. The female has

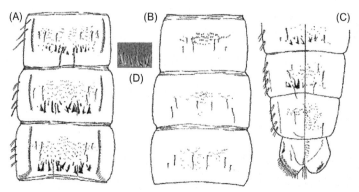

Figure 3.125 *Corynoneura sisbiota* Wiedenbrug et al., 2012; Pupa imago (From Wiedenbrug, S., Lamas, C. J. E., & Trivinho-Strixino, S. (2012). A review of the genus *Corynoneura* Winnertz (Diptera: Chironomidae) from the neotropical region. *Zootaxa, 3574*, 1–61). (A) Tergites II–IV. (B) Sternites II–IV. (C) Segments VI–IX; left is ventral view, right is dorsal view. (D) Detail of posterior shagreen of tergite IV.

sensilla on apical antennal apex as long as the last flagellomere, two seminal capsules subequal in size, both sclerotized, copulatory bursa oval and two separated labia. The pupa has rounded anal lobe, sternite II with elongate colorless spinules, tergites III–VI and sternites VI–VII with fine shagreen and a posterior row of about 10 strong spines with several smaller spines orally.

Male (Description from Wiedenbrug et al., 2012)

Total length 1.13 mm. Wing length 0.50–0.53 mm.

Coloration. Thorax brownish. Abdominal tergites I–II whitish, other tergites and genitalia brownish. Legs whitish.

Head. AR = 0.25–0.27. Antenna with 8 flagellomeres, apical flagellomere 52–55 μm. Flagellomeres with more than one row of setae each. Sensilla on antennal apex, grouped, longer than last flagellomere, curved (Fig. 3.123A). Eyes bare to pubescent.

Thorax. Antepronotal lobes dorsally tapering.

Wing. Clavus/wing length 0.22–0.24. Anal lobe absent (Fig. 3.123B).

Legs. Hind tibial scale 25–35 μm long, with one small S-seta.

Hypopygium (Fig. 3.123C and D). Tergite IX with 4 setae. Laterosternite IX with 1 seta. Superior volsella and inferior volsella absent. Sternapodeme as an inverted "V." Phallapodeme strongly curved, attached on posterolateral third of sternapodeme. Aedeagal lobe present.

Female (Description from Wiedenbrug et al., 2012)

Total length 1.06 mm. Wing length 0.54 mm.

Head. AR = 0.30–0.34. Antenna with 5 flagellomeres, apical flagellomere 27–30 μm. Flagellomeres with one row of setae each. Eyes bare to pubescent. Long sensilla grouped on antennal apex, length subequal to last flagellomere, curved.

Thorax. Antepronotal lobes dorsally tapering.

Legs. Hind tibial scale 27 μm long, with one small S-seta.

Genitalia (Fig. 3.124A and B). Tergite IX with 4 setae. Laterosternite IX with 1 seta. Two subequal seminal capsules 35 μm long; both spermathecal ducts with a loop, joint together shortly before seminal eminence, which has sclerotized outer borders. Notum 57 μm long. Apodeme lobe well sclerotized. Coxosternapodeme curved. Copulatory bursa apparently oval (not possible to discern). Labia membranous, bare, divided in two lobes. Gonocoxapodeme straight, gonapophyses median not well defined. Cercus 17 μm long.

Pupa (Description from Wiedenbrug et al., 2012)

Total length 1.32–1.47 mm.

Coloration (exuviae). Cephalothorax light brown abdomen almost transparent with grayish muscle markings.

Cephalothorax. Frontal apotome slightly rugose. Thorax suture slightly rugose except scutal tubercle region with spinules. Dc-setae thin taeniate, except Dc_3 wider and 102–117 μm long. Dc_1 displaced ventrally. Wing sheaths with two to three rows of pearls.

Abdomen (Fig. 3.125A–D). Sternite I and tergite I bare. Tergite II with fine shagreen and 12–14 small spines posteriorly. Tergites III–VI with fine shagreen and a posterior row of about 10 strong spines with several smaller spines orally. Tergites VII–IX with fine shagreen without strong spines. Sternite II with shagreen of elongate colorless spines. Sternites III–VIII with fine shagreen, sternite V additionally with stronger spinules posteriorly, sternites VI–VII additionally with about 10 strong spines posteriorly. Conjunctives without spinules. Segment I with 1, II with 3 L-setae and IV–VIII with 4 long taeniate L-setae. Anal lobe rounded (Fig. 3.125C). Anal lobe with fringe not complete, 3 taeniate macrosetae and inner setae taeniate.

Remarks. The male of *C. sisbiota* keys by Fu et al. (2009) near *C. latusatra* Fu, Sæther and Wang, *Corynoneura isigaheia* Sasa and Suzuki, *C. confidens* Fu, Sæther and Wang and *C. secunda* Makarchenko and Makarchenko. The antenna with 8 flagellomeres, and the length and position of the setae on the apical antennal flagellomeres distinguish the male of *C. sisbiota* from those species. Only *C. secunda* is known as pupa. According to the diagnosis and illustration given by Makarchenko and Makarchenko (2006b), this species is similar to *C. sisbiota*, however the species described here have longer posterior spines at the tergites.

Pupal exuviae of *C. sisbiota* were collected in small streams with alternate stony bottom and coarse sand of a Semideciduous Forest in Mato Grosso do Sul State and from a small mountain stream of the Atlantic Forest in Rio Grande do Sul State, Brazil.

Distribution. NT: Brazil.

3.84 *CORYNONEURA SORACHIBECEA* SASA & SUZUKI, 2001 (FIG. 3.126A−C)

Corynoneura sorachibecea Sasa & Suzuki, 2001: 7; Fu et al., 2009: 32.

Material examined. Japan: Hokkaido, Sorachi River, Sorachi, sweep net, male holotype (No. 404: 69), 5.ix.2000, H. Suzuki.

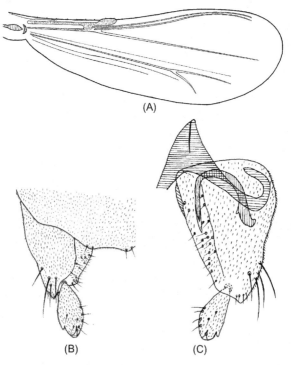

(A)

(B)   (C)

Figure 3.126 *Corynoneura sorachibecea* Sasa, Kitami, Suzuki, 2001, Male imago (From Fu, Y., Sæther, O. A., & Wang, X. (2009). *Corynoneura* Winnertz from East Asia, with a systematic review of the genus (Diptera: Chironomidae: Orthocladiinae). *Zootaxa, 2287*, 1−44). (A) Wing. (B) Hypopygium, dorsal view. (C) Hypopygium, ventral view.

Table 3.72 Lengths (in μm) and proportions of legs segments of male *Corynoneura sorachibecea* Sasa & Suzuki, 2001

	fe	ti	ta$_1$	ta$_2$	ta$_3$	ta$_4$	ta$_5$	LR	BV	SV	BR
p$_1$	270	325	190	100	55	23	38	0.58	3.7	3.1	1.4
p$_2$	380	355	205	90	50	23	40	0.58	4.6	3.6	1.9
p$_3$	315	330	185	105	40	20	38	0.56	4.1	3.5	1.8

Diagnostic characters. The species can be separated from other East Asian *Corynoneura* by having antenna with 10 flagellomeres, apically acute, AR 0.56−0.61, sternapodeme inverted V-shaped, phallapodeme strongly curved, no inferior volsella, gonostylus with small inner lobe in subapical part.

Additional description and corrections. Abdominal tergites with few setae, 0 on tergite Ⅰ, 2 on each of tergites Ⅱ−Ⅸ, while in the original description Sasa and Suzuki (2001) mentions abdominal tergites with small number of setae, 0 on Ⅰ, 1 on Ⅱ to Ⅳ, 2 on Ⅵ and Ⅶ, 1 on Ⅷ, and 2 on Ⅸ. The ninth tergite without an anal point like Y-shaped ridge on the ventral side. The ventral Y-shaped ridge mentioned by Sasa apparently is the phallapodeme. Some additional characters should be added: tentorium 125 μm long, 13 μm wide. Lengths and proportions of legs as in Table 3.72. Sternapodeme inverted V-shaped; coxapodeme 30 μm long; attachment point with phallapodeme placed caudal and ventrally directed; phallapodeme 75 μm long, strongly curved with projection for joint with sternapodeme placed prelateral. Gonocoxite 80 μm long with 4 long setae apically. Gonostylus 38 μm long; megaseta very short. HR 2.1, HV 3.3. The wing and hypopygium of holotype are redrawn (Fig. 3.126A−C).

Distribution. PA: Japan.

3.85 *CORYNONEURA SUNDUKOVI* MAKARCHENKO & MAKARCHENKO, 2010 (FIG. 3.127A−F)

Corynoneura sundukovi Makarchenko & Makarchenko, 2010: 365.

Diagnostic characters. Antenna with 12 flagellomeres; terminal flagellomere with long setae and with light sensitive hairs in 1/3 of apical part. t$_3$ with very slightly curve apical setae; $a/d = 1.5−1.6$; $b/d = 1.0−1.08$. Apex of inferior volsella nearest of gonostylus base and covered by setae in outer edge. Superior volsella large, roundish-triangular

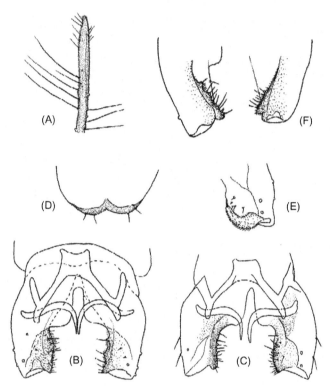

Figure 3.127 *Corynoneura sundukovi* Makarchenko & Makarchenko, 2010, Male imago (From Makarchenko, E. A., & Makarchenko, M. A. (2010). New data on the fauna and taxonomy of *Corynoneura* Winnertz (Diptera, Chironomidae, Orthocladiinae) for the Russian Far East and bordering territories. *Euroasian Entomological Journal, 9*(3), 353−370 + II). (A) Distal part of terminal antennal flagellomere. (B and C) Total view of hypopygium, from above (tergite IX deleted). (D) Posterior margin of tergite IX. (E) Distal part of gonocoxite and gonostylus. (F) Gonocoxites with inferior volsellae, from one side.

and bare. Sternapodeme inverted U-shaped. Phallapodemes low, in distal part is curved and directed down, is situated on trunnion.

Male (Description from Makarchenko & Makarchenko, 2010)

Total length 1.25−1.40 mm. Total length/wing length 1.20−1.34.

Head. Eyes bare. Antenna with 12 flagellomeres (Fig. 3.127A); AR 0.54−0.74. Lengths of palpomeres (in μm): 13−15, 15−18, 30, 40−45, 64−65.

Thorax. Brown. Dorsocentrals 5, prealars 2. Scutellum with 2 setae.

Wing. Wing length 0.93−1.1 mm.

Legs. Fore trochanter with dorsal keel. Spurs of fore tibia 23−28 μm long, of mid tibia 8 μm and 15 μm long, of hind tibia 13−15 μm and

Table 3.73 Lengths (in μm) and proportions of legs segments of male *Corynoneura sundukovi* Makarchenko & Makarchenko, 2010

	fe	ti	ta$_1$	ta$_2$	ta$_3$	ta$_4$
p$_1$	280−320	360−370	200−210	125−133	55−60	20−23
p$_2$	410−450	365−430	230−250	128−138	55	20−23
p$_3$	350−390	380−420	240−264	143−153	53−63	20−23

	ta$_5$	LR	BV	SV	BR
p$_1$	43−45	0.56−0.57	3.56	3.20−3.28	2.7−3.3
p$_2$	45−48	0.58−0.63	4.06−4.30	3.37−3.52	1.2−3.3
p$_3$	43−50	0.63	3.74−3.77	3.04−3.06	2.7−3.9

35−40 μm long. Hind tibia expanded, with comb of 13−16 setae, lengths and proportions of legs as in Table 3.73.

Hypopygium (Fig. 3.127B−F). The posterior margin of tergite IX medially concave, with 6−10 short setae. Gonocoxite 103−120 μm, superior volsella large rounded-triangular, bare. Inferior volsella apical hooked. Sternapodeme curved into U-shaped, transverse sternapodeme 28 μm. Gonostylus bent medially, 25−45 μm long.

Distribution. PA: Russia (Far East).

3.86 *CORYNONEURA TENUISTYLA* TOKUNAGA, 1936 (FIG. 3.128A AND B)

Corynoneura tenuistyla Tokunaga, 1936: 44; Schlee, 1968b: 51; Makarchenko & Makarchenko, 2006b: 157.

Diagnostic characters. Male antenna with 12 flagellomeres; AR 0.28−0.40 (Tokunaga, 1936), AR 0.43−0.44 (Schlee, 1968b), the apex of last flagellomere slightly rounded, and with approximately 15 setae "rosette-like" just before the distal end of last flagellomere.

Male (Description from Makarchenko & Makarchenko, 2006b)

Total length 0.80−1.20 mm.

Coloration. Head dark brown, thorax almost entirely black, dark brown at pleural membrane; scutum black, with shoulder parts dark brown; scutal vittae black, completely fused with each other, tergites I−V yellowish, VI−VIII black; legs pale brown.

Head. Male antenna with 12 flagellomeres; AR 0.28−0.44, the apex of last flagellomere slightly rounded, and with approximately 15 setae

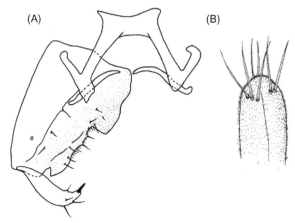

Figure 3.128 *Corynoneura tenuistyla* Tokunaga, 1936; Male imago (From Makarchenko, E. A., & Makarchenko, M. A. (2006b). Chironomids of the genera *Corynoneura* Winnertz, 1846 and *Thienemanniella* Kieffer, 1919 (Diptera, Chironomidae, Orthocladiinae) of the Russian Far East. *Euroasian Entomological Journal*, 5(2), 151–162). (A) Hypopygium, from above. (B) Apex of apical flagellomere.

"rosette-like" just before the distal end of last flagellomere, not distinctly swollen but bluntly ending at tip.

Legs. Fore trochanter with strong dorsal keel.

Hypopygium (Fig. 3.128A and B). Dark brown or black, inferior volsella developed, slightly hooked, placed caudally; gonostylus long, slender, distinctly curved at tip, with four small preapical setae, without excision.

Distribution. PA: Japan, Russia (Far East).

3.87 *CORYNONEURA TERTIA* MAKARCHENKO & MAKARCHENKO, 2010 (FIG. 3.129A–E)

Corynoneura tertia Makarchenko & Makarchenko, 2010: 367.

Diagnostic characters. Male antenna with 8–9 flagellomeres; terminal flagellomere in apex with group of 6–8 sensitive hairs. t_3 with 1 spur and S-shaped apical seta; $a/d = 2.0$; $b/d = 2.0$. Inferior volsella long and narrow, in basal part of outer edge with some setae; superior volsella long and narrow, angle shaped or roundish angle-shaped. Sternapodeme inverted U-shaped. Phallapodemes are crescent-shaped, situated on trunnion. Shagreen of tergites III–VI of pupa intensive and cover most part of surface. Shagreen of sternite VIII with spinules more or less the same size which evenly cover of sternite surface.

Male (Description from Makarchenko & Makarchenko, 2010)

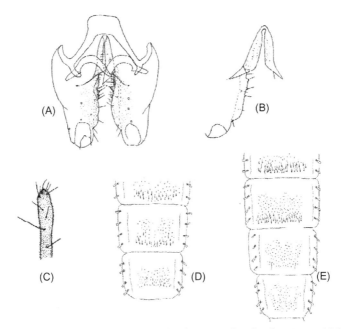

Figure 3.129 *Corynoneura tertia* Makarchenko & Makarchenko, 2010, Male imago (From Makarchenko, E. A., & Makarchenko, M. A. (2010). New data on the fauna and taxonomy of *Corynoneura* Winnertz (Diptera, Chironomidae, Orthocladiinae) for the Russian Far East and bordering territories. *Euroasian Entomological Journal*, *9*(3), 353−370 + II). (A) Total view of hypopygium, from above (tergite IX deleted). (B) Superior and inferior volsellae. (C) Distal part of terminal antennal flagellomere. (D) Sternites VI−VIII. (E) Tergites V −VIII.

Total length 1.55 mm. Total length/wing length 1.99.

Head. Eyes bare. Antenna with 8 or 9 flagellomeres (Fig. 3.129C); AR 0.64 (antenna with 8 flagellomeres), AR 0.45 (antenna 9 flagellomeres). Clypeus with 6−8 setae. Lengths of palpomeres (in μm): —, 12, 20, 22−24, 52−64.

Thorax. Brown. Dorsocentrals 4, prealars 2. Scutellum with 2 setae.

Wing. Wing length 0.78 mm.

Legs. Fore trochanter with dorsal keel. Spurs of fore tibia 20 μm long, of mid tibia 12 μm long, of hind tibia 32 μm and 16 μm long. Hind tibia expanded, with comb of 16 setae, lengths and proportions of legs as in Table 3.74.

Hypopygium (Fig. 3.129A and B). The posterior margin of tergite IX medially concave, with 4 short setae. Gonocoxite 72 μm, superior volsella long and narrow, angle or roundly angle. Inferior volsella long and

Table 3.74 Lengths (in μm) and proportions of legs segments of male *Corynoneura tertia* Makarchenko &Makarchenko, 2010

	fe	ti	ta₁	ta₂	ta₃	ta₄	ta₅	LR	BV	SV	BR
p₁	240	288	140	84	44	24	32	0.49	3.63	3.77	2.4
p₂	–	300	188	88	40	20	32	0.63	–	–	2.8
p₃	280	296	160	96	40	20	32	0.54	3.91	3.60	2.9

narrow. Sternapodeme curved into U-shaped, transverse sternapodeme 20 μm. Gonostylus slightly curved.

Pupa (Description from Makarchenko & Makarchenko, 2010)

Total length 1.4−1.6 mm.

Cephalothorax. Median antepronotals 2, 12 μm long. Wing sheath with 2−3 row of pearls.

Abdomen (Fig. 3.129D and E). Shagreen and chaetotaxy as illustrated. Tergite II with small spines in the distal half. Tergites III−IV with different sizes shagreen spines, median shagreen field separated from posterior shagreen field, last with spines increasing in size, posterior spines wider than longer. The surface of tergites VII−VIII uniformly covered with more or less the same spines (Fig. 3.129D). Sternite I naked. Sternite II covered colorless needle-shaped spines. Sternites III−VIII with shagreen of different sizes in the distal half, the largest of which are located closer to the posterior margin (Fig. 3.129E). Anal lobe distally 112−140 μm long, posterior with 23−29 taeniate setae 120−160 μm long; 3 taeniate macrosetae.

Distribution. PA: Russia (Far East).

3.88 *CORYNONEURA TOKARAPEQUEA* SASA & SUZUKI, 1995 (FIG. 3.130A AND B)

Corynoneura tokarapequea Sasa & Suzuki, 1995: 282; Fu et al., 2009: 33.

Material examined. Japan: Kagoshima Prefecture, Nakanoshima Island, sweep net, male holotype (No. 289: 36), 20.v.1994, H. Suzuki.

Diagnostic characters. The species can be separated from other East Asian *Corynoneura* by having antenna with 11 flagellomeres, AR 0.62; sternapodeme inverted U-shaped; phallapodeme short, almost straight; gonocoxite with well triangular superior volsella; and nearly rectangular inferior volsella.

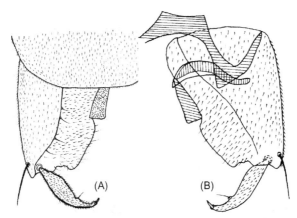

Figure 3.130 *Corynoneura tokarapequea* Sasa & Suzuki, 1995, Male imago (From Fu, Y., Sæther, O. A., & Wang, X. (2009). *Corynoneura* Winnertz from East Asia, with a systematic review of the genus (Diptera: Chironomidae: Orthocladiinae). *Zootaxa, 2287,* 1−44) A. Hypopygium, dorsal view. B. Hypopygium, ventral view.

Table 3.75 Lengths (in μm) and proportions of legs segments of male *Corynoneura tokarapequea* Sasa & Suzuki, 1995

	fe	ti	ta₁	ta₂	ta₃	ta₄	ta₅	LR	BV	SV	BR
p₁	240	285	153	80	45	20	35	0.54	3.8	3.4	1.7
p₂	315	285	175	75	43	23	33	0.61	4.5	3.4	1.5
p₃	260	275	150	85	35	25	33	0.55	3.6	3.9	1.6

Additional description and corrections. The anal point is absent and the gonocoxite has a well-developed superior volsella and a rectangular inferior volsella, while the original description by Sasa and Suzuki (1995) mentions a triangular anal point and a gonocoxite with two inner lobes. The original description mentions a midleg leg ratio of 0.57 while we found 0.61. Some additional characters should be added: Lengths and proportions of legs as in Table 3.75. Tergites I−V yellow, tergites VI−IX brown. Tentorium 125 μm long, 18 μm wide. Sternapodeme inverted U-shaped; coxapodeme 25 μm long; weak attachment point with phallapodeme placed caudal and caudally directed; phallapodeme 35 μm long, weakly curved. Gonocoxite 65 μm long, with 2 long setae apically. Gonostylus 33 μm long; megaseta 8 μm long. HR 2.0, HV 4.4. Wing damaged. Hypopygium of holotype redrawn (Fig. 3.130A and B).

Distribution. PA: Japan (Ryukyu Archipelago), Russia (Far East).

3.89 *CORYNONEURA TOKARAQUEREA* SASA & SUZUKI, 1995 (FIG. 3.131A–C)

Corynoneura tokaraquerea Sasa & Suzuki, 1995: 282; Fu et al., 2009: 34.

Material examined. Japan: Kagoshima Prefecture, Nakanoshima Island, sweep net, male holotype (No. 289: 36), 20.v.1994, H. Suzuki.

Diagnostic characters. The species can be separated from other East Asian members of the genus by having antenna with 11 flagellomeres, apically acute, AR of about 0.5, inferior volsella reduced, sternapodeme inverted U-shaped, and gonostylus with inner lobe.

Additional description and corrections. The antennal apex has short, subapical sensilla chaetica while the original description by Sasa and Suzuki (1995) that there are no sensory setae. The leg ratio of the mid leg is 0.63 while the original description mentions a leg ratio of 0.60. Some additional characters should be added: tentorium 130 μm long, 15 μm wide. VR 3.4. C length 190 μm, C/wing length 0.27. Cu length

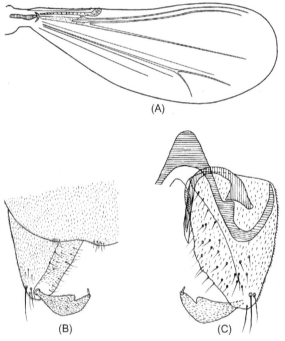

(A)

(B)   (C)

Figure 3.131 *Corynoneura tokaraquerea* Sasa & Suzuki, 1995, Male imago (From Fu, Y., Sæther, O. A., & Wang, X. (2009). *Corynoneura* Winnertz from East Asia, with a systematic review of the genus (Diptera: Chironomidae: Orthocladiinae). *Zootaxa, 2287*, 1–44). (A) Wing. (B) Hypopygium, dorsal view. (C) Hypopygium, ventral view.

Table 3.76 Lengths (in μm) and proportions of legs segments of male *Corynoneura tokaraquerea* Sasa & Suzuki, 1995

	fe	ti	ta₁	ta₂	ta₃	ta₄	ta₅	LR	BV	SV	BR
p₁	245	300	188	95	53	28	40	0.63	3.4	2.9	2.0
p₂	350	315	203	85	43	25	35	0.64	4.6	3.3	1.6
p₃	290	305	—	—	—	—	—	—	—	—	—

425 μm. Cu/wing length 0.60. Wing width/wing length 0.41. C with 5 setae. Lengths and proportions of legs as in Table 3.76. Sternapodeme inverted U-shaped; coxapodeme 45 μm long; attachment point with phallapodeme placed in caudal third of lateral sternapodeme and directed caudally; phallapodeme 75 μm long, strongly curved with projection for joint with sternapodeme placed prelateral. Gonocoxite 88 μm long, with 2 long setae apically. Gonostylus 38 μm long, megaseta 8 μm long. HR 2.3. The wing and hypopygium of holotype have been redrawn (Fig. 3.131A–C).

Distribution. PA: Japan (Ryukyu Archipelago).

3.90 *CORYNONEURA TRONDI* WIEDENBRUG ET AL., 2012 (FIGS. 3.132A–E AND 3.133A–C)

Corynoneura trondi Wiedenbrug et al., 2012: 50.

Diagnostic characters. Male of *Corynoneura trondi* differs from other Neotropical species with phallapodeme attached in the caudal apex of the sternapodeme by the curved small seta at the subapex of the hind tibia, transverse sternapodeme short and inferior volsella rounded. The pupa can be differentiated by the anal lobes laterally rounded, tergites III–V with larger spines of posterior row, on tergite III present on a mound.

Male (n = 1) (Description from Wiedenbrug et al., 2012)

Thorax length 0.45 mm.

Head. AR = 0.27. Antenna with 10 flagellomeres, apical flagellomere 72 μm (Fig. 3.133A). Flagellomeres, except first and last, with more than one row of setae each. Eyes pubescent.

Thorax. Antepronotal lobes dorsally tapering.

Wing. Not possible to measure.

Legs. Hind tibial scale with curved seta (Fig. 3.132B and C).

Hypopygium (Fig. 3.132D and E). Tergite IX with 4 setae. Laterosternite IX with 1 seta. Superior volsella absent. Inferior volsella

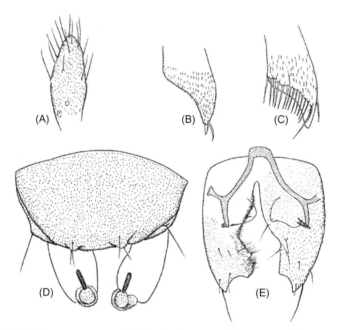

Figure 3.132 *Corynoneura trondi* Wiedenbrug et al., 2012; Male imago (From Wiedenbrug, S., Lamas, C. J. E., & Trivinho-Strixino, S. (2012). A review of the genus *Corynoneura* Winnertz (Diptera: Chironomidae) from the neotropical region. *Zootaxa, 3574*, 1−61). (A) Terminal flagellomeres. (B and C) Apex of hind tibia. (D and E) Hypopygium. (D) Tergite IX and gonostylus. (E) Tergite IX and gonostylus removed, sclerites hatched; right is ventral view, left is dorsal view.

rounded. Aedeagal lobe apparently absent. Transverse sternapodeme short. Phallapodeme with posterior margin sclerotized, attached in the caudal apex of the sternapodeme.

Pupa ($n = 1-2$) (Description from Wiedenbrug et al., 2012)

Total length 1.68 mm.

Coloration (exuviae). Cephalothorax and abdomen transparent.

Cephalothorax. Frontal apotome almost smooth. Thorax suture smooth except scutal tubercle region with small spines. All Dc-setae thin taeniate. Dc_1 displaced ventrally. Wing sheaths with two to three brows of pearls.

Abdomen (Fig. 3.133A−C). Sternite I and tergite I bare, tergite II with fine shagreen posteriorly. Tergites III−V with fine shagreen increasing in size posteriorly, with posterolateral teeth larger on a mold. Tergites VI−IX with more homogeneous shagreen. Sternites II−V with fine shagreen posteriorly. Sternites VI−VIII with shagreen points slightly increasing in size posteriorly. Conjunctives tergites II/III−VI/VII and Sternites V/VI−VII/VIII with

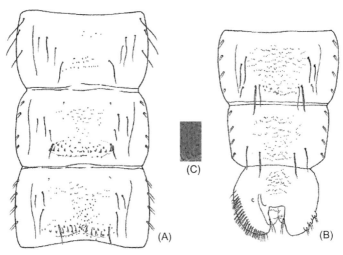

Figure 3.133 *Corynoneura trondi* Wiedenbrug et al., 2012; Pupa imago (From Wiedenbrug, S., Lamas, C. J. E., & Trivinho-Strixino, S. (2012). A review of the genus *Corynoneura* Winnertz (Diptera: Chironomidae) from the neotropical region. *Zootaxa, 3574,* 1—61). (A) Segments II — IV; dorsal view. (B) Tergites VII—IX and anal lobe, dorsal view; anal lobe, left is ventral view, right is dorsal view. (C) Detail of posterior shagreen of tergite IV.

spinules. Segment I with 1, II with 3 L-setae and IV—VIII with 4 long taeniate L-setae. Anal lobe rounded (Fig. 3.133B). Anal lobe with fringe almost complete, 3 taeniate macrosetae and inner setae taeniate.

Distribution. NT: Brazil.

3.91 *CORYNONEURA TYRRHENA* MOUBAYED-BREIL, 2015 (FIGS. 3.134A—J, 3.135A—D, AND 3.136A—E)

Corynoneura tyrrhena Moubayed-Breil, 2015: 1.

Diagnostic characters. Male adult. Hairs present on proximal half part on inner eye margin. Antenna 8-segmented, last flagellomere notched and slightly clubbed distally, longer than the 3 preceding segments. Palpomere 2 square-like, 3 globe-shaped, last segment bearing one single apical seta. Inner margin of tentorium swollen medially, anterior margin of the cibarial pump weakly concave. Hind tibia expanded, bearing a distinct S-shaped and relatively long apical seta. Phallapodeme scalpel-like with medial projection and attached caudally for joint on sternapodeme. Sternapodeme inverted U-shaped with oral projection, transverse sternapodeme relatively wide. Gonocoxite markedly truncate; superior volsella

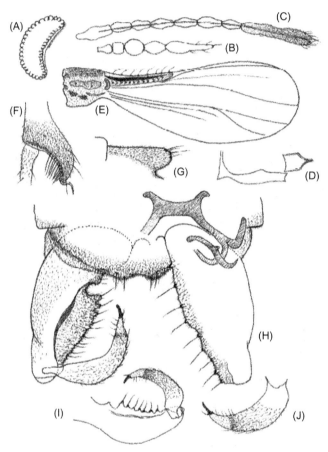

Figure 3.134 *Corynoneura tyrrhena* Moubayed-Breil, 2015; Male imago (From Moubayed-Breil, J. (2015). *Corynoneura tyrrhena* sp. n., a crenophilous species occurring in high mountain streams of Corsica [Diptera, Chironomidae, Orthocladiinae]. *Ephemera*, 2014(2015), *16*(1), 1−12). (A) Eyes. (B) Palp. (C) Antenna. (D) Tentorium and cibarial pump. (E) Wing. (F) Hind tibial apex. (G) Tergite IX in lateral view. (H) Hypopygium, dorsal (left) and ventral (right). (I) Gonocoxite in lateral view. (J) Left gonostylus in dorsal view.

with smoothly dented edge, bearing a characteristic semicircular notch; inferior volsella smooth. Gonostylus long, distinctly curved and gradually narrowed, megaseta long and slender. Pupal exuviae. Coloration brown except for the abdominal segments I−VIII, which are dark brown. Frontal apotome smooth. Cephalothorax distinctly wrinkled; median anteprono-tals long; pearl rows on antennal sheath nearly indistinct, composed of 5−6 rows. All L-setae on abdominal segments I−VIII bristle-like, taeniae

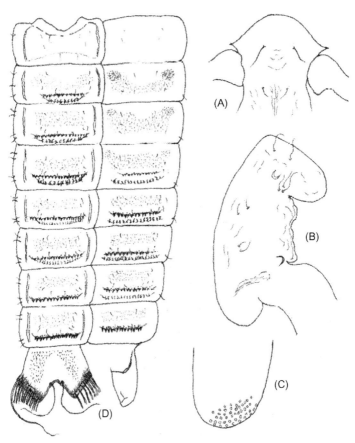

Figure 3.135 *Corynoneura tyrrhena* Moubayed-Breil, 2015; Pupa imago (From Moubayed-Breil, J. (2015). Corynoneura tyrrhena sp. n., a crenophilous species occurring in high mountain streams of Corsica [Diptera, Chironomidae, Orthocladiinae]. *Ephemera*, 2014(2015), *16*(1), 1–12). (A) Frontal apotome. (B) Cephalothorax. (C) Distribution pattern of pearl rows on wing sheath. (D) Armament pattern and chaetotaxy of abdominal segments, tergites and sternites Ⅰ–Ⅷ including anal segment.

are restricted to anal lobe. Posterior transverse rows of spines present on tergites Ⅱ–Ⅷ and sternites Ⅳ–Ⅷ; rows of hooks present on tergites Ⅱ–Ⅵ and sternites Ⅳ–Ⅶ. Anal lobe diamond-like, anterior side longer than posterior side, fringe restricted to the posterior part.

Male (Description from Moubayed-Breil, 2015)

Total length 1.85–1.90 mm. Wing length 0.67–0.70 mm.

Coloration. Contrasting brownish to dark brown especially in the thorax and abdomen. Head dark brown. Antennal and wing sheath brownish; antenna yellow-brown except for the half distal part of last

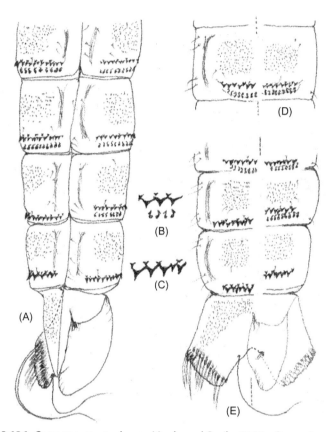

Figure 3.136 *Corynoneura tyrrhena* Moubayed-Breil, 2015; Pupa imago (From Moubayed-Breil, J. (2015). *Corynoneura tyrrhena* sp. n., a crenophilous species occurring in high mountain streams of Corsica [Diptera, Chironomidae, Orthocladiinae]. *Ephemera*, 2014(2015), *16*(1), 1–12). (A) Armament of abdominal segments V – IX in lateral view. (B and C) Details of rows of spines and hooks on sternites VII–VIII. (D) Abdominal segment IV, tergite and sternite. (E) Abdominal segments VI–VIII and anal segment, tergites and sternites.

flagellomere which is dark brown (Fig. 3.134C). Palpomere yellowish. Thorax dark brown with blackish scutum, scutellum, preepisternum, postnotum. Wing with yellowish veins and membrane, clavus brown dark to blackish, brachiolum brown dark with blackish areas anteriorly and posteriorly. Legs yellow brownish, third and fourth tarsomeres of all legs darker. Tergites I – IV brownish, VI–VIII dark brown. Anal segment brown.

Head. Eyes bare; hairs present on proximal half part on inner eye margin (Fig. 3.134A). Temporals absent. Clypeus with 10–11 setae in 3

Table 3.77 Lengths (in μm) and proportion of leg segments of male *Corynoneura tyrrhena* Moubayed-Breil, 2015

	fe	ti	ta$_1$	ta$_2$	ta$_3$	ta$_4$	ta$_5$	LR	BV	SV	BR
p$_1$	235	266	150	76	50	24	31	0.56	4.60	3.34	2.3
p$_2$	306	269	171	67	35	20	29	0.64	4.94	3.36	3.3
p$_3$	260	282	142	80	32	20	29	0.50	4.25	3.82	2.5

rows. Palp (Fig. 3.134B) 5-segmented, segment 2 square-like, 3 globular, 5 clubbed proximally and narrowed distally, bearing 1 apical seta; length (μm) of segments: 12, 9, 18, 21, 43. Antenna (Fig. 3.134C) 385−390 μm long, 8-segmented, antennal groove reaching segment 1; segment 1 long, segments 2−7 nearly subequal (33−39 μm long); ultimate flagellomere 120−125 μm long, longer than the 3 preceding segments, distal part club-shaped and distinctly notched, covered with blackish macrotrichia, apical club about 1/3 the length of last flagellomere. AR 0.47. Tentorium and cibarial pump as in Fig. 3.134D, tentorium swollen medially, anterior margin of cibarial pump slightly concave.

Thorax. Antepronotum with 5 setae, 2 placed near the median area and 3 laterally; dorsocentrals with 5 uniserial setae. Scutellum with 4 uniserial setae.

Wing (Fig. 3.134E). Width/wing length 0.46. Brachiolum with 2 setae, costa with 7−8 setae, clavus as in Fig. 3.134E.

Legs. Length (μm) of spurs on: fore tibia 25, mid tibia 15, hind tibia 22. Width (μm) at apex of: for tibia 14, mid tibia 18, hind tibia 41. Apex of hind tibia (Fig. 3.134F) well expanded, bearing one seta markedly S-shaped. Tarsomere 4 (ta$_4$) of fore leg, middle leg and hind leg bilobed apically. Length (μm) and proportions of legs as in Table 3.77.

Hypopygium (Fig. 3.134G−J). Tergite IX large at base, slightly narrowed posteriorly, posterior margin distinctly bilobed, 56 μm wide at posterior edge; anal point absent, lateral view of tergite IX as in Fig. 3.134G, each lobe bearing 4 setae (2 stout and long setae + 2 bristle-like setae). Laterosternite IX bare. Sternapodeme orally projecting, inverted U-shaped; transverse apodeme 15 μm long, 9 μm wide; lateral coxapodeme 28 μm long, coxapodeme 20 μm long, nearly straight; phallapodeme scalpel-like, with medial projection for joint with sternapodeme placed caudally. Gonocoxite 65 μm long, maximum width about 33 μm; distinctly truncate and strongly dented as it is figured in dorsal and lateral views (Fig. 3.134H and I); superior volsella bearing a characteristic semicircular notch; inferior volsella small and weakly swollen; ventral margin nearly straight, wider at base and bearing 9 stout setae. Gonostylus (Fig. 3.134H—J) 36 µm long, gradually narrowed, bent and curved inwards; distal 1/3 part covered with finest setae and macrotrichia; megaseta 7 µm long, slender and slightly bent.

Male pupal exuviae (Description from Moubayed-Breil, 2015)

Coloration. Brown to dark brown. Frontal apotome nearly smooth, cephalothorax distinctly wrinkled. Abdomen brown to dark brown, area delimited by the lateral muscle marks of segments I —VIII markedly dark brown. Total length 1.90—2.0 mm.

Frontal apotome (Fig. 3.135A). Frontal apotome with rounded anterior margin, frontal setae 25 µm long, relatively short. Cephalothorax as in Fig. 3.135B.

Thorax. Median antepronotal subequal, 61—63 µm long, lateral antepronotals absent; precorneal setae 49—51, 37 µm and 38 µm long. Dorsocentrals, all bristle-like. Length (µm) of dorsocentrals: Dc_1 17, Dc_2 25, Dc_3 17, Dc_4 32; distance (µm) between Dc_1 and Dc_2 25, Dc_2 and Dc_3 48, Dc_3 and Dc_4 12. Metanotal absent. Pearl rows on wing sheath (Fig. 3.135C) composed of dense characteristic 5—6 rows nearly indistinct.

Abdomen (tergites and sternites) and anal segment (Figs. 3.135D and 3.136A—E). Lateral view of segments V—IX and details of armament as in Fig. 3.136A—C. Armament and distribution pattern of shagreen, patches of spinules, hooks and points, chaetotaxy and lateral setation of abdominal segments as illustrated in Fig. 3.135D and Fig. 3.136A—E (tergites and sternites). Tergite I bare. Posterior transverse rows of spines present on tergites II—VIII and sternites IV—VIII; rows of hooks present on conjunctives of tergites II—VI and sternites IV—VII; tergite II armed with one row of spines and hooks occupying about 1/2 of segment width, row of hooks weak on II (7—9) becoming gradually more extensive (12—21) on tergites III—VII; row of hooks on sternites IV—VII nearly similar (15—22). Anteromedian patches of shagreen and small points present on both tergites and sternites II—VIII and tergite IX. Sternites II—IV armed anterolaterally with colorless patch of spinules, which is gradually decreasing from II to IV. Pedes spurii B absent. Apophyses absent. Lateral muscle marks well represented on tergites I—VIII. Anal segment diamond-shaped, maximum width 210 µm, minimum width 180 µm; anterior side (90 µm long) longer than posterior side (70 µm long); fringe composed of 12—14 taeniate setae; median setae bristle-like and relatively long (155−160 μm). Genital sac 120−125 μm long, rounded apically, overreaching apical margin of anal lobe by 7−10 μm. Anal macrosetae 215−225 μm long; taeniae 240−260 μm long.

Distribution. PA: France (Corsica).

3.92 *CORYNONEURA UNICAPSULATA* WIEDENBRUG & TRIVINHO-STRIXINO, 2011 (FIGS. 3.137A−G, 3.138A−D, AND 3.139A−L)

Corynoneura-group spec. 3: Wiedenbrug, 2000: 100. *Corynoneura unicapsulata* Wiedenbrug & Trivinho-Strixino, 2011: 30.

Diagnostic characters. The adult males are separable from other species by the antenna with 9−10 flagellomeres, AR 0.4−0.6, apical flagellomere composed of two or three fused flagellomeres. Eyes bare, attachment of the phallapodeme caudal on the lateral sternapodeme, phallapodeme posterior margin sclerotized and strongly curved to anterior,

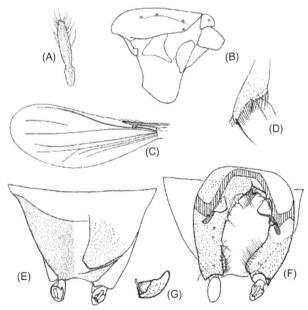

Figure 3.137 *Corynoneura unicapsulata* Wiedenbrug & Trivinho-Strixino, 2011 (From Wiedenbrug, S., & Trivinho-Strixino, S. (2011). New species of the genus *Corynoneura* Winnertz (Diptera, Chironomidae) from Brazil. *Zootaxa, 2822*, 1−40). (A−D) Male imago. (A) Terminal flagellomeres. (B) Thorax. (C) Wing. (D) Apex of hind tibia. (E−G) Hypopygium. (E) Tergite IX; left is dorsal view, right is ventral view. (F) Tergite IX removed, sclerites hatched; left is ventral view, right is dorsal view. (G) Gonostylus.

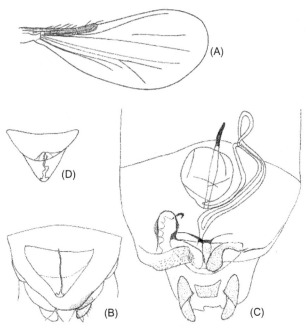

Figure 3.138 *Corynoneura unicapsulata* Wiedenbrug & Trivinho-Strixino, 2011 (From Wiedenbrug, S., & Trivinho-Strixino, S. (2011). New species of the genus *Corynoneura* Winnertz (Diptera, Chironomidae) from Brazil. *Zootaxa, 2822,* 1–40). (A–D) Female imago. (A) Wing. (B–D) Genitalia. (B) Dorsal with view of labia. (C) Ventral view of genitalia; left is gonapophysis removed. (D) Ventral view of labia.

small superior volsella bearing long seta and the gonostylus with prominent crista dorsalis. Adult females are separable from other species by the labia funnel shaped as *C. septadentata*, one large seminal capsule sclerotized about 47 μm and the second apparently absent, gonapophysis margin medially lobed. The pupae are distinguished from other species by the absence of pearl row on the wing sheath, the spines of shagreen on tergites IV and V increasing posteriorly in size, posterior spines as long as wide, 3 short lateral setae on segments III–VIII, anal lobe rectangular, anal lobe lateral margin without seta and posterior margin with fringe. The larvae can be recognized by the head integument without sculptures, the mentum with 3 median and 6 lateral teeth, antenna longer than postmentum, AR 0.9–1.1. Frontal apotome about 180–195 μm long, modified abdominal setae wider but not apically split, basal seta on posterior parapods with one pectinate side.

Male (Description from Wiedenbrug & Trivinho-Strixino, 2011)

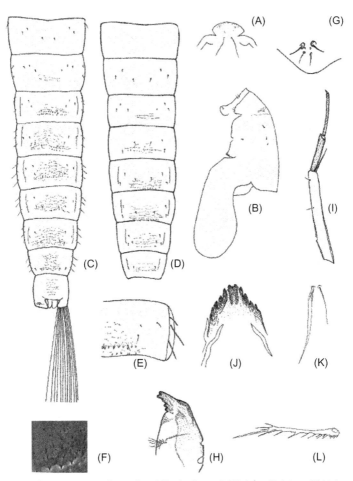

Figure 3.139 *Corynoneura unicapsulata* Wiedenbrug & Trivinho-Strixino, 2011 Immatures imago (From Wiedenbrug, S., & Trivinho-Strixino, S. (2011). New species of the genus *Corynoneura* Winnertz (Diptera, Chironomidae) from Brazil. *Zootaxa, 2822*, 1−40). (A−F) Pupa imago. (A) Frontal apotome. (B) Thorax, lateral view. (C) Tergites I − IX and anal lobe, left is without fringe. (D) Sternites I − VIII. (E) Detail of tergite IV. (F) Detail of posterior shagreen of tergite IV. (G−L) Larva imago. (G) Labrum. (H) Mandible. (I) Antenna. (J) Mentum. (K) Abdominal setae. (L) Subbasal seta of posterior parapod.

Total length 0.88−1.02 mm. Wing length 0.59−0.61 mm. Total length/wing length 1.51. Wing length/length of profemur 3.28−3.45.

Coloration. Thorax brownish; legs whitish; tergite I whitish, II−IX brownish.

Head. AR = 0.38−0.63. Antenna with 9−10 flagellomeres, ultimate flagellomere 95−145 μm long, composed of two or three fused

Table 3.78 Lengths (in μm) and proportions of legs segments of male *Corynoneura unicapsulata* Wiedenbrug & Trivinho-Strixino, 2011

	fe	ti	ta_1	ta_2	ta_3	ta_4
p_1	147−185	225−245	97−100	57−65	32−35	15−17
p_2	225−250	225−243	140−157	57−68	30−33	15
p_3	175−190	212−215	115−122	65−70	27−30	15

	ta_5	LR	BV	SV	BR	
p_1	22−25	0.41−0.43	3.72−3.79	3.97−4.30	1.75−2.00	
p_2	22−25	0.62−0.66	4.67−4.73	3.05−3.25	3.00−3.66	
p_3	22−25	0.53−0.58	3.71−3.97	3.24−3.39	2.50−2.75	

flagellomeres; one sensilla chaetica at the second and third and two at the last flagellomere. Flagellomeres with more than one row of setae each (Fig. 3.137A). Temporal setae absent. Clypeus 30−45 μm long, with 8 seta. Tentorium 102−107 μm long, 10−12 μm wide. Palpomere lengths (in μm): 7−10, 10−12, 12−15, 15−20 (17), 45. Third palpomere with one sensilla clavata. Eyes pubescent.

Thorax (Fig. 3.137B). Antepronotum with 1 lateral seta. Dorsocentrals 3, prealars 2. Scutellars 2. Antepronotal lobes dorsally reduced, tapering.

Wing (Fig. 3.137C). VR 3.30−3.41. Clavus 25−30 μm wide, ending 127−135 μm from arculus. Clavus/wing length 0.21−0.23. Anal lobe absent.

Legs. Spur of front tibia 17−22 μm long; spur of middle tibia 10 μm long; spurs of hind tibia 30−42 μm long, second spur S-shaped. Width of apex of front tibia 15−20 μm, of middle tibia 12−15 μm, of hind tibia 30−37 μm. Hind tibial scale 30−35, 32 μm long (Fig. 3.137D). Leg measurements (in μm) and ratios as in Table 3.78.

Abdomen. Without seta.

Hypopygium (Fig. 3.137E−G). Tergite IX with 2 setae. Laterosternite IX with 1 seta. Superior volsella small, rounded with long setae medially directed. Inferior volsella prominent with longer microtrichia medially directed. Phallapodeme strongly curved to anterior, sclerotized on posterior margin, 17−22 μm long; transverse sternapodeme 17−20 μm long. Gonocoxite 60−65 μm long. Gonostylus with a strong crista dorsalis 17 μm long; megaseta 10 μm long. HR 3.42−3.71; HV 5.06.

Female (Description from Wiedenbrug & Trivinho-Strixino, 2011)

Thorax 0.36 mm ($n = 1$). Abdomen 0.71 mm length ($n = 1$). Wing length 0.58−0.68 mm. Wing length/profemur 3.54−3.98.

Table 3.79 Lengths (in μm) and proportions of legs segments of female *Corynoneura unicapsulata* Wiedenbrug & Trivinho-Strixino, 2011

	fe	ti	ta₁	ta₂	ta₃	ta₄
p₁	162−170	170−245	87−95	42−63	22−33	17
p₂	225−280	205−255	135−165	52−70	27−35	15
p₃	162−205	200−230	100−128	62−70	25−30	15

	ta₅	LR	BV	SV	BR	
p₁	15−25	0.39−0.51	3.71−4.31	3.80−4.37	2.0−2.50	
p₂	22−25	0.62−0.66	4.79−4.81	3.18−3.34	2.33−2.75	
p₃	22−25	0.50−0.55	3.70−4.02	3.41−3.63	1.5−2.50	

Coloration. Thorax brownish; legs whitish; tergite I whitish, II −IX brownish.

Head. AR 0.44−0.48. Antenna with 5 flagellomeres, apical flagellomere 57 μm long; one sensilla chaetica on flagellomeres 2−4 and two on flagellomeres 5. Temporal setae absent. Clypeus 35−42 μm long, with 6−8 setae. Tentorium 45−75 μm long, 5 μm wide. Palpomere lengths (in μm): 10−12, 12, 15−17, 17−20, 45−47. One sensilla clavata on third palpomere. Eyes pubescent.

Thorax. Antepronotals 1, dorsocentrals 3, prealars 2. Scutellars not possible to see. Antepronotal lobes dorsally tapering.

Wing (Fig. 3.138A). VR 2.17−2.26. Clavus 17−27 μm wide, ending 192−235 μm from arculus. Clavus/wing length 0.33−0.34.

Legs. Spur of front tibia 7−12 μm long; spur of middle tibia 12 μm long; spurs of hind tibia 37−40 μm long, second spur small S-shaped. Hind tibial scale 35−45 μm long. Width of apex of front tibia 15 μm, of middle tibia 12−15 μm, of hind tibia 25−35 μm. Leg measurements (in μm) and ratios as in Table 3.79.

Abdomen. Setae on abdomen absent.

Genitalia (Fig. 3.138B−D). Gonocoxite IX with 1 seta. Tergite IX with 4 setae. Seminal capsule 47 μm, second seminal capsule apparently absent, both seminal ducts with a loop joined together shortly before seminal eminence, which has sclerotized outer borders. Notum 30−35 μm long, membrane median well sclerotized. Apodeme lobe anterior margin in the same shape as membrane, apically pointed. Coxosternapodeme strongly curved with well-developed lateral lamellae, coxosternapodeme with one end stronger sclerotized at the roof of copulatory bursa, which is semicircle-shaped. Labia membranous, bare, separated, triangle-shaped, ventrally longer than copulatory bursa posterior margin, apically funnel shaped, median apical margin with small curves building the accessory gonopore, labia dorsally with straight median margin, building the funnel posterior to the posterior margin of the copulatory burse. Gonocoxapodeme straight, gonapophyses strongly lobed medially. Postgenital plate quadrangular, with microtrichia. Cercus 35−42 μm long.

Pupa (Description from Wiedenbrug & Trivinho-Strixino, 2011) Total length 1.24−1.29 mm.

Cephalothorax. Frontal apotome with few granulation (Fig. 3.139A). Frontal setae 40−112 μm long. Postorbitals 20−42 μm and 25−42 μm. Median antepronotals 25−45 μm. Lateral antepronotals 17−35 μm. Pc_2 37 μm. Dorsocentrals Dc_1 22−27 μm long; Dc_2 15−27 μm long; Dc_3 20 μm long; Dc_4 15 μm long. Distance between Dc_1 and Dc_2 75−90 μm; Dc_2 and Dc_3 45−52 μm; Dc_3 and Dc_4 20−22 μm. Dc_1 displaced ventrally. Wing sheaths without row of pearls (Fig. 3.139B).

Abdomen (Fig. 3.139C−F). Tergite I and sternites I−II without shagreen. Tergite II with few shagreen points. Tergites III−VIII with fine shagreen grading to a row of stronger points posteriorly (Fig. 3.139E and F). Tergite IX with fine shagreen. Sternites III−IV, VIII with fine shagreen posteriorly, V−VII with fine shagreen. Sternite VIII on female bare. Conjunctive tergite II/III with 4−11 hooklets; tergite III/IV with 12−21 hooklets; tergite IV/V with 13−21 hooklets; tergite V/VI with 12−19 spinules; tergite VI/VII with 25−32 spinules; tergite VII−VIII with a row of 30 small spinules; sternite IV/V with 9−12 spinules; sternite V/VI with 12−13 hooklets; sternite VI/VII with 9−12 hooklets, sternite VII/VIII with 6−11 hooklets.

Segment I with 3 D-setae, 1 L-seta, without V-setae; segments II−III 3 D-setae, 3 V-setae and 3 L-setae; segments IV−VII with 4 D-setae, 3 V-setae and 3 very thin and short taeniate L-setae; segment VIII with 2 D-setae, 1 V-seta and 3 very thin and short taeniate L-setae.

Anal lobe rectangular 105 μm long. Anal lobe fringe without lateral setae, posterior with 13−15 taeniate setae 250−325 μm long; 3 macrosetae thin and taeniate, 20−42 μm long; inner setae taeniate 50 μm long.

Larva (Description from Wiedenbrug & Trivinho-Strixino, 2011) *Head.* Head capsule integument smooth. Frontal apotome 182−192 μm long; head width 147−165 μm; postmentum 140−147 μm long; postmentum/head width 0.89−0.95. Sternite I simple (Fig. 3.139G). Premandible with an outer lateral lamellae with a brush. Mentum with three median teeth, intermediate teeth minute and six adjacent teeth (Fig. 3.139J). Distance between setae submenti 50−57 μm. Mandible (Fig. 3.139H) length 47−50 μm with outer sclerotized protuberance. Antennae (Fig. 3.139I), AR 0.95−1.02. Length of segment I 95−110 μm, II 45−47 μm, III 50−55 μm, IV 5 μm; basal segment width 12 μm; antennal blade 30−37 μm long; ring organ at 20−25 μm from the base of first antennal segment. Antennal segments two and three brown.

Abdomen. Ventral setae modified, wide and longer (Fig. 3.139K). Anal setae length 75 μm. Subbasal seta on posterior parapod 135−137 μm long, with a serrate side (Fig. 3.139L).

Remarks. All larvae of *Corynoneura unicapsulata* were collected from litter laying near the water surface from small mountain streams. This species was found in Minas Gerais, São Paulo, Paraná and Rio Grande do Sul States.

Distribution. NT: Brazil.

3.93 *CORYNONEURA VIDIAPODEME* WIEDENBRUG ET AL., 2012 (FIG. 3.140A−D)

Corynoneura vidiapodeme Wiedenbrug et al., 2012: 53.

Diagnostic characters. The male of *Corynoneura vidiapodeme* is differentiated from other species with phallapodeme attachment placed in the caudal apex of sternapodeme and apex of hind tibia with a S-shaped seta by the antenna plumose with 7 flagellomeres and inferior volsella narrow and large. *C. vidiapodeme* is similar to *C. renata* and can be distinguish from this species due to the narrower aedeagal lobe and by the by shape of the phallapodeme.

Male (Description from Wiedenbrug et al., 2012)

Wing length 0.52−0.57 mm.

Coloration. Thorax brownish. Abdominal tergites I − V whitish, rest brownish. Legs whitish.

Head. AR = 0.62. Antenna with 7 flagellomeres, apical flagellomere 105 μm (Fig. 3.140A). Flagellomeres with more than one row of setae each. Eyes pubescent.

Thorax. Antepronotal lobes dorsally tapering.

Wing. Clavus/wing length 0.24−0.25. Anal lobe absent (Fig. 3.140B).

Legs. Hind tibial scale 30 μm long, with one S-seta.

Hypopygium (Fig. 3.140C and D). Tergite IX with 4 setae. Laterosternite IX with 1 seta. Superior volsella low with short setae. Inferior volsella large and narrow. Aedeagal lobe with narrow base well

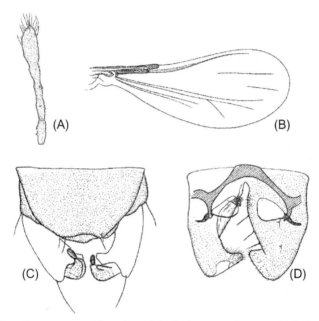

Figure 3.140 *Corynoneura vidiapodeme* Wiedenbrug et al., 2012; Male imago (From Wiedenbrug, S., Lamas, C. J. E., & Trivinho-Strixino, S. (2012). A review of the genus *Corynoneura* Winnertz (Diptera: Chironomidae) from the neotropical region. *Zootaxa, 3574,* 1–61). (A) Terminal flagellomere. (B) Wing. (C and D) Hypopygium. (C) Tergite IX and gonostylus. (D) Tergite IX and gonostylus removed, sclerites hatched; right is ventral view, left is dorsal view.

sclerotized. Sternapodeme rounded, anteriorly straight, phallapodeme caudal attached. Phallapodeme oar-shaped.

Distribution. NT: Brazil.

3.94 *CORYNONEURA VITTALIS* TOKUNAGA, 1936

Corynoneura vittalis Tokunaga, 1936: 45.

Diagnostic characters. Antenna with 12 flagellomeres, AR 0.2, antennae with apical rosettes of hairs at extreme tip.

Male (Description from Tokunaga, 1936)

Total length 1.0 mm.

Coloration. Head yellowish; thorax yellow, while lateral are black; scutellum pale brown or yellow; postscutellum brown; wing with yellowish clava; tergites of first two segments pale yellow; other segments yellowish; legs uniformly yellowish.

Head. Antenna with 12 flagellomeres, AR 0.2; antenna slightly expanded apically.

Hypopygium. As in *Corynoneura celtica.*

Distribution. PA: Japan.

3.95 *CORYNONEURA YOSHIMURAI* TOKUNAGA, 1936 (FIG. 3.141A−C)

Corynoneura yoshimurai Tokunaga, 1936: 46; Sasa & Suzuki, 2000b: 99; not *Corynoneura yoshimurai* Wang, 2000: 635 (= C. *inefligiata* Fu et al., 2009, by present designation); Fu et al., 2009: 36.

Material examined. P.R. China: Sichuan Province, Emeishan County, Medicine School, 29°21′N, 103°17′E, alt. 1500 m, light trap, 6 males (BDN No. 1178, 1179, 05364, 05365, 05366, 05367), 17.v.1986, X. Wang; Sichuan Province, Emeishan County, Jinding, 29°21′N, 103° 17′E, alt. 3, 077 m, light trap, 1 male (BDN No. 05362), 18.v.1985, X.

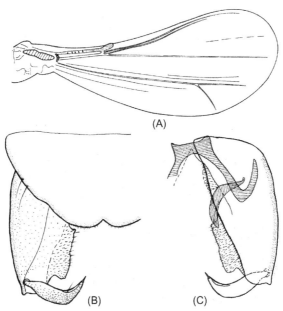

(A)

(B)   (C)

Figure 3.141 *Corynoneura yoshimurai* Tokunaga, 1936; Male imago (From Fu, Y., Sæther, O. A., & Wang, X. (2009). *Corynoneura* Winnertz from East Asia, with a systematic review of the genus (Diptera: Chironomidae: Orthocladiinae). *Zootaxa, 2287,* 1−44). (A) Wing. (B) Hypopygium, dorsal view. (C) Hypopygium, ventral view.

Wang; Tibet Autonomous Region, Chayu County, 29°N, 90°E, alt. 2500 m, light trap, 1 male (BDN No. 07845), 13.vi.1988, X. Wang.

Diagnostic characters. The species can be separated from other East Asian members of the genus by having antenna with 9 or 10 flagellomeres; an AR of about 0.3−0.5; sternapodeme inverted U-shaped; and gonostylus slender and nearly straight.

Male (*n* = 8)

Total length 0.87−1.15, 0.98 mm. Wing length 0.57−0.69, 0.62 mm. Total length/wing length 1.24−1.8, 1.52. Wing length/length of profemur 2.7−3.1, 2.9

Coloration. Head brown. Thorax yellow with brown vittae, scutum, scutellum, preepisternum, postnotum. Abdominal segments brown. Legs yellow. Wings transparent and yellowish, with pale yellow clava.

Head. Eyes bare, reniform. Antenna with 10 flagellomeres; ultimate flagellomere 93−113, 99 μm long, apex with rosette of short sensilla chaetica, apically expanded, thick 65−75, 70 μm long, with maximum width 18−23, 21 μm; AR 0.34−0.39, 0.36. Clypeus with 5−9, 7 setae. Temporal setae lacking. Tentorium 110−125, 120 μm long, 13−20, 16 μm wide. Stipes: 50−55, 52 μm long. Palpomeres length (in μm): 10−15, 13(7); 10−13, 12(7); 15−18, 17(7); 18−25, 23(7); 48−53, 49(7). Palpomere 2 and 3 ellipsoid, 4 rectangular, 5 long and slender. Palpomere 5/3 ratio 2.7−3.5, 2.9.

Thorax. Antepronotum without lateral setae. Dorsocentrals 1−7, 4, prealars 1−4, 2(4). Scutellum with 1−2 setae.

Wing (Fig. 3.141A). VR 3.0−3.5, 3.2(6). C length 140−170, 153 μm, C/wing length 0.23−0.26, 0.25. Cu length 350−440, 386(6) μm. Cu/wing length 0.59−0.66, 0.6. Wing width/wing length 0.4−0.43, 0.41. C with 4−6 setae.

Legs. Fore trochanter with keel. Spur of front tibia 20−25, 21 μm and 10−15, 13 μm long, spurs of middle tibia 10−18, 11 μm long, of hind tibia 28−33, 30 μm long. Width at apex of fore tibia 15−30, 20 μm, of mid tibia 15−28, 18 μm, of hind tibia 25−38, 30 μm. Tip of hind tibia expanded, with comb of 12−16, 15 setae and 1 seta near spur developed as hook. Lengths and proportions of legs as in Table 3.80.

Hypopygium (Fig. 3.141B and C). Posterior margin weakly bilobed, and with many short setae, laterosternite IX with 1 long seta. Anal point absent. Inferior volsellae large, between rectangular and triangular, sternapodeme inverted U-shaped, coxapodeme 20−25, 22 μm long. Phallapodeme curved not extending beyond tergite IX, 30−35, 33 μm

Table 3.80 Lengths (in μm) and proportions of legs segments of male *Corynoneura yoshimurai* Tokunaga, 1936 (based on Chinese specimens)

	fe	ti	ta_1	ta_2	ta_3	ta_4
p_1	200–250, 223	235–280, 260	115–140, 127	60–85, 69	33–43, 39	15–23, 19
p_2	285–305, 297	245–280, 260	130–160, 151	63–75, 66	30–43, 34	13–20, 16
p_3	220–260, 241	240–285, 263	120–150, 133	60–80, 73	25–30, 29	15–18, 16

	ta_5	LR	BV	SV	BR	
p_1	25–35, 29	0.48–0.52, 0.49	3.9–4.1, 4.0	3.6–3.8, 3.7	1.8–2.1, 2.0	
p_2	23–33, 26	0.57–0.62, 0.58	4.2–6.3, 5.2	3.4–3.8, 3.6	2.0–2.3, 2.1	
p_3	25–30, 27	0.46–0.57, 0.51	4.3–4.6, 4.5	3.4–4.1, 3.8	1.7–2.6, 1.9	

long. Gonocoxite 68−80, 73 μm long with 2 setae apically. Gonostylus slightly curved apically, 20−30, 26 μm long. Megaseta 5−8, 6 μm long. HR 2.5−3.6, 2.6, HV 2.9−5.1, 3.9.

Remarks. Tokunaga (1936) described the species based on Japan specimen. When we examined the specimens from Sichuan Province, we found that *C. celtica* listed in Wang (2000) was wrongly identified; the males from Sichuan all belong to *C. yoshimurai*.

Distribution. PA: China (Tibet), Japan. **OR**: China (Sichuan).

3.96 *CORYNONEURA ZEMPOALA* WIEDENBRUG ET AL., 2012 (FIG. 3.142A−E)

Corynoneura zempoala Wiedenbrug et al., 2012: 55.

Diagnostic characters. This species is similar to the *C. scutellata* group of species as defined by Hirvenoja and Hirvenoja (1988). This species has gonostylus with a basal lobe as *C. scutellata*. The antenna with 10 flagellomeres, AR about 0.72−0.76, the median pointed sternapodeme, the shape of the aedeagal lobus and gonocoxite without inferior volsella segregate this species from other species of the group.

Male (Description from Wiedenbrug et al., 2012)

Total length 1.46−1.58 mm. Wing length 1.05−1.07 mm.

Coloration. Thorax brownish. Abdominal tergites brownish, tergite VII usually with posteromedian white area. Legs light brown.

Head. AR = 0.72−0.76. Antenna with 10 flagellomeres, apical flagellomere 240−255 μm (Fig. 3.142A). Flagellomeres with more than one row of setae each. Eyes pubescent.

Thorax. Antepronotal lobes dorsal slightly tapering.

Wing. Clavus/wing length 0.29−0.31. Anal lobe absent (Fig. 3.142C).

Legs. Hind tibial scale 25−37 μm long, with one small seta curved (Fig. 3.142B).

Hypopygium (Fig. 3.142D and E). Tergite IX with about 10 small setae. Laterosternite IX with 3 setae. Superior and inferior volsella absent. Aedeagal lobe triangular, rounded apically. Sternapodeme triangular pointed. Phallapodeme strongly curved, attached on posterolateral third to sternapodeme. Gonostylus apically thin and curved, with a hyaline basal lobe, sometimes difficult to see.

Distribution. NT: Mexico.

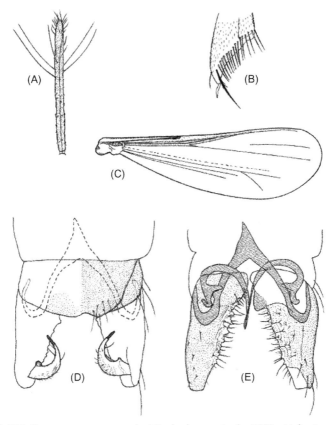

Figure 3.142 *Corynoneura zempoala* Wiedenbrug et al., 2012; Male imago (From Wiedenbrug, S., Lamas, C. J. E., & Trivinho-Strixino, S. (2012). A review of the genus *Corynoneura* Winnertz (Diptera: Chironomidae) from the neotropical region. *Zootaxa*, *3574*, 1–61). (A) Terminal flagellomere. (B) Apex of hind tibia. (C) Wing. (D and E) Hypopygium. (D) Tergite IX and gonostylus. (E) Tergite IX and gonostylus removed, sclerites hatched; right is dorsal view, left is ventral view.

3.97 *CORYNONEURA* SP. *CIRCULIMENTUM* WIEDENBRUG & TRIVINHO-STRIXINO, 2011 (FIGS. 3.143A–C AND 3.144A–H)

Corynoneura sp. *circulimentum* Wiedenbrug & Trivinho-Strixino, 2011: 36.

    Diagnostic characters. Adult females are separable from other species by the eyes pubescent, labia funnel shaped, one large and one small seminal capsule, both sclerotized and gonapophysis with posterior median margin straight (Fig. 3.143B and C). Pupa are difficult to separable from *C. sertaodaquina*, but the shagreen is more similar to *C. septadentata*, having

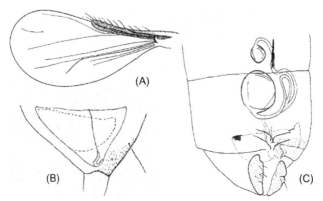

Figure 3.143 *Corynoneura* sp. *circulimentum* Wiedenbrug & Trivinho-Strixino, 2011, Female imago (From Wiedenbrug, S., & Trivinho-Strixino, S. (2011). New species of the genus *Corynoneura* Winnertz (Diptera, Chironomidae) from Brazil. *Zootaxa, 2822*, 1−40). (A) Wing. (B) Dorsal with view of labia. (C) Ventral view of genitalia; left is gonapophysis removed.

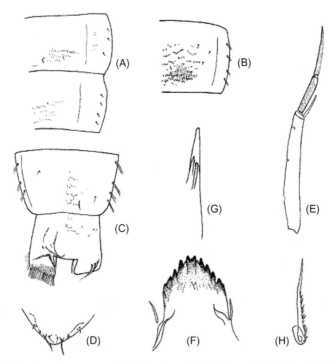

Figure 3.144 *Corynoneura* sp. *circulimentum* Wiedenbrug & Trivinho-Strixino, 2011, Immatures imago (From Wiedenbrug, S., & Trivinho-Strixino, S. (2011). New species of the genus *Corynoneura* Winnertz (Diptera, Chironomidae) from Brazil. *Zootaxa, 2822*, 1−40). (A−C) Pupa imago. (A) Sternites II−III. (B) Tergite IV. (C) Segment VIII and anal lobe, left is ventral view, right is dorsal view, without fringe. (D−H) Larva imago. (D) Labrum. (E) Antenna. (F) Mentum. (G) Abdominal setae. (H) Subbasal seta of posterior parapod.

shorter spines, not elongated as in *C. sertaodaquina* (Fig. 3.144A—C). Larva
are distinguished from other species due to antennal length subequal to
postmentum length (Fig. 3.144E). Frontal apotome about 170 μm long,
mentum with three median teeth (Fig. 3.144F), ventral setae on abdomen
apically split (Fig. 3.144G), basal seta on posterior parapod with one pectinate side (Fig. 3.144H).

Remarks. Unfortunately the male of this morphotype is still
unknown. Pupa and larva are similar to *C. septadentata* and *C. sertaodaquina*. Not only the male but more material necessary to define this morphotype as a valid species.

3.98 *CORYNONEURA* SP. "HEADSPINES" WIEDENBRUG ET AL., 2012 (FIG. 3.145A—D)

Corynoneura sp. "headspines" Wiedenbrug et al., 2012: 55.

Diagnostic characters. This larva can be differentiated from other
larvae of the genus by the presence of two spines on the head, anterior to
the eyes spots and two anal setae on each procercus instead of the usual
four found on the other species of the genus.

Larva (*n* = 2) (Description from Wiedenbrug et al., 2012)

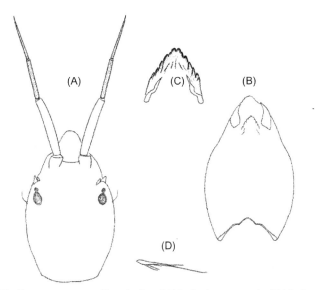

Figure 3.145 *Corynoneura* sp. "headspines" Wiedenbrug et al., 2012; Larva imago
(From Wiedenbrug, S., Lamas, C. J. E., & Trivinho-Strixino, S. (2012). A review of the
genus *Corynoneura* Winnertz (Diptera: Chironomidae) from the neotropical region.
Zootaxa, 3574, 1—61). (A) Head and antenna, dorsal view. (B) Head ventral view. (C)
Mentum. (D) Subbasal seta of posterior parapod.

Head (Fig. 3.145A and B). Postmentum 160—163 μm long. Head capsule integument of larva apparently smooth. Two spines on the head anterior to eyes spots (Fig. 3.145A). Mentum with two median teeth, first lateral teeth adpress to the median (Fig. 3.145C). Antenna 225 μm long, segments two and three darker. First segment shorter than postmentum length.

Abdomen. Ventral setae modified, taeniate and longer than the half of the segment width. Subbasal seta on posterior parapod split from the base with one spine longer (Fig. 3.145D). Only two anal setae present on each procercus.

Remarks. Larval morphotype *Corynoneura* sp. "headspines" was collected in artificial substrate of stones and leaves at slow flowing, lowland streams.

3.99 *CORYNONEURA* SP. NEAR *LACUSTRIS* EDWARDS, 1924 (FIG. 3.146A—F)

Corynoneura n. sp. 11 Bolton, 2007: 30; Fu & Sæther, 2012: 45.

Material examined. USA, Ohio, Ashtabula County, Ashtabula River at Root Rd. Ohio, 1 pharate male, 3 larval exuviae, 30.viii.1995, J. T. Freda & M. J. Bolton; Ashtabula County, Mill Creek, RM 25. 7 Clay Rd., 2 prepupal larval exuviae, 22.ix. & 7.x.2003, J. T. Freda (MJB).

Diagnostic characters. The pharate male is similar to *C. lacustris* Edwards in having the rectangular inferior volsella and in the shape of the sternapodeme, but can be separated by having antenna with 11 flagellomeres, AR 0.39, and rectangular superior volsella. The larvae are separable by having unsculptured head capsule, mentum with three median teeth, lateral tooth adjacent to median tooth smaller, all antennal segments same colored, and length of antenna/length of head 1.70—1.87, 1.77.

Pharate male ($n = 1$) (Description from Fu & Sæther, 2012)

Coloration. Body entirely pale yellow. The limited characters which can be observed are: Antenna (Fig. 3.146A) with 11 flagellomeres, ultimate flagellomere 103 μm, AR 0.39; antenna apically expanded, with about 8 apical sensilla chaetica. Hind tibia distinctly expanded, with comb of 15 setae, 1 seta near spur S-shaped. Posterior margin of tergite IX slightly concave, with 4 long setae. Laterosternite IX with 1 long seta. Superior volsella small, rectangular, anteriomedially separated. Inferior volsella rectangular. Phallapodeme broad apically and strongly curved, 41 μm long, placed in caudal position of sternapodeme. Transverse

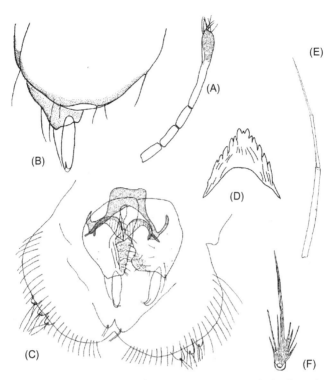

Figure 3.146 *Corynoneura* sp. near *lacustris* (From Fu, Y., & Sæther, O. A. (2012). *Corynoneura* Winnertz and *Thienemanniella* Kieffer from the Nearctic region (Diptera: Chironomidae: Orthocladiinae). *Zootaxa*, *3536*, 1−61. www.mapress.com/j/zt). (A−C) Male imago. (A) Antenna. (B) Hypopygium, dorsal view. (C) Hypopygium, ventral view. (D−F) Larva. (D) Mentum. (E) Antenna. (F) Subbasal seta of posterior parapod.

sternapodeme very broad longitudinally, 28 μm wide. Gonocoxite 63 μm long. Gonostylus apically slightly curved, 36 μm long; megaseta 4 μm long. HR 1.75 (Fig. 3.146B and C).

Larva (*n* = 3) (Description from Fu & Sæther, 2012)

Coloration. Entirely yellowish.

Head. Capsule length 224−248, 236 μm; width 140−148, 145 μm. Postmentum 200−220, 209 μm long. Sternite II obvious, rising from small tubercle, I simple, and III invisible. Premandible 24−28, 26 μm long. Mentum as in Fig. 3.147D. Mandibles 53−59, 56 μm long. Antenna (Fig. 3.146E), AR 0.86−0.88, 0.87. Lengths of flagellomeres I−IV (in μm): 192−198, 195; 105−109, 107; 111−113, 112; 4−6, 5. Basal segment width 16−18, 17 μm; length of blade at apex of basal segment 29 μm. Length of antenna/length of head 1.70−1.87, 1.77.

Abdomen. Length of anal setae 141—172, 145 µm. Procercus 10—12, 11 µm long, 6—8, 7 µm wide. Subbasal setae of posterior parapods split as in Fig. 3.147F, 48—55, 51 µm long. Posterior parapods 141—145, 142 µm long.

3.100 *CORYNONEURA* SP. *SEXADENTATA* WIEDENBRUG & TRIVINHO-STRIXINO, 2011 (FIGS. 3.147A—C AND 3.148A—I)

Corynoneura sp. *sexadentata* Wiedenbrug & Trivinho-Strixino, 2011:37.

Diagnostic characters. Female (Fig. 3.147A—C) and pupa (Fig. 3.148A—C) are not separable from *C. septadentata*, the only difference between both species is the larval mentum of sp. *sexadentata* that has 6 median teeth and instead of 7 (Fig. 3.148E).

Remarks. Unfortunately the male of this morphotype is still unknown. All females of *C.* sp. *sexadentata* were collected in Ubatuba (0 m alt.), and specimens of *C. septadentata* in the Parque Estadual do Jaraguá (about 1000 m alt.). It is possible that the difference on the

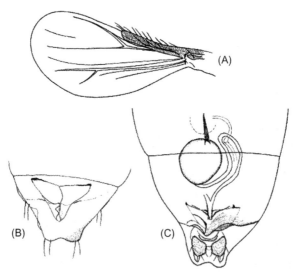

Figure 3.147 *Corynoneura* sp. *sexadentata* Wiedenbrug & Trivinho-Strixino, 2011, Female imago (From Wiedenbrug, S., & Trivinho-Strixino, S. (2011). New species of the genus *Corynoneura* Winnertz (Diptera, Chironomidae) from Brazil. *Zootaxa, 2822*, 1—40). (A) Wing. (B) Dorsal view of genitalia with labia, right is dorsal view of labia, left is ventral view of labia. (C) Ventral view of genitalia; left is gonapophysis removed.

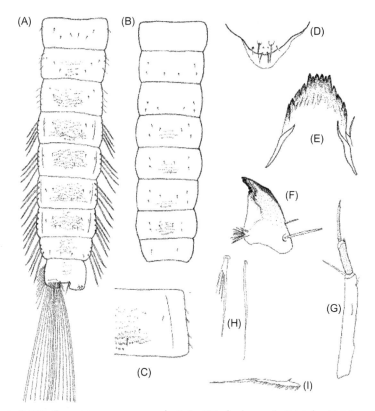

Figure 3.148 *Corynoneura* sp. *sexadentata* Wiedenbrug & Trivinho-Strixino 2011, Immatures imago (From Wiedenbrug, S., & Trivinho-Strixino, S. (2011). New species of the genus *Corynoneura* Winnertz (Diptera, Chironomidae) from Brazil. *Zootaxa, 2822,* 1–40). (A–C) Pupa imago. (A) Tergites Ⅰ–Ⅸ and anal lobe, right is dorsal view without fringe, left is ventral view. (B) Sternites Ⅰ–Ⅷ. (C) Detail of tergite Ⅳ. (D–I) Larva imago. (D) Dorsal view of labrum. (E) Mentum. (F) Mandible. (G) Antenna. (H) Abdominal setae. (I) Subbasal seta of posterior parapod.

mentum is a population variation, but only with knowledge of the male and more material it would be possible to affirm that.

3.101 *CORYNONEURA* SP. 6 (BOLTON) (FIG. 3.149A–F)

Corynoneura sp. 6 Bolton, 2007: 30; Fu & Sæther, 2012: 46.

Material examined. USA, Ohio, Wayne County, Brown's Lake, bog, 1 female, 1 larval and 1 pupal exuviae, 21.iv.1988, M. J. Bolton (MJB).

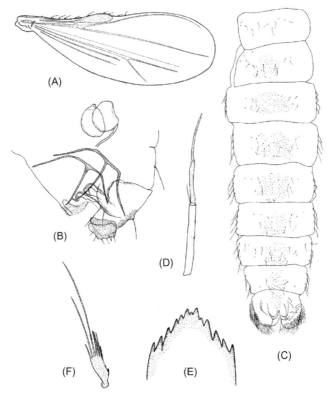

Figure 3.149 *Corynoneura* sp. 6 (Bolton) (From Fu, Y., & Sæther, O. A. (2012). *Corynoneura* Winnertz and *Thienemanniella* Kieffer from the Nearctic region (Diptera: Chironomidae: Orthocladiinae). *Zootaxa, 3536*, 1−61. www.mapress.com/j/zt). (A and B) Female imago. (A) Wing. (B) Genitalia, ventral view. (C) Pupa: tergites I−IX. (D−F) Larva. (D) Antenna. (E) Mentum. (F) Subbasal seta of posterior parapod.

Diagnostic characters. The female has an AR of 0.44, hind tibia only slightly elongate and without S-shaped spur, gonocoxite IX with 3 setae, and coxosternapodeme with 4−5 lateral lamellae. The pupa has no taeniate L-setae on tergites I−II, 34 taeniae in fringe of anal lob, and tergites IV−VII with only several thick hooklets. The larvae are separable by AR 0.82, mentum with three median teeth, median one smaller than the lateral median tooth, and first lateral teeth a little smaller than adjacent ones.

Female (*n* = 1)

Total length 2.48 mm. Wing length 0.71 mm. Wing width/wing length 0.42. Total length/wing length 3.49. Wing length/profemur length 3.3.

Coloration. Head light brown. Thorax brown with median part yellowish. Antenna and legs yellowish, abdomen yellowish brown.

Head. AR 0.44. Lengths of flagellomeres (in μm): 35, 28, 29, 29, 53. Ultimate flagellomere with 9 sensilla chaetica. Clypeus with 7 setae. Palpomere lengths (in μm): 14, 17, 21, 29, 53. Palpomere 5/3 ratio 2.6.

Thorax. Dorsocentrals 7, prealars 2. Scutellum with 2 setae.

Wing (Fig. 3.149A). VR 2.0, two anal veins present, Cu 368 μm long, Cu/wing length 0.52, C 280 μm long, C/wing length 0.39. Wing width/wing length 0.42. Costa with 11 setae.

Legs. Fore trochanter with keel. Spurs of fore tibia 8 μm long, of mid tibia 12 μm and 10 μm long, of hind tibia 32 μm and 15 μm long. Width at apex of fore tibia 18 μm wide, of mid tibia 19 μm, of hind tibia (a) 29 μm. Width of hind tibia 1/3 from apex (d) 21 μm, elongation length (b) 23 μm, length of maximum thickening (c_1) 55 μm, total length of thickening (c_2) 72 μm, a/d 1.4, b/d 1.1, c_1/d 2.6, c_2/d 3.4. Apex of hind tibia slightly expanded, with comb of 15 setae, without S-shaped spur. Lengths and proportions of legs as in Table 3.81.

Abdomen. Number of setae on tergites not visible.

Genitalia (Fig. 3.149B). Tergite IX with 4 long caudal setae. Cercus 44 μm long, 26 μm wide. Notum length 79 μm. Gonocoxite with 3 long setae. Coxosternapodeme with 4−5 transparent fused lateral lamellae on inner side of coxosternapodeme. Seminal capsule 61 μm long. Neck 10 μm long, 8 μm wide.

Pupa ($n = 1-2$)

Total length 1.69 mm. Exuviae yellowish including cephalothorax.

Cephalothorax. Frontal setae 44 μm long. Median antepronotals 6 μm and 10 μm long. Lateral antepronotals not visible. Anterior precorneal seta (PcS_1) 14 μm long, PcS_2 17 μm long, PcS_3 21 μm long. PcS_{1-3} almost in a line; PcS_1 4 μm from PcS_2; PcS_2 7 μm from PcS_3; PcS_1 11 μm from PcS_3; PcS_3 19 μm from thoracic horn. Anterior dorsocentral (Dc_1) 6 μm long; Dc_2 7 μm long; Dc_3 4 μm long; Dc_4 not visible. Dc_1 located 12 μm

Table 3.81 Lengths (in μm) and proportions of legs segments of female *Corynoneura.* sp. 6 (Bolton) (n = 1)

	fe	ti	ta₁	ta₂	ta₃	ta₄	ta₅	LR	BV	SV	BR
p₁	212	260	128	69	36	18	30	0.49	3.9	3.7	2.2
p₂	268	280	139	71	34	12	34	0.50	4.5	3.9	2.1
p₃	240	264	147	87	32	16	34	0.56	3.9	3.4	1.9

from Dc_2, Dc_2 located 51 μm from Dc_3. Wing sheath with 4 rows of pearls.

Abdomen (Fig. 3.149C). Shagreen and chaetotaxy as illustrated. No taeniate L-setae on tergites I−II. Anal lobe 125 μm long. Anal lobe fringe with 34 setae, 360 μm long. Length of three taeniate anal macrosetae hard to measure; median setae about 61 μm long.

Larva ($n = 1-2$)

Coloration. Head brown. Antenna with basal segment yellowish, other segments dark brown. Abdomen yellowish.

Head. Capsule length 272 μm, width 152 μm. Postmentum 240 μm long. Sternite II obvious, rising from small tubercle, I and III not visible. Premandible 21 μm long. Mentum as in Fig. 3.150E. Mandible 50 μm long. Antenna (Fig. 3.149D), AR 0.82. Lengths of flagellomeres I−IV (in μm): 160, 93, 95, 8. Basal segment width 15 μm, length of blade at apex of basal segment 44 μm. Length of antenna/length of head 1.31.

Abdomen. Length of anal setae 240 μm. Procercus 12 μm long, 10 μm wide. Subbasal seta of posterior parapods split as in Fig. 3.149F, 73 μm long. Posterior parapods 97 μm long, 46 μm wide.

3.102 *CORYNONEURA* SP. 10 (BOLTON) (FIG. 3.150A−D)

Corynoneura sp. 10 Bolton, 2007: 31; Fu & Sæther, 2012: 48.

Material examined. USA, Ohio, Hocking County, Deep Woods, 1.3 miles east-southeast of South Bloomingville, tributary to east fork of Queer Creek, 3 larval and 1 pupal exuviae, 25.iii. & 5.v.2004, M. J. Bolton (MJB).

Diagnostic characters. The pupa has no taeniate L-setae on tergite I, 29 taeniae in fringe of anal lobe, and tergites III−VI with strong shagreen posteriorly. The larvae are separable by unicolorous antenna, AR 0.71−0.79, length of antenna/length of head 1.6; mentum with three median teeth, median one smaller than lateral median tooth; and first lateral teeth smaller than adjacent ones.

Pupa ($n = 1$)

Total length 2.25 mm. Exuviae yellowish including cephalothorax.

Cephalothorax. Frontal setae 26 μm long. Median antepronotals 12 μm and 14 μm long. Lateral antepronotals 8 μm long. Anterior precorneal seta (PcS_1) 10 μm long, PcS_2 12 μm long, PcS_3 20 μm long. PcS_{1-3} in a line, PcS_1 6 μm from PcS_2, PcS_2 4 μm from PcS_3, PcS_1 12 μm from PcS_3, PcS_3 61 μm from thoracic horn. Anterior dorsocentral (Dc_1) not measurable (only base retained), Dc_2 7 μm long, Dc_3 8 μm long, Dc_4 7 μm long.

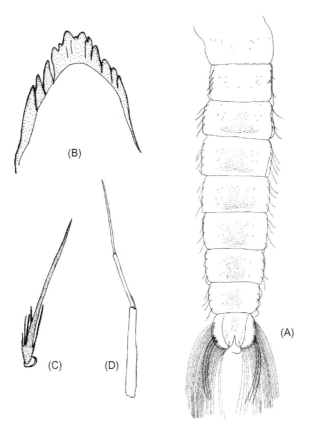

Figure 3.150 *Corynoneura* sp. 10 (Bolton) (From Fu, Y., & Sæther, O. A. (2012). *Corynoneura* Winnertz and *Thienemanniella* Kieffer from the Nearctic region (Diptera: Chironomidae: Orthocladiinae). *Zootaxa, 3536,* 1−61. www.mapress.com/j/zt). (A) Pupa: tergites I −IX. (B−D) Larva. (B) Mentum. (C) Subbasal seta of posterior parapod. (D) Antenna.

Dc_1 located 21 μm from Dc_2, Dc_2 located 48 μm from Dc_3, Dc_3 located 19 μm from Dc_4. Wing sheath with 4 rows of pearls.

Abdomen (Fig. 3.150A). Shagreen and chaetotaxy as illustrated. No taeniate L-setae on tergite I . Anal lobe 145 μm long. Anal lobe fringe with 29 setae, 380 μm long; three taeniate anal macrosetae about 240 μm long; median setae about 121 μm long.

Larva ($n = 3$)

Coloration. Body entirely pale yellow.

Head. Capsule length 220−240, 233 μm; width 156−164, 160 μm. Postmentum 200−208, 205 μm long. Sternite II obvious, rising from small tubercle, I and III not visible. Premandible 14—17, 16 μm long. Mentum as in Fig. 3.150B. Mandible 50—55, 52 μm long. Antenna (Fig. 3.150D), AR 0.71—0.79. Lengths of flagellomeres I—IV (in μm): 162—172, 93—97, 121, 4. Basal segment width 18—22 μm; length of blade at apex of basal segment 28—30 μm. Length of antenna/length of head 1.6.

Abdomen. Length of anal setae 158—165 μm. Procercus 12—14 μm long, 10—12 μm wide. Subbasal setae of posterior parapods split as in Fig. 3.150C, 67 μm long.

3.103 *CORYNONEURA* SP. 12 (BOLTON) (FIG. 3.151A—C)

Corynoneura sp. 12 Bolton, 2007: 28; Fu & Sæther, 2012: 50.

Material examined. USA, Ohio, Delaware County, tributary to Big Run Tributary, 1 prepupal exuviae, 1.vi.2005, M. J. Bolton; Delaware County, Olentangy River at Williams Road, 1 larval exuviae, 13.ix.2006, M. J. Bolton; Highland County, northern fork of White Oak Creek at Sicily Road, 3 larval exuviae, 21.viii.2006, M. J. Bolton (MJB).

Diagnostic characters. The larvae are separable by having the mentum with two median teeth, and smaller lateral tooth adjacent to median tooth; basal antennal segment yellowish, other segments dark brown;

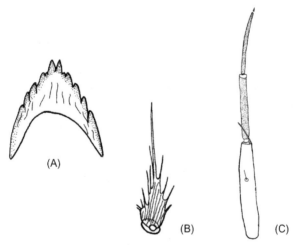

Figure 3.151 *Corynoneura* sp. 12 (Bolton), Larva (From Fu, Y., & Sæther, O. A. (2012). *Corynoneura* Winnertz and *Thienemanniella* Kieffer from the Nearctic region (Diptera: Chironomidae: Orthocladiinae). *Zootaxa, 3536*, 1—61. www.mapress.com/j/zt). (A) Mentum. (B) Subbasal seta of posterior parapod. (C) Antenna.

length of antenna/length of head 1.47−1.50, 1.48, long posterior shagreen on abdominal tergites II −VIII in prepupa.

Larva (*n* = 3)

Coloration. Head brown. Antenna with basal segment yellowish, other
segments dark brown. Abdomen yellowish.

Head. Capsule length 172−182, 178 μm, width 132−140, 136 μm.
Postmentum 152−170, 161 μm long. Sternite II obvious, rising from
small tubercle, I and III not visible. Premandible 20−25, 22 μm long.
Mentum as in Fig. 3.151A. Mandibles 30−36, 33 μm long. Antenna as in
Fig. 3.151C, AR 0.73−0.78, 0.75. Lengths of flagellomeres I−IV
(in μm): 107−119, 112; 69−73, 72; 73−76, 75; 3−5, 4. Basal segment
width 18−24, 22 μm; length of blade at apex of basal segment
23−24 μm. Length of antenna/length of head 1.47−1.50, 1.48.

Abdomen. Length of anal setae 141−162, 150 μm. Procercus 10−12,
11 μm long, 8−12, 10 μm wide. Subbasal setae of posterior parapods split
as in Fig. 3.151B, 36−40, 39 μm long. Posterior parapods 121−131,
126 μm long.

CHAPTER 4

Zoogeography

Based on the data provided in this research, the zoogeographic features of *Corynoneura* found worldwide is reviewed and summarized in this chapter.

Zoogeography is a science which studies, describes, and tries to explain the past and present distribution of animals from a temporal and spatial view. In recent decades, the development of the theory of plate tectonics and continental drift have provided zoogeography with a historical aspect. The science of biogeography has evolved from the traditionally descriptive to quantitative. As we know, zoogeographic studies are characterized by their regionalism, which means that regional faunistic surveys promote the development of zoogeographic science.

The abbreviations of the zoogeography used in this book:

AF Afrotropical region (Ethiopian)
AU Australian region
NE Nearctic region
NT Neotropical
O Oriental region
P Palearctic region
W Worldwide

4.1 DIVERSITY OF THE *CORYNONEURA* GENERIC GROUP WORLDWIDE

The recorded species of *Corynoneura* generic group in the world amounts to 206 species belonging to 9 genera (Table 4.1)

According to Table 4.1, the genera *Corynoneura* and *Thienemanniella* have a worldwide distribution, other genera are only distributed in one or two geographical regions. From the total number of the species, Table 4.1 shows most of the species of *Corynoneura* generic group occurred in Palearctic region (78 species), Neotropical region (49 species), and Oriental region (32 species), which account for 37.9%, 23.8%, and 15.5% of all species, respectively. In addition, 29 species in the Nearctic region, take up 14.1%, 12 species are found in the Afrotropical region (5.8%), and 6 species occur in the Australian region (2.9%). The result shows that the distribution in Palearctic region is predominant.

Taxonomy of Corynoneura *Winnertz (Diptera: Chironomidae)*
DOI: https://doi.org/10.1016/B978-0-12-815263-8.00004-0

Table 4.1 Distribution of the *Corynoneura* generic group and species number in the worldwide

Generic group	Genus	P	NE	AF	NT	O	AU	Total
Corynoneura generic group	*Corynoneura*	46	19	4	25	17	5	98
	Thienemanniella	31	9	6	8	15	1	53
	Ichthyocladius	0	0	0	3	0	0	3
	Physoneura	0	0	0	2	0	0	2
	Tempisquitoneura	0	1	0	1	0	0	2
	Onconeura	0	a	0	9	0	0	9
	Ubatubaneura	0	0	0	1	0	0	1
Corynoneura generic group	*Notocladius*	0	0	1	0	0	0	1
	Corynoneurella	1	0	1	0	0	0	2
	Total genera	3	3	4	7	2	2	9
	Total species	78	29	12	49	32	6	206
	Percentage	37.9	14.1	5.8	23.8	15.5	2.9	—

[a]Only recorded larva.

Table 4.2 Distribution of *Corynoneura* species

Region	Species
Palearctic (46 species)	*C. arctica* Kieffer, *C. aurora* Makarchenko and Makarchenko, *C. brundini* Hirvenoja and Hirvenoja, *C. carriana* Edwards, *C. celeripes* Winnertz, *C. celtica* Edwards, *C. collaris* Makarchenko and Makarchenko, *C. coronata* Edwards, *C. cuspis* Tokunaga, *C. cylindricauda* Fu, Sæther and Wang, *C. doriceni* Makarchenko and Makarchenko, *C. edwardsi* Brundin, *C. ferelobata* Sublette and Sasa, *C. fittkaui* Schlee, *C. fujiundecima* Sasa, *C. gratias* Schlee, *C. gynocera* Tuiskunen, *C. inawapequea* Sasa, Kitami and Suzuki, *C. inefligiata* Fu, Sæther and Wang, *C. isigaheia* Sasa and Suzuki, *C. kadalinka* Makarchenko and Makarchenko, *C. kedrovaya* Makarchenko and Makarchenko, *C. kibunelata* Sasa, *C. kisogawa* Sasa *et* Kondo, *C. kibunespinosa* Sasa, *C. lacustris* Edwards, *C. longipennis* Tokunaga, *C. lobata* Edwards, *C. magna* Brundin, *C. marina* Kieffer, *C. makarchenkorum* Krasheninnikov, *C. nankaiensis* Fu, Sæther and Wang, *C. prima* Makarchenko and Makarchenko, *C. schleei* Makarchenko and Makarchenko, *C. scutellata* Winnertz, *C. secunda* Makarchenko and Makarchenko, *C. seiryuresea* Sasa, Suzuki and Sakai, *C. sorachibecea* Sasa, Kitami and Suzuki, *C. sundukovi* Makarchenko and Makarchenko, *C. tenuistyla* Tokunaga, *C. tertia* Makarchenko and Makarchenko, *C. tokarapequea* Sasa and Suzuki, *C. tokaraquerea* Sasa and Suzuki, *C. tyrrhena* Moubayed-Breil, *C. vittalis* Tokunaga, *C. yoshimurai* Tokunaga

(Continued)

Table 4.2 (Continued)

Region	Species
Oriental (17 species)	*C. carinata* Singh and Maheshwari, *C. centromedia* Hazra, Nath and Chaudhuri, *C. chandertali* Singh and Maheshwari, *C. confidens* Fu, Sæther and Wang, *C. ecphora* Fang et al., *C. ferelobata* Sublette and Sasa, *C. incidera* Hazra, Nath and Chaudhuri, *C. korema* Fu, Sæther and Wang, *C. lahuli* Singh and Maheshwari, *C. latusatra* Fu, Sæther and Wang, *C. lobata* Edwards, *C. macdonaldi* Fu, Sæther and Wang, *C. nasuticeps* Hazra, Nath and Chaudhuri, *C. prominens* Fu, Sæther and Wang, *C. scutellata* Winnertz, *C. yoshimurai* Tokunaga, *C. postcinctura* (Tokunaga)
Nearctic (19 species)	*C. acuminata* Fu and Sæther, *C. arctica* Kieffer, *C. ascensa* Fu and Sæther, *C. capitanea* Fu and Sæther, *C. caudicula* Fu and Sæther, *C. celeripes* Winnertz, *C. collaris* Makarchenko and Makarchenko, *C. diara* Roback, *C. doriceni* Makarchenko and Makarchenko, *C. disinflata* Fu and Sæther, *C. edwardsi* Brundin, *C. hortonensis* Fu and Sæther, *C. fittkaui* Schlee, *C. floridaensis* Fu and Sæther, *C. lacustris* Edwards, *C. lobata* Edwards, *C. macula* Fu and Sæther, *C. porrecta* Fu and Sæther, *C. scutellata* Winnertz
Afrotropical (4 species)	*C. cristata* Freeman, *C. dewulfi* Goetghebuer, *C. elongata* Freeman, *C. lobata* Edwards
Australian (5 species)	*C. australiensis* Freeman, *C. imperfecta* (Skuse), *C. lobata* Edwards, *C. scutellata* Winnertz, *C. postcinctura* (Tokunaga)
Neotropical (25 species)	*C. bodoquena* Wiedenbrug et al., *C. boraceia* Wiedenbrug et al., *C. canchim* Wiedenbrug et al., *C. diogo* Wiedenbrug et al., *C. espraiado* Wiedenbrug et al., *C. ferelobata* Sublette *et* Sasa, *C. fortispicula* Wiedenbrug *et* Trivinho-Strixino, *C. franciscoi* Wiedenbrug et al., *C. guanacaste* Wiedenbrug et al., *C. hermanni* Wiedenbrug *et* Trivinho-Strixino, *C. hirvenojai* Sublette *et* Sasa, *C. humbertoi* Wiedenbrug et al., *C. longiantenna* Wiedenbrug et al., *C. mediaspicula* Wiedenbrug *et* Trivinho-Strixino, *C. mineira* Wiedenbrug *et* Trivinho-Strixino, *C. renata* Wiedenbrug et al., *C. salviniatilis* Wiedenbrug et al., *C. septadentata* Wiedenbrug *et* Trivinho-Strixino, *C. sertaodaquina* Wiedenbrug *et* Trivinho-Strixino, *C. scutellata* Winnertz, *C. sisbiota* Wiedenbrug et al., *C. trondi* Wiedenbrug et al., *C. unicapsulata* Wiedenbrug *et* Trivinho-Strixino, *C. vidiapodeme* Wiedenbrug et al., *C. zempoala* Wiedenbrug et al., *C.* sp. *circulimentum*, *C.* sp. *sexadentata*, *C.* sp. "headspines"

4.2 DISTRIBUTION OF *CORYNONEURA* SPECIES

With reference to the genus *Corynoneura* Winnertz, 98 species have been described. Forty-six species are recorded from the Palearctic region, 19 from the Nearctic region, 25 from the Neotropical region, 17 from the Oriental region, 4 from the Afrotropical region, and 5 from the Australasian region (Ashe & O'Connor, 2012; Fu & Sæther, 2012; Fu, Sæther, & Wang, 2009; Makarchenko & Makarchenko, 2006b, 2010; Makarchenko, Makarchenko, Zorina, & Sergeeva, 2005; Schlee, 1968b; Wang, 2000; Wiedenbrug, Lamas, & Trivinho-Strixino, 2012; Wiedenbrug & Trivinho-Strixino, 2011; Yamamoto, 2004). The distribution information of *Corynoneura* species research work are listed in Table 4.2.

REFERENCES

Allegrucci, G., Carchini, G., Todisco, V., Convey, P., & Sbordoni, V. (2006). A molecular phylogeny of antarctic Chironomidae and its implications for biogeographical history. *Polar Biology*, *29*, 320−326.

Andersen, T., & Sæther, O. A. (2005). *Onconeura*, a new Neotropical orthoclad genus (Chironomidae, Orthocladiinae). *Zootaxa*, *957*, 1−16.

Ashe, P., & Cranston, P. S. (1990). Family Chironomidae. In Á. Soós, & L. Pap (Eds.), *Catalogue of Palaearctic Diptera Psychodidae-Chironomidae* (Vol. 2, pp. 113−355). Budapest: Akadémiai Kiadó.

Ashe, P., & O'Connor, J. P. (2012). *A world catalogue of Chironomidae (Diptera). Part 2. Orthocladiinae* (p. 968) Dublin: Irish Biogeographical Society, National Museum of Ireland.

Boesel, M. W., & Winner, R. W. (1980). Corynoneurinae of northeastern United States, with a key to adults and observations on their occurrence in Ohio (Diptera: Chironomidae). *Journal of the Kansas Entomological Society*, *53*(3), 501−508.

Bolton, M. J. (2007). *Ohio EPA Supplemental keys to the larval Chironominae (Diptera) of Ohio and Ohio Chironomidae checklist.* Unpublished report from Ohio EPA. <http://www.epa.state.oh.us/portals/35/documents/MidgeLarvaeKeyJune2007.pdf> [2016-5-3].

Brodin, Y., Ejdung, G., Strandberg, J., & Lyrholm, T. (2012). Improving environmental and biodiversity monitoring in the Baltic Sea using DNA barcoding of Chironomidae (Diptera). *Molecular Ecology Resources*, *13*(6), 996−1004.

Brundin L. (1949). Chironomiden und andere Bodentiere der südschwedischen Urgebirgsseen. Ein Beitrag zur Kenntnis der bodenfaunistischen Charakterzüge schwedischer oligotropher Seen (Vol. 30, pp. 1−914). Report Institute of Freshwater Research Drottningholm.

Brundin L. (1956). Zur Systematik der Orthocladiinae (Dipt. Chironomidae) (Vol. 37, pp. 5−185). Report Institute Freshwater Research Drottningholm.

Carew, M. E., Marshal, S. E., & Hoffman, A. A. (2011). A combination of molecular and morphological approaches resolves species in the taxonomically difficult genus *Procladius* Skuse (Diptera: Chironomidae) despite high intraspecific morphological variation. *Bulletin of Entomological Research*, *101*, 505−519.

Coffman, W. P. (1979). Neglected characters in pupal morphology as tools in taxonomy and phylogeny of Chironomidae (Diptera). In O. A. Sæther (Ed.), *Recent development in Chironomid studies (Diptera: Chironomidae)* (Vol. 10, pp. 37−46). Entomologica Scandinavica Supplement.

Coffman, W. P., Cranston, P. S., Oliver, D. R., & Sæther, O. A. (1986). The pupae of Orthocladiinae (Diptera: Chironomidae) of the Holarctic region. Keys and diagnoses. In T. Wiederholm (Ed.), *Chironomidae of the Holarctic Region. Keys and Diagnoses. Part 2. Pupae*. Entomologica Scandinavica Supplement (Vol. 28, pp. 1−482).

Cranston, P. S. (1995). Systematics. In P. D. Armitage, P. S. Cranston, & L. V. C. Pinder (Eds.), *The Chironomidae: Biology and ecology of non-biting midges.* (pp. 31−61). London: Chapman & Hall.

Cranston, P. S., Hardy, N. B., & Morse, G. E. (2012). A dated molecular phylogeny for the Chironomidae (Diptera). *Systematic Entomology*, *37*, 172−188.

Cranston, P. S., & Oliver, D. R. (1988). Additions and corrections to the Nearctic Orthocladiinae (Diptera: Chironomidae). *The Canadian Entomologist*, *120*, 425−462.

Cranston, P. S., Oliver, D. R., & Sæther, O. A. (1983). The larvae of Orthocladiinae (Diptera: Chironomidae) of the Holarctic region. Keys and diagnoses. In T. Wiederholm (Ed.), *Chironomidae of the Holarctic Region. Keys and diagnoses. Part 1. Larvae.* Entomologica Scandinavica Supplement (Vol. 19, pp. 149–292).

Cranston, P. S., Oliver, D. R., & Sæther, O. A. (1989). The adult males of Orthocladiinae (Diptera: Chironomidae) of the Holarctic region. Keys and diagnoses. In T. Wiederholm (Ed.), *Chironomidae of the Holarctic Region. Part 3. Adult males.* Entomologica Scandinavica Supplement (Vol. 34, pp. 165–352).

Edwards, F. W. (1924). Some British species of *Corynoneura* (Diptera—Chironomidae). *Entomologist's Monthly Magazine, 60,* 182–189.

Edwards, F. W. (1929). British non-biting midges (Diptera, Chironomidae). *Transactions of the Royal Entomological Society of London, 77,* 279–430.

Ekrem, T., Stur, E., & Hebert, P. D. N. (2010). Females do count: documenting Chironomidae (Diptera) species diversity using DNA barcoding. *Organisms Diversity and Evolution, 10,* 397–408.

Epler, J. H. (2001). *Identification Manual for the larval Chironomidae (Diptera) of North and South Carolina. A guide to the taxonomy of the midges of the southeastern United States, including Florida.* North Carolina Department of Environment and Natural Resources, Raleigh, NC, and St. Johns River Water Management District, Palatka, FL. Special Publication SJ2001-SP13. 526 pp.

Epler, J. H., & de la Rosa, C. L. (1995). *Tempisquitoneura,* a new genus of Neotropical Orthocladiinae (Diptera: Chironomidae) symphoretic on *Corydalus* (Megaloptera: Corydalidae). *Journal of the North American Benthological Society, 14,* 50–60.

Fang, X. L., Wang, X., & Fu, Y. (2014). A new species of *Corynoneura* Winnertz from oriental China (Diptera, Chironomidae, Orthocladiinae). *Zootaxa, 3884*(6), 567–572.

Ferrington, L. C., Jr, & Sæther, O. A. (1995). *Physoneura,* a new genus of Orthocladiinae from Patagonia and South Chile (Diptera: Chironomidae). *Aquatic Insects, 17,* 57–63.

Fittkau, E. J. (1974). *Ichthyocladius* n. gen., eine neotropische Gattung der Orthocladiinae (Chironomidae, Diptera) deren Larven epizoisch auf Welsen (Astroblepidae und Loricariidae) leben. *Entomologisk Tidskrift Supplement, 95,* 91–106.

Freeman, P. (1953). Chironomidae from Western Cape Province—II. *Proceedings of the Royal Entomological Society of London (B), 22,* 201–253.

Freeman, P. (1956). A study of the Chironomidae (Diptera) of Africa South of Sahara, Part 2. Bulletin of the British Museum (Natural History). *Entomology, 4,* 287–368.

Freeman, P. (1961). The Chironomidae (Diptera) of Australia. *Australian Journal of Zoology, 9,* 611–738.

Fu, Y., Hestenes, T. C., & Sæther, O. A. (2010). Review of Afrotropical *Thienemanniella* Kieffer (Diptera: Chironomidae: Orthocladiinae). *Zootaxa, 2338,* 1–22.

Fu, Y., & Sæther, O. A. (2012). *Corynoneura* Winnertz and *Thienemanniella* Kieffer from the Nearctic region (Diptera: Chironomidae: Orthocladiinae). *Zootaxa, 3536,* 1–61.

Fu, Y., Sæther, O. A., & Wang, X. (2009). *Corynoneura* Winnertz from East Asia, with a systematic review of the genus (Diptera: Chironomidae: Orthocladiinae). *Zootaxa, 2287,* 1–44.

Fu, Y., Sæther, O. A., & Wang, X. (2010). *Thienemanniella* Kieffer from East Asia, with a systematic review of the genus (Diptera: Chironomidae: Orthocladiinae). *Zootaxa, 2431,* 1–42.

Galindo-Leal, C., & Câmara, I. G. (2003). *The Atlantic Forest of South America: Biodiversity status, threats, and outlook.* Washington: Island Press.

Goetghebuer, M. (1919). Observations sur les larves et les nymphes de quelques Chironomides de Belgique. *Annales de Biologie Lacustre, 9,* 51–78.

Goetghebuer, M. (1935). Chironomides du Congo belge (1). *Revue de Zoologie et de Botanique Africaines, 27,* 364–366.

Goetghebuer, M. (1939). Tendipedidae (Chironomidae). e) Subfamilie Corynoneurinae. A. Die Imagines. In E. Lindner (Ed.), *Die Fliegen der palaearktischen Region* (Vol. 13f, pp. 1−14).

Harrison, A. D. (1992). Chironomidae from Ethiopia, Part 2. Orthocladiinae with two new species and a key to *Thienemanniella* Kieffer (Insecta, Diptera). *Spixiana, 15,* 149−195.

Harrison, A. D. (1997). Two small Orthocladiinae (Chironomidae, Diptera) from the Western Cape Province, South Africa. *Annals of the Cape Provincial Museums (Natural History), 19,* 375−386.

Hazra, N., Nath, S., & Chaudhuri, P. K. (2003). The genus *Corynoneura* Winnertz (Diptera. Chironomidae) from the Darjeeling-Sikkim Himalayas of India, with description of three new species. *Entomologist's Monthly Magazine, 139,* 69−82.

Henriques-Oliveira, A. L., Dorvillé, L. F. M., & Nessimian, J. L. (2003). Distribution of Chironomidae larvae fauna (Insecta: Diptera) on different substrates in a stream of Floresta da Tijuca, RJ, Brazil. *Acta Limnologica Brasiliense, 15*(2), 69−84.

Hestenes, T. C., & Sæther, O. A. (2000). Three new Nearctic *Thienemanniella* Kieffer species with a review of the Nearctic species. In O. Hoffrichter (Ed.), *Late 20th century research on Chironomidae: An Anthology from the 13th international symposium on Chironomidae* (pp. 103−127). Aachen: Shaker Verlag.

Hirvenoja, M., & Hirvenoja, E. (1988). Corynoneura brundini spec. nov. Ein Beitrag zur Systematik der Gattung Corynoneura (Diptera, Chironomidae). *Spixiana Supplement, 14,* 213−232.

Kieffer, J. J. (1911). Nouveaux Tendipédides du groupe *Orthocladius* (Dipt.). 2. note. *Bulletin de la Société entomologique de France, 8,* 199−202.

Kieffer, J. J. (1915). Neue Chironomiden aus Mitteleuropa. *Broteria Serie Zoologica, 13,* 65−87.

Kieffer, J. J. (1923). *Nouvelle contribution à l'étude des Chironomides de la Nouvelle-Zemble.* Report of the scientific results of the Norwegian Expedition to Novaya Zemlya (Vol. 9, pp. 3−11).

Kieffer, J. J. (1924). Chironomides nouveaux ou rares de l'Europe centrale. *Bulletin de la Société d'Histoire Naturelle de Metz, 30,* 11−110.

Kieffer, J. J. (1925). Deux genres nouveaux et plusieurs espèces nouvelles du groupe des Orthocladiariae (Dipteres, Chironomides). *Annales de la Societe Scientifique de Bruxelles, 44,* 555−566.

Krasheninnikov, A. B. (2012). A new and little known Chironomidae of subfamily Orthocladiinae (Diptera, Chironomidae) from the Urals. *Eurasian Entomological Journal, 11*(1), 83−86.

Krasheninnikov, A. B., & Makarchenko, E. A. (2009). On the chironomid fauna of subfamilies Podonominae, Diamesinae and Orthocladiinae (Diptera, Chironomidae) of the Vishersky Nature Reserve and bordering territories (North Urals). *Euroasian Entomological Journal, 8*(3), 335−340.

Langton, P. H., & Pinder, L. C. V. (2007). *Keys to the adult male Chironomidae of Britain and Ireland* (64). Freshwater Biological Association, Scientific Publication, 239 + 168 figs.

Langton, P., & Visser, H. (2003). *Chironomidae exuviae. A key to pupal exuviae of the West Palaearctic Region* (64). Freshwater Biological Association, Scientific Publication, 239 + 168 figs.

Lehmann, J. (1979). Chironomidae (Diptera) aus Fließgewässern Zentralafricas (Systematik, Ökologie, Verbreitung und Produktionsbiologie), Teil I: Kivu Gebiet, Ostzaïre. Spixiana, Supplement (Vol. 3, pp. 1−144).

Makarchenko, E. A., & Makarchenko, M. A. (2006a). Subfamily Orthocladiinae. In A. Lelej (Ed.), *Key to the insects of Russian Far East. Vol. 6. Diptera and Siphonaptera.* Pt 4. Vladivostok, Dal'nauka (pp. 280–372, 482–530, 623–671).

Makarchenko, E. A., & Makarchenko, M. A. (2006b). Chironomids of the genera *Corynoneura* Winnertz, 1846 and *Thienemanniella* Kieffer, 1919 (Diptera, Chironomidae, Orthocladiinae) of the Russian Far East. *Euroasian Entomological Journal,* *5*(2), 151–162.

Makarchenko, E. A., & Makarchenko, M. A. (2010). New data on the fauna and taxonomy of *Corynoneura* Winnertz (Diptera, Chironomidae, Orthocladiinae) for the Russian Far East and bordering territories. *Euroasian Entomological Journal,* *9*(3), 353–370, + II.

Makarchenko, E. A., Makarchenko, M. A., Zorina, O. V., & Sergeeva, I. V. (2005). Preliminary data on fauna and taxonomy of Chironomids (Diptera, Chironomidae) of the Russian Far East. Vladimir Ya. *Levanidov's Biennial Memorial Meetings* (pp. 394–420).

Martin, J., Blinov, A., Alieva, E., & Hirabayashi, K. (2007). A molecular phylogenetic investigation of the genera closely related to *Chironomus* Meigen (Diptera: Chironomidae). In T. Andersen (Ed.), *Contributions to the systematics and ecology of aquatic Diptera: A tribute to Ole A. Sæther.* (pp. 193–203). Ohio: The Caddis Press.

Mendes, H., Andersen, T., & Sæther, O. A. (2004). New species of *Ichthyocladius*, a member of the *Corynoneura*-group (Diptera: Chironomidae: Orthocladiinae), with a review of the genus. *Studies of Neotropical Fauna and Environment,* *39*, 15–35.

Moubayed-Breil, J. (2015). *Corynoneura tyrrhena* sp. n., a crenophilous species occurring in high mountain streams of Corsica [Diptera, Chironomidae, Orthocladiinae]. *Ephemera,* *16*(1), 1–12, 2014 (2015).

Myers, N., Mittermeier, R. A., Mittermeier, C. G., Fonseca, G. A. B., & Kent, J. (2000). Biodiversity hotspots for conservation priorities. *Nature, 403*, 853–858.

Oliver, D. R., Dillon, M. E., & Cranston, P. S. (1990). *A catalog of Nearctic Chironomidae* (pp. 1–89). Research Branch Agriculture Canada, Publication, 1857/B.

Ouyang, Y., Lu, K., & Yan, J. (1984). The new records of the larvae of chironomid for China in Zhoushan archipelago. *Journal of the Zhejiang College of Fisheries, 3*, 29–42.

Roback, S. S. (1957a). The immature tendipedids of the Philadelphia area (Diptera: Tendipedidae). *Monographs of the Academy of Natural Sciences of Philadelphia, 9*, 1–152.

Roback, S. S. (1957b). Some Tendipedidae from Utah. *Proceedings of the Academy of Natural Sciences of Philadelphia, 109*, 1–24.

Sæther, O. A. (1969). Some Nearctic Podonominae, Diamesinae and Orthocladiinae (Diptera: Chironomidae). *Bulletin of the Fisheries Research Board of Canada, 170*, 1–154.

Sæther, O. A. (1977). Female genitalia in Chironomidae and other Nematocera: morphology, phylogenies, keys. *Bulletin of the Fisheries Research Board of Canada, 197*, 1–209.

Sæther, O. A. (1980). Glossary of chironomid morphology terminology (Diptera: Chironomidae). *Entomologica Scandinavica Supplement, 14*, 1–51.

Sæther, O. A. (1981). Orthocladiinae (Diptera: Chironomidae) from British West Indies, with descriptions of *Antillocladius* n. gen., *Lipurometriocnemus* n. gen., *Compterosmittia* n. gen. and *Diplosmittia* n. gen. *Entomologica Scandinavica Supplement, 16*, 1–46.

Sæther, O. A. (1982). Orthocladiinae (Diptera: Chironomidae) from SE U.S.A., with descriptions of *Plhudsonia*, *Unniella* and *Platysmittia* n. genera and *Atelopodella* n. subgen. *Entomologica Scandinavica, 13*, 465–510.

Sæther, O. A., & Kristoffersen, L. (1996). Chironomids with "M-fork". A reevaluation of the *Corynoneura* group (Insecta, Diptera, Chironomidae). *Spixiana Supplement, 19*(2), 229–232.

Sæther, O. A., & Spies, M. (2004). *Fauna Europaea: Chironomidae.* Fauna Europaea version 1.1. Available from <http://www.faunaeur.org> [2010-1-14].

Sanseverino, A. M., & Nessimian, J. L. (2001). Hábitats de larvas de Chironomidae (Insecta, Diptera) em riachos de Mata Atlântica no Estado do Rio de Janeiro. *Acta Limnologica Brasiliensia, 13*(1), 29–38.

Sasa M. (1985). Studies on Chironomid midges of some lakes in Japan, Part III. Studies on the Chironomids collected from lakes in the Mount Fuji area (Diptera, Chironomidae) (Vol. 83, pp. 130–131). Research Report from the National Institute for Environmental Studies.

Sasa M. (1988). Studies on the chironomid midges collected from lakes and streams in the southern region of Hokkaido, Japan. In M. Sasa, Y. Sugaya, & M. Yasuno (Eds.) *Studies on the chironomid Midges of Lakes in southern Hokkaido* (Vol. 121, pp. 9–76). Research Report from the National Institute for Environmental Studies.

Sasa, M. (1989). *Some characteristics of nature conservation within the chief rivers in Toyama Prefecture. The Upper Reach of Shou River* (pp. 60–62). Toyama Prefectural Environmental Pollution Research Centre.

Sasa, M. (1997). *Part 1. Studies on the chironomids collected throughout the year in the Shofuku Garden, near the mouth of Kurobe River, Toyama* (1997, pp. 14–69). Bulletin of Toyama Prefectural Environmental Science Research Centre.

Sasa, M., & Kondo, S. (1993). *Some characteristics of water quality and aquatic organism in the chief lakes in Toyama Prefecture, . Lake Nawagaike. Part 7. Additional chironomids recorded from the middle reaches of Kiso River* (1993, pp. 1–102). Toyama Prefectural Environmental Pollution Research Centre.

Sasa, M., Kitami, K., & Suzuki, H. (1999). *Studies on the chironomid midges collected with light traps and by sweeping on the shore of Lake Inawashiro, Fukishima Prefecture* (1999, pp. 1–37). Research Report of the Dr. Noguchi Memorial Hall.

Sasa, M., & Suzuki, H. (1995). The chironomid species collected on the Tokara Islands, Kagoshima (Diptera). *Japanese Journal of Sanitary and Zoology, 46*(3), 255–288.

Sasa, M., & Suzuki, H. (1998). Studies on the chironomid midges collected in Hokkaido and Northern Honshu. *Tropical Medicine, 40*(1), 9–43.

Sasa, M., & Suzuki, H. (1999). Studies on the chironomid midges of Tsushima and Iki Islands, western Japan: Part 2. Species of Orthocladiinae and Tanypodinae collected on Tsushima. *Tropical Medicine, 41*, 75–132.

Sasa, M., & Suzuki, H. (2000a). Studies on the chironomid species collected on Ishigaki and Iriomote Islands, Southwestern Japan. *Tropical Medicine, 42*(1), 1–37.

Sasa, M., & Suzuki, H. (2000b). Studies on the chironomid midges collected on Yakushima Island, Southwestern Japan. *Tropical Medicine, 42*(2), 53–134.

Sasa, M., & Suzuki, H. (2001). Studies on the chironomid species collected in Hokkaido in September, 2000. *Tropical Medicine, 43*(1/2), 1–38.

Sasa, M., Suzuki, H., & Sakai, T. (1998). Studies on the chironomid midges collected on the shore of Shimanto River in April 1998. Part 2. Description of additional species belonging to Orthocladiinae, Diamesinae and Tanypodinae. *Tropical Medicine, 40*(3), 99–147.

Schlee, D. (1968a). Phylogenetic studies on Chironomiden (Diptera), with special reference to the *Corynoneura* group. *Annales Zoologici Fennici, 5*, 130–138.

Schlee, D. (1968b). Vergleichende Merkmalsanalyse zur Morphologie und Phylogenie der *Corynoneura*-Group (Diptera, Chironomidae). *Stuttgarter Beiträge zur Naturkunde, 180*, 1–150.

Silva, F. L., Ekrem, T., & Fonseca-Gessner, A. A. (2013). DNA barcodes for species delimitation in Chironomidae (Diptera): a case study on genus Labrundinia. *Canadian Entomologist, 145*(6), 589–602.

Silva, F. L., & Wiedenbrug, S. (2014). Integrating DNA barcodes and morphology for species delimitation in the *Corynoneura* group (Diptera: Chironomidae: Orthocladiinae). *Bulletin of Entomological Research, 104*, 65–78.

Simpson, K. W., & Bode, R. W. (1980). Common larvae of Chironomidae (Diptera) from New York state streams and rivers: with particular reference to the fauna of artificial substrates. *New York State Museum Bulletin, 439*, 1—105.

Singh, S., & Maheshwari, G. (1987). Chironomidae (Diptera) of Chandertal Lake, Lahul Valley (Northwest Himalaya). *Annals of Entomology, 5*(2), 11—20.

Stur, E., & Ekrem, T. (2011). Exploring unknown life stages of Arctic Tanytarsini (Diptera: Chironomidae) with DNA barcoding. *Zootaxa, 2743*, 27—39.

Sublette, J. E. (1970). Type specimens of Chironomidae (Diptera) in the Illinois Natural History Survey Collection, Urbana. *Journal of the Kansas Entomological Society, 43*, 44—95.

Sublette, J. E., & Sasa, M. (1994). Chironomidae collected in Onchocerciasis endemic areas of Guatemala (Insecta, Diptera). *Spixiana Supplement, 20*, 1—60.

Swofford, D. L. (2002). *PAUP: phylogenetic analysis using parsimony (*and other methods). Version 4.0b10.* Sunderland, MA: Sinauer Associates.

Tokunaga, M. (1936). Japanese *Cricotopus* and *Corynoneura* species (Diptera: Chironomidae). *Tenthredo, 1*, 9—52.

Tokunaga, M. (1964). Diptera, Chironomidae. *Insects of Micronesia, 12*(5), 485—628.

Tuiskunen, J. (1983). A description of *Corynoneura gynocera* sp. n. (Diptera, Chironomidae) from Finland. *Annales entomologica Fennici, 49*, 100—102.

Wiedenbrug, S. (2000). *Studie zur Chironomidenfauna aus Bergbächen von Rio Grande do Sul, Brasilien.* Dissertation, Faculty of Biology, Ludwigs-Maximilian-Universität-München, Munich, Germany (p. 444).

Wang, X. (2000). A revised checklist of Chironomidae from China (Diptera). In O. Hoffrichter (Ed.), *Late 20th century research on Chironomidae. An anthology from the 13th international symposium on Chironomidae.* (pp. 629—652). Achen: Shaker Verlag.

Wiedenbrug, S., Lamas, C. J. E., & Trivinho-Strixino, S. (2012). A review of the genus *Corynoneura* Winnertz (Diptera: Chironomidae) from the neotropical region. *Zootaxa, 3574*, 1—61.

Wiedenbrug, S., Lamas, C. J. E., & Trivinho-Strixino, S. (2013). A review of neotropical species in *Thienemanniella* Kieffer (Diptera, Chironomidae). *Zootaxa, 3670*, 215—237.

Wiedenbrug, S., Mendes, H. F., Pepinelli, M., & Trivinho-Strixino, S. (2009). Review of the genus *Onconeura* Andersen et Sæther (Diptera: Chironomidae), with the description of four new species from Brazil. *Zootaxa, 2265*, 1—26.

Wiedenbrug, S., & Trivinho-Strixino, S. (2009). Ubatubaneura, a new genus of the *Corynoneura* group (Diptera: Chironomidae: Orthocladiinae) from the Brazilian Atlantic Forest. *Zootaxa, 1993*, 41—52.

Wiedenbrug, S., & Trivinho-Strixino, S. (2011). New species of the genus *Corynoneura* Winnertz (Diptera, Chironomidae) from Brazil. *Zootaxa, 2822*, 1—40.

Winnertz, J. (1846). Beschreibung einiger neuer Gattungen aus der Ordnung der Zweiflügler. *Stettiner entomologische Zeitung, 7*, 11—20.

Yamamoto, M. (2004). A catalog of Japanese Orthocladiinae (Diptera: Chironomidae). *Acta Dipterologica, 21*, 1—121.

Zaher, H., Barbo, F. E., Martínez, P. S., Nogueira, C., Rodrigues, M. T., & Sawaya, R. J. (2011). Répteis do Estado de São Paulo: Conhecimento Atual e Perspectivas. *Biota Neotropica, 11*, 1—15.

FURTHER READING

Boothroyd, I. K. G. (1999). Description of *Kaniwhaniwhanus* n. gen. (Diptera: Chironomidae: Orthocladiinae) from New Zealand. *New Zealand Journal of Marine and Freshwater Research, 33*, 341—350.

Cranston, P. S., & Martin, J. (1989). Family Chironomidae. In N. L. Evenhuis (Ed.), *Catalog of the Diptera of the Australasian and Oceanian Regions* (pp. 252–274). Honolulu: Bishop Museum.

Freeman, P., & Cranston, P. S. (1980). Family Chironomidae. In R. W. Crosskey (Ed.), *Catalogue of the Diptera of the Afrotropical Region* (pp. 175–202). London: British Museum (Natural History).

Sæther, O. A., Ashe, P., & Murray, D. A. (2000). Family Chironomidae. In L. Papp, & B. Darvas (Eds.), *Contributions to a manual of Palaearctic Diptera (with special reference to the flies of economic importance)* (Vol. 4. A.6, pp. 113–334). Budapest: Science Herald.

Spies, M., & Reis, F. (1996). Catalog and bibliography of Neotropical and Mexican Chironomidae (Insecta, Diptera). *Spixiana Supplement, 22*, 61–119.

Wiedenbrug, S., & Fittkau, E. J. (1997). *Oliveiriella almeidai* (Oliveira, 1946), gen. nov., comb. nov., from South America with description of the pupae (Insecta, Diptera, Chiromidae, Orthocladiinae). *Spixiana, 20*, 167–172.

INDEX

Note: Page numbers followed by "*f*" and "*t*" refer to figures and tables, respectively.